5.95

D1696352

Sequence Alignment

Sequence Alignment

*Methods, Models, Concepts,
and Strategies*

Edited by Michael S. Rosenberg

UNIVERSITY OF CALIFORNIA PRESS
Berkeley · Los Angeles · London

University of California Press, one of the most
distinguished university presses in the United States,
enriches lives around the world by advancing
scholarship in the humanities, social sciences, and
natural sciences. Its activities are supported by the UC
Press Foundation and by philanthropic contributions
from individuals and institutions. For more information,
visit www.ucpress.edu.

University of California Press
Berkeley and Los Angeles, California

University of California Press, Ltd.
London, England

© 2009 by The Regents of the University of California

Library of Congress Cataloging-in-Publication Data

Sequence alignment: methods, models, concepts, and
strategies/edited by Michael S. Rosenberg.
    p.; cm.
  Includes bibliographical references and index.
  ISBN 978-0-520-25697-2 (cloth: alk. paper)
  1. Bioinformatics. 2. Computational biology.
  I. Rosenberg, Michael S., 1972-
  [DNLM: 1. Sequence Alignment—methods.
  QU 450 S479 2009]
  QH324.2.S47 2009
  572.80285—dc22

2008029222

Manufactured in the United States

16 15 14 13 12 11 10 09 08
10 9 8 7 6 5 4 3 2 1

The paper used in this publication meets the minimum
requirements of ANSI/NISO Z39.48-1992 (R 1997)
(*Permanence of Paper*).

Contents

Contributors	vii
Preface	xi
1. Sequence Alignment: Concepts and History	1
2. Insertion and Deletion Events, Their Molecular Mechanisms, and Their Impact on Sequence Alignments	23
3. Local versus Global Alignments	39
4. Computing Multiple Sequence Alignment with Template-Based Methods	55
5. Sequence Evolution Models for Simultaneous Alignment and Phylogeny Reconstruction	71
6. Phylogenetic Hypotheses and the Utility of Multiple Sequence Alignment	95
7. Structural and Evolutionary Considerations for Multiple Sequence Alignment of RNA, and the Challenges for Algorithms That Ignore Them	105
8. Constructing Alignment Benchmarks	151
9. Simulation Approaches to Evaluating Alignment Error and Methods for Comparing Alternate Alignments	179

10. Robust Inferences from Ambiguous Alignments   209
11. Strategies for Efficient Exploitation of the Informational Content of Protein Multiple Alignments   271

References   297

Index   333

Contributors

Roland Fleißner
Center for Integrative Bioinformatics, Vienna, Austria
Roland.Fleissner@campus.lmu.de

Anne Friedrich
*Institut de Génétique et de Biologie
   Moléculaire et Cellulaire, France*
friedric@igbmc.fr

Joseph J. Gillespie
*University of Maryland, Baltimore County
Virginia Bioinformatics Institute, Virginia Tech*
jgille@vbi.vt.edu

Gonzalo Giribet
Harvard University
ggiribet@oeb.harvard.edu

Karl Kjer
Rutgers University
kjer@aesop.rutgers.edu

Liam J. McGuffin
University of Reading, United Kingdom
l.j.mcguffin@reading.ac.uk

Dirk Metzler
Ludwig-Maximilians-Universität, Munich, Germany
metzler@bio.lmu.de

Burkhard Morgenstern
University of Göttingen, Germany
bmorgen@gwdg.de

Luc Moulinier
Institut de Génétique et de Biologie Moléculaire et Cellulaire, France
luc.moulinier@igbmc.fr

Cédric Notredame
Centre for Genomic Regulation, Spain
cedric.notredame@crg.es

T. Heath Ogden
Idaho State University
ogdet@isu.edu

Olivier Poch
Institut de Génétique et de Biologie Moléculaire et Cellulaire, France
poch@igbmc.fr

Benjamin Redelings
North Carolina State University
benjamin_redelings@ncsu.edu

Michael S. Rosenberg
Arizona State University
msr@asu.edu

Usman Roshan
New Jersey Institute of Technology
usman@oak.njit.edu

Marc A. Suchard
University of California, Los Angeles
msuchard@ucla.edu

Contributors

Julie D. Thompson
*Institut de Génétique et de Biologie
   Moléculaire et Cellulaire, France*
julie@igbmc.fr

Ward Wheeler
American Museum of Natural History
wheeler@amnh.org

Preface

Alignment is a vastly underappreciated aspect of comparative genomics and bioinformatics, in part because alignment tools have become so good. As a biologist who occasionally writes software (although, so far, no alignment software), I have become very aware of the tradeoff between ease of use and potential for misuse and abuse. If software is difficult to use, the implemented algorithms may not be widely applied, but those who do apply them will often have a greater understanding of what they are actually doing. When software is easy, many more people will use the algorithms, but the average level of understanding will drop, and the potential for users to use the algorithms in ways which they should not increases. For the most part (there are certainly exceptions), alignment software is easy and fast and has become an almost trivial part of bioinformatics. The danger is the assumption of triviality. Aligned sequences are used as the "raw" input for a wide array of genome studies, and people thus forget that the alignment is itself a hypothesis of homology—a hypothesis that can be wrong. More precisely, each pair of aligned sites is itself a hypothesis. Thus the alignment is actually a set of hypotheses, some of which may be correct and some incorrect—meaning that the alignment as a whole may be neither right nor wrong, but somewhere in the middle.

My own interest in alignment started when I casually tried aligning the upstream regions of a large number of genes from a single species just to see what would happen. While examining the results, I started to wonder what the expectation would be if the data were completely random (which, given the expected homology of the upstream region of a large set of unrelated genes, is more or less what I had). All of my work on alignment stems from my horror upon discovering the answer (when a pair of completely random DNA sequences, which should have 25% identity due to simple random chance, are aligned using common algorithms and parameters, the resulting aligned sequences are identical at over 40% of the sites) and wondering what the consequence of this and similar issues was on pretty much everything we do in bioinformatics.

This volume had its genesis over an encounter at the 2005 joint meeting of the Society for the Study of Evolution, the Society of Systematic Biologists, and the American Society of Naturalists in Fairbanks, Alaska. Having just finished a presentation about some underappreciated aspects of sequence alignment, I was approached by Chuck Crumly from the University of California Press, who had become interested in finding someone to put together a broad volume on just those sorts of topics. As planned, this volume contains a range of opinions and input from alignment researchers and users in a wide variety of disciplines, including biology, genomics, bioinformatics, computer science, and mathematics. There are two general underlying themes: First, sequence alignments should not be taken for granted; one way or another, they are extremely important for comparative sequence analysis in evolutionary and functional genomics and bioinformatics. Second, the sequence alignment problem is not solved; there are still many challenges and issues that need to be overcome. This book is a dialectic meant to encourage discussion addressing these challenges.

The eleven chapters roughly fall into four (unlabeled) sections: introduction (Chapter 1), biological mechanisms (Chapter 2), algorithms (Chapters 3–5), and broader issues (Chapters 6–11), although in some sense we never escape from either algorithms or broader issues, which are discussed in greater or lesser detail through most of the book.

I begin the book with an introduction to the concepts and history of sequence alignment by describing the dynamic programming approach and the basic algorithms that have been fundamental to the development of sequence alignment software. Emphasis is on the biological concept of homology and the contrast between the biological motivation for aligning data and the computational goals for which algorithms are generally designed.

Liam McGuffin next summarizes the current state of knowledge about the molecular mechanisms leading to indel (insertion and deletion) mutations and explores the root causes of indel events that allow for better approaches to context-dependent alignment.

There follows Burkhard Morgenstern's in-depth comparison of global and local alignment, including alignment tools that combine both local and global procedures into single algorithms. His discussion of benchmarks highlights methods designed to test the sensitivity of alignment (correctly aligning homologous regions), which should also be evaluated on specificity (not aligning nonhomologous regions). He ends with a discussion of the challenges and tools developed for full genome alignment.

In the next chapter, Cédric Notredame examines the state of the art in multiple sequence alignment by summarizing the major approaches for multiple sequence alignment, including matrix- and consistency-based approaches. He discusses recent methods for combining alternate alignments into a single meta-alignment, and he concludes by examining the importance of using additional data sources (such as structural information) in guiding multiple sequence alignments using template-based approaches.

Dirk Metzler and Roland Fleißner follow with a review of statistical approaches for simultaneously estimating alignments and phylogeny. They focus especially on the modeling of insertion and deletion events in a phylogenetic framework and how advances in maximum-likelihood and Bayesian approaches enable these advanced statistical procedures to be used to align sequences.

In the next contribution, Ward Wheeler and Gonzalo Giribet view alignments as inferential objects rather than data and maintain that alignments should be treated thus in phylogenetic analysis. They then criticize the traditional approach to alignment and phylogenetic analysis (Ogden and Rosenberg 2007a) with a specific implementation of simultaneous phylogeny and alignment construction (De Laet and Wheeler 2003). This highlights the difference between a computational goal (finding the phylogeny by minimizing the number of steps necessary to create an observed set of sequences) and a biological goal (constructing an alignment that best represents the true positional homologies of the sequences or finding the phylogeny that best represents the true evolutionary history of the sequences). The chapters in this book and the two references cited in this paragraph are intended to help readers to draw their own conclusions.

Karl Kjer and colleagues, in Chapter 7, delve into the biological motivations for aligning data and explore the limitations of strict algorithmic approaches. They emphasize, with simple examples, the importance of including both structural and evolutionary information in postalgorithmic manual curation of alignments. They include a detailed "how-to" explanation for the manual structural alignment of rRNA sequences.

Next, Julie Thompson follows by discussing the current state of benchmark databases for evaluating alignment algorithms. These databases are critical to algorithm development because most new algorithms use such benchmark databases to evaluate performance. There may be a danger of algorithms being overoptimized for the specific characteristics of these databases. She concludes by examining recent benchmark tests for a variety of alignment programs, focusing on recent algorithmic advances that have generally improved alignment quality, but also identifying areas where there is still room for improvement.

Heath Ogden and I team up to examine the increasing role of computer simulation in alignment evaluation, not just for the benchmarking of alignment algorithms but also for exploring the consequences of alignment errors (or use of alternate alignments) in bioinformatic analysis. We describe, compare, and contrast the strengths and weaknesses of a number of approaches for comparing alternate alignments.

In the penultimate chapter, Benjamin Redelings and Marc Suchard take a general look at certainty and uncertainty in sequence alignment, including the root causes for ambiguity, and they explore a variety of approaches for including (or excluding) ambiguously aligned sites in an analysis. They detail how recent advances in Bayesian statistical approaches permit the estimation of alignment uncertainty and how uncertainty can be used in other bioinformatic analyses, including, in particular, phylogeny construction.

In the concluding chapter, Anne Friedrich and colleagues put alignments to use in further evolutionary, structural, functional, and mutational studies of proteins. They also summarize many of the programs and packages currently available.

"WHAT ALIGNMENT PROGRAM SHOULD I USE?"

The question of what software to use is the most common question anyone who works on alignment receives. A definitive answer will not be found in this book. The best alignment program may depend on the specific circumstances of the data being aligned, including the nature of the sequences (e.g., DNA, RNA, protein), the number of sequences to be aligned, the lengths of the sequences, the evolutionary divergence of the sequences, whether structural information is available, the type of structures (e.g., globular or disordered), and perhaps even the specific purpose for constructing the alignment (i.e., what will the alignment be used for). For large-scale bioinformatic studies, the speed of an algorithm is very important, although I personally feel it can be an overrated factor for smaller comparative studies of the kind encountered in the average molecular lab. If waiting a few extra hours (or even days) will produce a better result, one should take the time to get the best possible answer (in large-scale bioinformatics, where thousands to millions of alignments may be produced as part of a single study, the time difference may scale to months or even years, at which point speed becomes of greater concern).

For a long time, ClustalW/ClustalX (Thompson et al. 1997; Thompson et al. 1994) has been the alignment program of choice for many users because of the general quality of its alignments, its wide implementation, and its ease of use (as of this writing, ClustalW and ClustalX have been cited over 33,000 times [data from ISI Web of Knowledge]). Over the past few years, recent programs such as MUSCLE (Edgar 2004a, b), MAFFT (Katoh et al. 2005), and ProbCons (Do et al. 2005) have consistently received high marks in a variety of benchmark tests (e.g., Blackshields et al. 2006; Pollard et al. 2004; Rosenberg, unpublished data). Given the constant and continued development of alignment programs (see Chapter 1), one would expect that the answer to the question of which is the best program for a specific circumstance may change through time. So, although this book will not provide a definitive answer to the question, it will hopefully help guide your decision as to the factors and issues to consider when thinking about how best to align and interpret your data.

I wish to thank all of the people who contributed to the production of this volume. First and foremost, all of the authors who responded to requests for a contribution and followed through with a manuscript: without you, this all would have come to nothing. As previously mentioned, Chuck Crumly at UC Press deserves much of the credit for getting this volume started; he also deserves thanks for his patience in dealing with the delays and foibles of the editor. Sudhir Kumar offered support and encouragement throughout the creation of this book, including critical commentary on much of my own work. The postdocs and students in my lab deserve thanks for helping with analyses as well as patience and understanding when the needs of the book may have occasionally taken priority over their own: Heath Ogden, Corey Anderson, Ahmet

Kurdoglu, Loretta Goldberg, Meraj Aziz, Virginia Earl-Mirowski, and Amy Harris. During the preparation of this volume I received financial support from the National Library of Medicine of the National Institutes of Health, the National Science Foundation, and the Center for Evolutionary Functional Genomics of the Biodesign Institute and the School of Life Sciences at Arizona State University. Finally, I must thank my wife, Maureen Olmsted, for her continual support, love, and understanding.

Michael S. Rosenberg
January 2008

CHAPTER 1
Sequence Alignment
Concepts and History

MICHAEL S. ROSENBERG
Arizona State University

Pairwise Alignment and Dynamic Programming 3
Global Alignment vs. Local Alignment 11
    Local Alignment vs. Database Searching 14
Importance of the Cost Function 14
Multiple Alignments 16
Statistical Approaches to Sequence Alignment 19
Homology 19
Challenges for the Future 21

Sequence alignment is a fundamental procedure (implicitly or explicitly) conducted in any biological study that compares two or more biological sequences (whether DNA, RNA, or protein). It is the procedure by which one attempts to infer which positions (sites) within sequences are homologous, that is, which sites share a common evolutionary history (see the section "Homology" in this chapter for more detail). For the majority of scientists, alignment is a task whose automated solution was solved years ago; the alignment is of little direct interest but is rather a necessary step that allows one to study deeper questions, such as the identification and quantification of conserved regions or functional motifs (Kirkness et al. 2003; Thomas et al. 2003), profiling of genetic disease (Miller and Kumar 2001; Miller et al. 2003), phylogenetic analysis (Felsenstein 2004), and ancestral sequence profiling and prediction (Cai et al. 2004; Hall 2006). For other scientists, alignment is an active area of research, where basic questions on how one should construct and evaluate an alignment are under heavy scrutiny and debate. Because alignment is the first step in many complex, high-throughput studies (Lecompte et al. 2001), it is important to remember that alignment algorithms produce a hypothesis of homology (just as a phylogenetic tree is a hypothesis of evolutionary history). Like other hypotheses, these alignments may contain more or less error depending on the nature of the data, some of which may have huge downstream effects on other analyses (Kumar and Filipski 2007; Ogden and Rosenberg 2006; Rosenberg 2005a, b).

In a casual survey, most researchers guess that two or three dozen alignment programs and algorithms have been published. The true number is actually in the hundreds, with the numbers increasing each year. Figure 1.1 shows a summary of the number of named alignment programs released over the 20-year period from 1986 to 2005 (alignment algorithms go back to 1970, but prior to the mid-1980s and the advent of the personal computer, most were simply published as logical descriptions or as source code rather than as compiled executables). Just counting named programs and not papers describing algorithmic

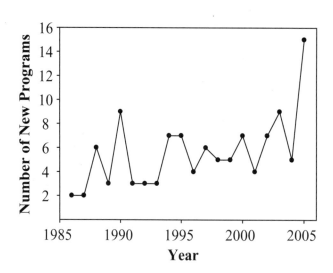

Figure 1.1. Number of new named alignment programs released each year from 1986 to 2005.

Alignment Concepts and History

advances or unnamed code or software, there has been an average of over five programs released per year, with a generally increasing trend through time.

This chapter provides a brief historical overview of sequence alignment with descriptions of the common basic algorithms, methods, and approaches that underlie most of the way sequence alignments are performed today. More thorough treatments of many of these topics are discussed throughout the text.

PAIRWISE ALIGNMENT AND DYNAMIC PROGRAMMING

To be able to compare potential sequence alignments, one needs to be able to determine a value (or score) that estimates the quality of each alignment. The formulas behind an alignment score are generally known as *objective functions*; they range from simple cost-benefit sums to complex maximum-likelihood values. This introduction will use a simple cost-benefit approach, but much more advanced scoring algorithms and mechanisms are available, many of which are discussed in detail throughout this book.

When using a cost-benefit approach to evaluating a pairwise alignment, one must specify scores for the various ways in which a pair of sites can be compared. In the simplest case, three scores are specified: (1) the benefit of aligning a pair of sites that contain the same character (state) in both sequences; (2) the cost of aligning a pair of sites that contain different characters in the sequences; and (3) the cost of aligning a character in one sequence with a gap in the other sequence. Depending on how one defines the scores, the eventual goal could be to find the alignment that maximizes the benefit or to find the alignment that minimizes the cost. This is essentially an arbitrary choice based on how one chooses to define the costs and benefits. Many of the popular alignment programs in use today (e.g., ClustalW) find the maximum score, a convention that will be retained throughout this description. In computer science, one simple, cost-based scoring function is the *edit distance*, that is, the minimum number of changes necessary to convert one sequence into another. In this case, the goal of alignment is to minimize the edit distance.

Given a scoring function, one can compare any set of alignments for the same set of initial sequences. The one with the best score would be considered the best alignment. For example, let us set the benefit of a match to +1, the cost of a mismatch to −3, and the cost of aligning

```
ACCTGATCCG        ACCTGATCCG
| |  | | | | |  |    | |       | | |
AC-TGATCAG        ACTGA-TCAG
S=8-4-3=1         S=5-4-12=-11
```

Figure 1.2. Alternate alignments of a pair of sequences illustrating a simple scoring function with matches = +1, mismatches = −3, and gaps = −4. The alignment on the left is better than the alignment on the right because its overall score is larger (1 vs. −11).

a character to a gap to −4. Figure 1.2 shows two possible alignments of sequences ACCTGATCCG and ACTGATCAG. In the first potential alignment, there are 8 matches (8 × +1 = +8), one mismatch (1 × −3 = −3), and one site aligned with a gap (1 × −4 = −4), for a total score of +1. In the second potential alignment, there are 5 matches (+5), 4 mismatches (−12), and one site aligned with a gap (−4), for a total score of −11. The first alignment has a higher score than the second and would be considered a better alignment. But how do we know that it is the best possible alignment?

It is impossible to evaluate all possible alignments. Take the simple case where a sequence of 100 characters is being aligned with a sequence of 95 characters. If all we do is add 5 gaps to the second sequence (to bring it to 100 total sites), there are approximately 55 million possible alignments (Krane and Raymer 2003). Because we may need to add gaps to both sequences, the actual number of possible alignments is significantly greater.

The number of potential alignments made automated procedures for aligning sequences a critical aspect of molecular sequence comparison (but see Chapter 7 for a discussion of why one should not rely solely on automated methods). The first attempts at developing a computational method for the alignment of sequences were undertaken in the mid-1960s with studies such as those of Fitch (1966) and Needleman and Blair (1969), but it was not until 1970 that the first elegant solution to the alignment problem was produced (Needleman and Wunsch 1970). It is this solution, using dynamic programming, that has made their procedure the grandfather of all alignment algorithms.

Dynamic programming is a computational approach to problem solving that essentially works the problem backwards. Dynamic programming is best illustrated with a simple mathematical example, say calculating the nth value of a Fibonacci sequence. The Fibonacci sequence is a series of numbers in which each value is equal to the sum of the two values preceding it, $F_n = F_{n-1} + F_{n-2}$ (by definition, the first two values of the sequence must be specified). Thus, to calculate the 10th value of the Fibonacci sequence, one needs to know the 8th and 9th values; to calculate the 9th value, one needs to know the 7th and 8th values; to calculate the 8th value one needs to know the 6th and 7th values; and so on. Although it is not particularly difficult to solve this problem with a straightforward recursive algorithm, finding the nth value by a recursive approach is very inefficient. Note that if each value is determined independently, one needs to compute the same value more than one time (e.g., calculation of the 9th and 8th values both require calculating the 7th value). Using a standard recursive algorithm, determination of the 10th value of the Fibonacci sequence would require 109 steps (Figure 1.3A); the formal complexity of the recursive approach to the Fibonacci sequence is exponential. A dynamic programming approach, on the other hand, is simpler and more efficient. It works the problem in the opposite direction from the recursive approach by starting with the first value in the sequence rather than the nth value. In the Fibonacci sequence, the first two values must be predefined; for this example we will set the first value to 0 and the second value to 1. Given the first two values, we can easily determine the 3rd value, $0 + 1 = 1$. The 4th value is the sum of the already determined 2nd and 3rd values and therefore is $1 + 1 = 2$. At this point it is trivial to keep moving forward until one reaches the 10th value (5th value = $1 + 2 = 3$; 6th value = $2 + 3 = 5$; 7th value = $3 + 5 = 8$; 8th value = $5 + 8 = 13$; 9th value = $8 + 13 = 21$, and 10th value = $13 + 21 = 34$). By avoiding the redundant determination of the lower values, the dynamic programming approach takes only 10 steps. That is, the complexity is linear, requiring only n steps (Figure 1.3B).

Pairwise sequence alignment is more complicated than calculating the Fibonacci sequence, but the same principle is involved. The alignment score for a pair of sequences can be determined recursively by breaking the problem into the combination of single sites at the end of the sequences and their optimally aligned subsequences (Eddy 2004). If sequences x and y have m and n sites, respectively, the last position of their alignment can have three possibilities: x_m and y_n are aligned, x_m is aligned with a gap with y_n somewhere upstream, or y_n is aligned with a gap with x_m somewhere upstream. The alignment score for each case is the score for that final position plus the score of the optimal alignment

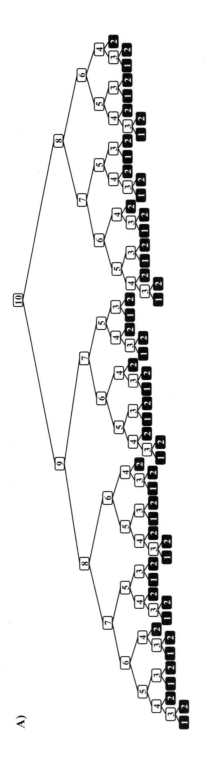

Figure 1.3. Calculating the 10th value of the Fibonacci sequence. The numbers in the cells represent the nth number in the sequence. The actual value of the nth number is equal to the sum of the previous two values ($f_n = f_{n-1} + f_{n-2}$). The first two values of the sequence are predefined (represented by black cells). (A) The top-down recursive approach has duplication of effort since many values have to be determined multiple times (e.g., the 3rd value is determined 21 times). It requires 109 steps, including looking up the predefined values 55 times. (B) The bottom-up dynamic programming approach starts with the first two values and works forward to the desired 10th value. There is no duplication of effort, and the 10th value can be determined in just 10 steps.

of the upstream subsequence (each site is scored independently, so the score of nonoverlapping alignment segments can be added for a total score). The overall optimal score is the maximum of the score for the three cases. Critically, the optimal score for each of the aligned subsequences can be determined the same way: three possible cases for the final position plus the optimal alignment of the upstream subsequence for each case. Thus, a formula for the optimal alignment score can be written in a simple recursive format:

$$S(i, j) = \max \begin{vmatrix} S(i-1, j-1) + \sigma(x_i, y_j) \\ S(i-1, j) + \gamma \\ S(i, j-1) + \gamma \end{vmatrix},$$

where $S(i, j)$ is the score for the optimal alignment from position 1 to i in sequence x and 1 to j in sequence y, γ is the gap cost, and $\sigma(x_i, y_j)$ is the mismatch/match score for the pair of states at positions x_i and y_j (Eddy 2004).

Although simple to write, this formula is very inefficient to solve and, as for the Fibonacci sequence, leads to redundant determination of the optimal score for subsequence alignments that will show up over and over again. The dynamic programming solution works by starting with the optimal alignment of the smallest possible subsequences (nothing in sequence x aligned to nothing in sequence y) and progressively determining the optimal score for longer and longer sequences by adding sites one at a time. By keeping track of the optimal score for each possible aligned subsequence in a matrix, the optimal score and alignment of the full sequences can easily and efficiently be determined.

This method is illustrated in Figure 1.4 with the alignment of a pair of short sequences: ATG and GGAATGG, using a match score of +1, a mismatch score of −1, and a gap score of −2 (example adapted from Lesk 2002). The first step is to construct a matrix that contains each sequence along an axis, with an extra empty row and column at the top and left sides of the matrix. Each cell in the matrix will be filled with the maximum of three possible values: (1) the sum of the score of the cell that is diagonally to its upper left and the match or mismatch score (depending on whether the characters in the row/column of the cell match or mismatch); (2) the sum of the score of the cell that is directly above the cell and the gap score; or (3) the sum of the score of the cell directly to the left of the cell and the gap score. (There is essentially a hidden rule that allows one to skip any rule which does not apply, for

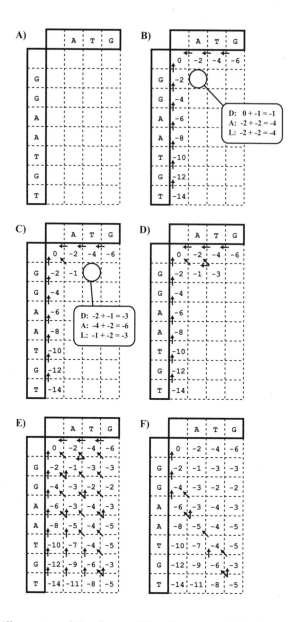

Figure 1.4. Illustration of Needleman–Wunsch (1970) global alignment algorithm. (A) Setting up the matrix. (B) The first row and column are filled with increasing multiples of the gap cost. The first cell will be given the maximum of three possible values. (C) The value for the first cell is entered along with the path that led to the value. The possible values for the second cell are illustrated. (D) The value for the second cell is entered; multiple paths are recorded since multiple paths led to the maximum score. (E) The completed matrix. (F) The completed matrix with all suboptimal paths removed. Tracing the arrows from the bottom right corner to the upper left leads to four possible paths and (therefore) four equally optimal alignments.

example, the cells in the top row do not have any cells above them, so rule 2 cannot apply). The practical upshot of these rules is that the first step is to place a zero in the upper left corner of the matrix and then to fill the first row and column with increasing multiples of the gap cost. The rest of the cells in the matrix are filled in one by one. An important point is that one needs to keep track of which rule was applied, that is, which neighboring cell (diagonal, above, or left) led to the value that was filled in. Computationally this is usually stored in a second matrix called a trace-back matrix; in Figure 1.4 this is illustrated with arrows for simplicity. A second important point is that it is possible for multiple rules to produce the identical maximum value; in such a case the trace-back matrix should properly include *all* possible paths to the maximum value.

In our example, the first cell (A vs. G) can have three values: (1) the score of the cell to the upper left, 0, plus the mismatch cost (A/G is a mismatch), $-1 = -1$; (2) the score of the cell above it, -2, plus the gap score, $-2, = -4$; or (3) the score of the cell to the left, -2, plus the gap score, $-2, = -4$. The maximum of these is the first score, -1, which is entered in the cell, along with an arrow to the upper left to remind us that it is that path from which the score was derived. We repeat for the next cell (to the right). The possible values for this cell are $-2 + -1 = -3$, $-4 + -2 = -6$, or $-1 + -2 = -3$. The maximum value is -3, but this can be achieved from two separate paths, so we record both paths using our trace-back arrows.

This procedure is repeated until the entire matrix is filled with values. The value in the last cell, the lower right (-5 in our example), represents the score of the best alignment (given the score function). To find this alignment, one starts with this cell and follows the arrows back to the upper left corner of the matrix. Following a diagonal arrow indicates that the sites represented by that row and column of the matrix should be aligned. Following a vertical arrow indicates that the character in the sequence along the vertical axis (the character represented by the row of the matrix) should be aligned with a gap in the sequence represented by the horizontal axis. Following a horizontal arrow indicates that the character in the sequence along the horizontal axis (the character represented by the column of the matrix) should be aligned with a gap in the sequence represented by the vertical axis.

If there is more than one possible path back to the top of the matrix, this indicates that multiple pairwise alignments lead to the identical score and are equally optimal. In our example, there are four possible

```
GGAATGG      GGAATGG      GGAATGG      GGAATGG
---ATG-      ---AT-G      --A-TG-      --A-T-G
```

Figure 1.5. Four equally optimal global alignments of sequences GGAATGG and ATG derived from the alignment matrix shown in Figure 1.2.

paths, leading to four possible alignments of these sequences, shown in Figure 1.5.

In this specific case, one might argue that the first of these appears to be a subjectively better alignment than the others, but, based solely on the specified cost function, all four of these alignments are equally good (each of these alignments contains three matches and four gaps). Changing the cost function may change the result (see below). Very few alignment programs produce more than one alignment, even if there are multiple equally optimal alignments; the sim algorithms of Huang et al. (1990) and Huang and Miller (1991) are a notable exception. How the single resultant alignment produced by most programs is chosen from the universe of possible optimal alignments is usually not clear.

This is the simplest approach to pairwise sequence alignment. Obvious enhancements include the use of more complicated scoring functions. Not all mismatches are necessarily equal, and different types of mismatches could be given different scores depending on the properties of the characters. For DNA sequences, these differential scores might be based on standard models of sequence evolution. For example, it is well known that transitional substitutions occur more often than transversional substitutions, therefore a transversional mismatch might be given a higher cost than a transitional mismatch. For protein sequences, empirically derived substitution matrices are usually used to determine relative costs of various mismatches; these matrices include the PAM (Dayhoff et al. 1978), JTT (Jones et al. 1992), and BLOSUM (Henikoff and Henikoff 1992) matrices. These matrices usually include estimated biological factors such as the conservation, frequency, and evolutionary patterns of individual amino acids. In principle, one could imagine giving different benefits to different matches (e.g., perhaps a larger benefit should come from aligning a pair of cysteine residues because of the extreme conservation and structural constraints of this amino acid).

Another enhancement to the scoring function has to do with the gap costs. As described above, all gaps are treated as identical single position events. Biologically, we recognize that single insertion-deletion events may (and often do) cover multiple sites. We therefore may not want the cost of a gap that covers three sites to be triple the cost of a gap that covers only one site (a linear gap cost). The general solution is to use one cost score for opening (starting) a gap and a second score for extending (lengthening) a gap (an affine gap cost) (Altschul and Erickson 1986; Gotoh 1982, 1986; Taylor 1984). In this case the total cost of the gap is $O + nE$ where O is the gap opening cost, E is the gap extension cost, and n is the length of the gap (or the length of the extension, depending on how the algorithm is defined). Much more complicated schemes for scoring gaps of varying lengths are possible, although the majority of modern alignment programs appear to use some form of an affine cost structure. Some algorithms will also vary in how they treat terminal gaps (that is, gaps that occur at the very beginning or ends of a sequence); some algorithms will give these reduced cost (even zero) since they are not inferred to occur between observed characters (this is sometimes known as semi-global alignment).

In our previous example, the use of an affine gap cost would eliminate the third and fourth alignments in Figure 1.5 from the optimal set, since these each have three putative independent gaps (of length 2, 1, and 1) while the first two alignments have only two gaps (of length 3 and 1). Lowering the cost of terminal gaps would then eliminate the second alignment, since one of its gaps is internal to the observed characters, while the first alignment contains only terminal gaps.

Beyond changes in how alignments are scored, there have been numerous improvements in efficiency of the basic dynamic programming approach described above, including, for example, decreasing memory requirements (Myers and Miller 1988) and the number of computational steps (Gotoh 1982).

GLOBAL ALIGNMENT VS. LOCAL ALIGNMENT

The procedure described thus far is a global alignment algorithm; that is, it assumes that the entirety of the sequences are sequentially homologous and tries to align all of the sites optimally within the sequences. This assumption may be incorrect as a result of large-scale sequence rearrangement and genome shuffling. In such a case, only subsections of the sequences may be homologous, or the homologous sections may be in a different order. For example, a long sequence may be ordered ABCDEF (where each letter represents a section of sequence and not an

```
AB--CDEF        ABCDEF       ABCDE--F
ABEDC--F        ABEDCF       AB--EDCF
```

Figure 1.6. Illustration of global alignment problem. Sequences ABCDEF and ABEDCF cannot be properly aligned because the homologous sections of the sequences are not in the same order.

individual site). A sequence inversion of section CDE may change the sequence in another species to ABEDCF. Although each section of the first sequence is homologous with a section of the second sequence, they cannot be globally aligned, because of the rearrangement. Figure 1.6 shows possible global alignments if section C, D, or E is aligned. In each case, the other two sections cannot be aligned properly.

An alternative approach to global alignment is local alignment. In a local alignment, subsections of the sequences are aligned without reference to global patterns. This allows the algorithm to align regions separately regardless of overall order within the sequence and to align similar regions while allowing highly divergent regions to remain unaligned.

Early approaches for local alignment were developed by Sankoff (1972) and Sellers (1979, 1980), but the basic local alignment procedure most widely used was proposed by Smith and Waterman (1981b). It is a simple adaptation to the standard Needleman–Wunsch algorithm. The first difference is in the determination of values for a cell. In addition to the three possible values described by the Needleman–Wunsch algorithm, the local alignment algorithm allows for a fourth possible value: zero. This prevents the alignment score from ever becoming negative; if this rule is invoked, no trace-back arrow is stored for the cell. Figure 1.7 shows the score matrix for a local alignment of the same sequences that were globally aligned in Figure 1.4. The addition of the fourth rule substantially changes the structure of the scores and the trace-back arrows.

Once the matrix has been filled, the second change in the local alignment algorithm is that rather than starting in the lower right corner of the matrix, one uses the cell with the largest value as the starting position for the trace-back. In our example, this is the cell directly above the lower right cell, with a score of 3. The final difference in this method is that the trace-back does not continue to the upper-left corner, but rather terminates when the trace-back arrows end. The result of this procedure is that only a portion of the sequences may be aligned; the remainder

Alignment Concepts and History

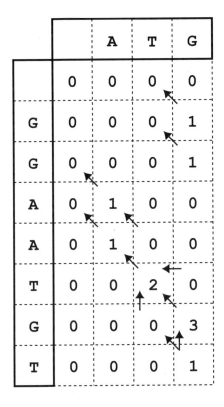

Figure 1.7. Completed score and trace-back matrix for local alignment using the Smith and Waterman (1981b) algorithm.

of the sequences are left as unaligned. For this example, the local alignment is simply that shown in Figure 1.8.

Only the aligned parts of the sequences are reported. Of course, additional local alignments of the sequences could be found if there are multiple cells with the same maximal score or by choosing submaximal starting points. As with global alignment, there have been major advances in approaches for local alignment; major local alignment programs and algorithms in use today include DIALIGN (Morgenstern 1999; Morgenstern, Frech, et al. 1998) and CHAOS (Brudno, Chapman, et al. 2003; Brudno and Morgenstern 2002).

With the recent sequencing revolution, one bioinformatic challenge has been the comparison of full genome sequences. Because genomes are so readily rearranged, alignment of entire genomes is a specialized

ATG
ATG

Figure 1.8. The local alignment of sequences GGAATGG and ATG derived from the alignment matrix shown in Figure 1.2.

case of local alignment applied on a very large scale. Many specialized programs for producing local alignments of entire genomes have been produced recently, include BlastZ (Schwartz et al. 2003), MUMMER (Delcher et al. 1999; Delcher et al. 2002), GLASS (Batzoglou et al. 2000), WABA (Kent and Zahler 2000), MAUVE (Darling et al. 2004), GRAT (Kindlund et al. 2007), MAP2 (Ye and Huang 2005), and AuberGene (Szklarczyk and Heringa 2006).

Local Alignment vs. Database Searching

Much of the work on local alignment has focused on database searching rather than simple sequence comparison. It was recognized very early on that algorithms would be necessary to retrieve sequences from a database with a pattern similar to that of a query sequence (e.g., Korn et al. 1977). Comparing sequences for similar patterns requires, in some form, local alignment, and local alignment methods form the basis of all database searching algorithms. The most famous sequence search algorithm, BLAST (Altschul et al. 1990), contains this phrase in its name: Basic *Local Alignment* Search Tool. Although not discussed in any detail within this book, must of the major work on local alignment derives from interests in database searching, particularly in the development of both the BLAST and FASTx (Lipman and Pearson 1985; Pearson and Lipman 1988) families of algorithms.

IMPORTANCE OF THE COST FUNCTION

The algorithms described above allow one to search the population of possible alignments efficiently for the most optimal alignment(s), but it is the cost function that is most important in determining the actual best alignment. As one would expect, changing the cost function may change which alignment is considered to be most optimal. Take for example, the pair of sequences CAGCCTCGCTTAG and AATGCCATTGACGG. If we perform global and local alignments of these sequences using the parameters

Alignment Concepts and History

A)
```
CA-GCC-TCGCTTAG    CA-GCC-TCGCTTAG    CA-GCC-TCGCTTAG    CA-GCC-TCGCTTAG
AATGCCATTGACG-G    AATGCCATTGAC-GG    AATGCCATTGA-CGG    AATGCCATTG-ACGG
```

B)
```
            GCC
            GCC
```

Figure 1.9. Optimal alignments of sequences CAGCCTCGCTTAG and AATGCCATTGACGG with a cost function with matches = +1, mismatches = −1, and gaps = −2. (A) Four equally optimal global alignments. (B) The single optimal local alignment.

as before (+1 match, −1 mismatch, −2 gap), we find four equally optimal global alignments, shown in Figure 1.9A (the only difference is the position of the gap in the second sequence), and one local alignment (Figure 1.9B).

If we change the parameters of our cost function so that the match benefit is still +1, but the mismatch cost is now −0.3 and the gap cost is −1.3, we would find the same four global alignments, but our local alignment would change to that shown in Figure 1.10. Different sets of scoring values may lead to different optimal alignments; the best alignment is not only dependent on the algorithm (global vs. local) but on parameter choices.

How does one determine what values should be used for the cost function? Most users tend to use program defaults, in which case the problem of determining the proper values is just left to the program authors rather than the end user. It should be noted that the absolute magnitudes of the values are unimportant; it is the relative values that matter (that is, multiplying all of the scores by a constant will not affect the resultant best alignment). As mentioned previously, relative match and mismatch values are usually determined from empirical substitution matrices (for proteins) or models of sequence evolution for DNA.

```
            GCC-TCG
            GCCATTG
```

Figure 1.10. The optimal local alignment of sequences CAGCCTCGCTTAG and AATGCCATTGACGG with a cost function with matches = +1, mismatches = −0.3, and gaps = −1.3. Contrast with the local alignment in Figure 1.9B.

However, the most important value in scoring alignments is often considered to be the ratio of the mismatch cost and gap cost (a more complex function may include different mismatch costs and affine gap costs, but the general principle still holds). The values in the above example indicate that a gap is twice as costly as a mismatch. One could take this to mean that point mutations occur twice as often as insertion/deletion events; however, biologically we know that indels tend to be much rarer than that, relative to point mutations. There is remarkably little data available on the observed ratio of indel events to point mutations. For DNA, indels appear to be about 12 to 70 times less common than point mutations, depending on the specific taxa and evolutionary divergences examined (Mills, Luttig, et al. 2006; Ophir and Graur 1997; Sundström et al. 2003); one would expect protein sequences to have a different ratio. It is easy to imagine how one can force an optimal alignment to be more or less "gappy" by manipulating the gap cost: A very high gap cost will increase the number of mismatches and decrease the number of gaps in the optimal alignment, while a lower gap cost will decrease the number of mismatches and increase the number of gaps.

Additionally, there appears to be a bit of a discrepancy between the biological ratio of point mutations and indels and the actual cost structure used in the alignment. Subjective evaluation of alignments produced with different cost ratios, as well as objective examination of both empirical and simulated benchmarks (unpublished) tend to find that the gap costs that produce the best alignments (those that most closely resemble the true alignment) are usually less than those that would be predicted from a straightforward evaluation of the actual mutation rates. Methods for optimizing gap costs (as well as other aspects of the cost function) and their effects are an understudied aspect of sequence alignment.

MULTIPLE ALIGNMENTS

Up to this point, the described algorithms have been for comparing pairs of sequences. In general, we are often interested in aligning more than two sequences (in general, called multiple alignment). In principle, one could use a Needleman–Wunsch approach for more than two sequences (for example, constructing a three-dimensional cubic matrix for three sequences) (Jue et al. 1980; Murata et al. 1985), but this quickly becomes computationally intractable and inefficient as a result of constraints in computational power and memory.

Early alternate approaches for multiple alignment required a known phylogenetic tree. Sankoff and colleagues (Sankoff 1975; Sankoff et al. 1976; Sankoff et al. 1973) developed parsimony-based approaches. Waterman and Perlwitz (1984) suggested an alternate approach that used weighted averaging. Instead of using a known tree, Hogeweg and Hesper (1984) suggested an iterative method where one starts with a putative tree, aligns the data, uses the alignment to estimate a new tree, and then uses the new tree to realign the data, and so forth.

The approach for multiple sequence alignment that eventually really caught on is known as progressive alignment (Feng and Doolittle 1987, 1990). In progressive alignment, one generally starts by constructing all possible pairwise alignments (for n sequences, there are $n \times (n-1)/2$ pairs). These pairwise alignments are used to estimate a phylogenetic tree using a distance-based algorithm such as the unweighted pair group method with arithmetic mean (UPGMA) or neighbor joining. Using the tree as a guide, the most similar sequences are aligned to each other using a pairwise algorithm. One then progressively adds sequences to the alignment, one sequence at a time, based on the structure of the phylogenetic tree. Numerous multiple alignment programs have been based on a progressive alignment adaptation of the Needleman–Wunsch algorithm, including ClustalW (Thompson et al. 1994), perhaps the most widely used global multiple alignment program.

Unlike the pairwise algorithm, multiple alignment algorithms are heuristic rather than exact solutions; by searching only a subset of the population of alignments, they efficiently find an alignment that is approximately optimal but is not guaranteed to be the most optimal alignment possible for the given cost function. For example, a general disadvantage of the progressive alignment approach is that it is what is known as a greedy algorithm; any mistakes that are made in early steps of the procedure cannot be corrected by later steps. For example, take the case (adapted from Duret and Abdeddaim 2000), with three short sequences whose optimal alignment is as shown in Figure 1.11A. Assuming the guide tree indicates we should start by aligning sequences 1 and 2, there are three possible alignments with the same score (one transversional mismatch and one gap), shown in Figure 1.11B. When adding sequence 3, the position of the gap cannot be changed. Thus, adding sequence 3 could lead to three possible multiple alignments, shown in Figure 1.11C, only the first of which is optimal. At the first step, only one of the three alignments can be used for the next step, and if the wrong one is chosen, the end results will not be the most optimal solution.

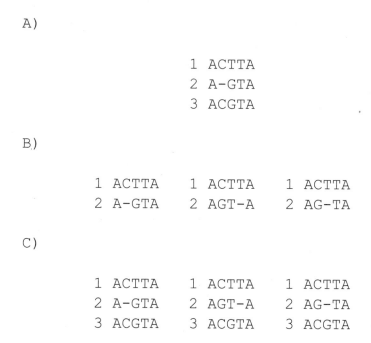

Figure 1.11. Illustration of the progressive alignment problem. (A) The optimal multiple alignment of three sequences. (B) The three possible optimal alignments that would be constructed in the first step of the alignment, depending on which pair were chosen to be aligned first. (C) The multiple alignments resulting from each of the three starting points from part B. Only one of these is equal to the actual optimal alignment illustrated in A. Example adapted from Duret and Abdeddaim (2000).

A number of less-greedy algorithms have been designed to try to get around this problem. For example, T-Coffee (Notredame et al. 2000) starts by using pairwise local alignments to find high scoring regions of similarity. Although a basic greedy progressive alignment is used to produce the global multiple alignment, these local alignments have an effect on the relative scoring at the early progressive stages and may help avoid errors that would otherwise be introduced into the multiple alignment. Another approach is to use iterative methods in which the alignment generated from one pass of an algorithm is used to construct a new guide tree, which can then be used to form a new alignment. Some of the better known iterative alignment programs include MultAlin (Corpet 1988), PRRP (Gotoh 1996), and DIALIGN (Morgenstern 1999; Morgenstern, Frech, et al. 1998).

STATISTICAL APPROACHES TO SEQUENCE ALIGNMENT

The methods described thus far could be considered "character-based" alignments in which the algorithms optimize a cost/benefit function. An alternate approach to alignment uses statistical estimation based on either maximum-likelihood or Bayesian methods. Although a statistical approach to alignment was suggested as far back as the mid-1980s (Bishop and Thompson 1986), this approach did not really begin to gain traction for another 15 years because of the computational complexity of the problem. The most influential contribution to statistical alignment has certainly been a pair of papers by Thorne et al. (1991, 1992) that describe the first tractable stochastic models for the insertion-deletion process. Although these models suffer from issues of realism (the first paper requires all insertion-deletion events to be a single character in length), the models suggested in these papers have formed the basis of most statistical alignment procedures in use today.

An important early advance was the formal development of hidden Markov models to describe the insertion-deletion process (Baldi et al. 1993, 1994; Krogh et al. 1994), including the release of the software HMMER (Eddy 1995) for statistical alignment. These were followed by a number of advances by Hein and colleagues for maximum-likelihood solutions to the multiple alignment problem (Hein 2001; Hein et al. 2003; Hein et al. 2000; Steel and Hein 2001).

Allison and colleagues (Allison and Wallace 1994; Allison et al. 1992a, b; Allison and Yee 1990) set the stage for Bayesian approaches for alignment through the formal modeling of and comparison of sequences. Mitchison (1999) describes one of the first statistical approaches to simultaneously estimating an alignment and a phylogeny, which helped lead to Handel (Holmes and Bruno 2001), one of the first software packages for Bayesian alignment.

In recent years, there has been an explosion of development in statistical alignment, in general, and simultaneous estimation of alignment and phylogeny specifically (Fleißner et al. 2005; Lunter et al. 2005; Redelings and Suchard 2005). Modern advances in statistical alignment are discussed in more detail in Chapters 5 and 10.

HOMOLOGY

The biological goal of alignment is the inference of site *homology*. Homology is similarity in a character or trait due to inheritance from a common ancestor. With respect to comparing biological sequences, homology can have three different interpretations: (1) The sequences can be homologous; (2) the sites within homologous sequences can be homologous; (3) the observed characters at a homologous site can be homologous. Sequence alignment (as discussed in this book) is mostly concerned with the second of these. The general purpose of alignment is to identify positions in homologous sequences that are descended from a common ancestral sequence, that is, to identify which sites in a pair (or more) of sequences are themselves homologous. A pair of sites is "homologous" if the position in both sequences corresponds to the identical position in the common ancestral sequence. A pair of sites is "identical" if both sequences contain the same nucleotide (or amino acid for protein sequences); identity could be due to homology (i.e., the specific nucleotide was inherited by both sequences from the common ancestral sequence with no substitutions) but may often be due to convergent or parallel substitutions or by misalignment (when two sites are thought to be homologous but are not). Thus, character homology is first dependent on site homology (homologous characters can be found only at homologous sites) and secondarily on inheritance of the character from the common ancestor. We assume that the sequences we wish to align are homologous (in the first sense), although local alignment is often used as part of the inference of sequence homology using database search algorithms such as BLAST (Altschul et al. 1990) and FASTA (Pearson and Lipman 1988).

An additional level of homology that plays an important role with respect to sequence alignment is structural homology. Inferred homology of the secondary and/or tertiary structure of a protein or RNA is often used to guide the inferred sequence alignment. This is often done because structural homology should be conserved more than sequence similarity, making it easier to infer than direct sequence homology. Very little discussion has been raised about whether structural homology and sequence homology need to be congruent; it seems logically possible to have structural homologs whose underlying sequences are not themselves homologous. How common this incongruence may be is very difficult to determine.

An extremely important issue to remember is that there is a fundamental difference between the biological and computational goals of alignment algorithms. The biological goal is the inference of homology. The computational goal is the (efficient) optimization of an objective function. Many investigators forget that the fact that a solution is computationally optimal does not mean it is biologically correct. This is a fundamental problem throughout computational biology, not just alignment. For example, phylogenetic methods attempt to find the tree that is most optimal based on a given criterion (such as parsimony, distance, or likelihood). It has been shown through simulation (Kumar 1996; Nei et al. 1998; Takahashi and Nei 2000; many unpublished studies) that the true tree is often not the most optimal tree for a given dataset, although this fact does not appear to be widely appreciated by the phylogenetics community at large. The same holds true for alignment.

CHALLENGES FOR THE FUTURE

Despite all of the progress that has been made in sequence alignment over the last four decades, there are still many challenges facing the sequence alignment community (both users and developers). In some sense, the primary purpose of this book is to highlight these challenges, and thus many of these challenges are discussed in detail throughout the remaining chapters. The following serves as a summary of some of these major challenges, as well as what to look for in the future.

- Generating objective functions (and parameters) that lead to the most biologically realistic homology predictions
- Generally improving methods for aligning highly diverged sequences
- Developing better methods for aligning larger data sets, both more and longer sequences, particularly at the scale of entire genomes
- Developing a better understanding of the molecular mechanisms leading to indel formation in the first place and find better ways to integrate these into context-dependent alignment
- Developing more realistic and computationally tractable models of indel mutation for use in statistical (maximum-likelihood or Bayesian) alignment and analysis
- Developing more efficient methods for statistical approaches to alignment so that they will become more useful and practical for larger data sets
- Further exploring the relationship between sequence homology and structural (or functional) morphology
- Developing better methods for the automatic incorporation of structural information into alignment

- Developing better methods for avoiding the circularity problem inherent in progressive multiple sequence alignment and phylogeny reconstruction (most likely by improving methods for simultaneous alignment and phylogeny recovery)
- Developing better and broader benchmarks for testing alignment algorithms, both in general and for specific alignment problems and situations
- Developing better approaches for comparing and contrasting alternate alignments
- Developing better methods for recognizing and displaying ambiguities in an alignment
- Developing better methods for incorporating alignment ambiguity into other analyses
- Exploring in much greater detail the effects of alignment error on downstream analyses in bioinformatics and genomics

CHAPTER 2

Insertion and Deletion Events, Their Molecular Mechanisms, and Their Impact on Sequence Alignments

LIAM J. MCGUFFIN
University of Reading, United Kingdom

Mutations..26
   Point Mutations.................................26
   Insertions and Deletions (Indels)..................27
Proposed Mechanisms of Indel Formation..............29
   Errors in DNA Replication........................30
   Transposons...................................30
   Alternative Splicing.............................32
   Unequal Crossover and Chromosomal Translocation..........33
Impact of Indels on Sequence Alignments..............35
   Global and Local Alignments.....................36
   Gap Penalties..................................36
   Improving Gap Penalties by Understanding the Context of Indel Events..........................37
Conclusion..38

The alignment of biological sequences allows us to infer the evolutionary relationships between different genes and proteins. Most new genes and proteins will evolve either through insertions or deletions of sets of subsequences, or through point mutations, where one amino acid is replaced with another. Therefore, we can judge the evolutionary

A) Yeast: FTKENVRILESWFAKNIENPYLDTKGLENLMKNTSLSRIQIKNWVSNRRRKEK
    Human: YSKGQLRELEREYAAN---KFITKDKRRKISAATSLSERQITIWFQNRRVKEK

B)

Figure 2.1. (A) A sequence alignment between the mat alpha2 Homeodomain protein from yeast (Protein Data Bank (PDB) code *1k61*) and Homeobox protein Hox-B13 from Human (PDB code *2cra*). (B) A structural alignment between *1k61* and *2cra* carried out using TM-align (Zhang and Skolnick 2005).

distance between related organisms by scoring the differences occurring between their protein and DNA sequences. In this chapter, we will focus on insertion and deletion events and explain how these events affect how we carry out and score sequence alignments.

We will begin by looking at a sequence alignment in order to illustrate the problem of handling insertions and deletions in related sequences. Figure 2.1A shows a local alignment carried out for similar proteins from two different eukaryotic organisms separated by more than a billion years of evolution.

It is apparent that despite the fact that these proteins have a similar DNA binding function and similar structure (Figure 2.1B), their respective amino acid sequences are quite different. Many mutations have occurred to produce the changes in sequences, most of which may be point mutations; however, the two proteins are also different in length, which means that a gap has been introduced into the alignment.

Given further evidence that the yeast sequence was closer to the common ancestral sequence, one might argue that a deletion event (or a number of deletion events) had probably occurred. However, without such information, we are unable to make assumptions about whether an insertion event had occurred in the yeast sequence or whether a deletion event had occurred in the human sequence. For this reason, when we carry out an alignment between sequences, we often treat insertion events and deletion events the same. Thus, insertions and deletions are often grouped together and collectively referred to as *indels*.

In fact, it could be said that all mutation events occurring to alter protein sequences could be explained by indel events. For example, the occurrence of point mutations may be accounted for by the deletion of one base or amino acid and the insertion of another. However, in the case of sequence alignments, the molecular mechanisms of how such an amino acid substitution occurs is less important. What is more important, for the scoring of protein sequence alignments in particular, is which type of substitution has occurred at a particular position in the sequence.

The effectiveness of the alignment of protein sequences can be improved by the use of an amino acid substitution matrix. A plethora of matrices has been developed, such as PAM (Dayhoff et al. 1978), GCB (Gonnet et al. 1992), JTT (Jones et al. 1992), BLOSUM62 (Henikoff and Henikoff 1992), and, more recently, OPTIMA (Kann et al. 2000), which are used to score the alignment of different pairs of amino acids with different weightings. These weightings account for the different physical, chemical, and structural properties shared by each pair of amino acids; a leucine–isoleucine match within an alignment may be scored higher than a leucine–tryptophan match. Similarly, when DNA sequences are aligned, there are various substitution models; for example, *transitions* (e.g., C-T, T-C, A-G, G-A) will often be weighted higher than *transversions* (e.g., T-A, A-T, C-G, G-C).

However, mutations considered as "true" indel events are those that cause changes in the lengths of biological sequences, and these have a more drastic individual effect on the scoring of sequence alignments. To account for the insertions or deletions of stretches of subsequences, gaps must be introduced into the alignment, and this incurs a penalty.

Traditionally, there are two types of penalty imposed when accounting for gaps within an alignment: gap opening penalties and gap extension penalties. The cost of opening a new gap is generally far weightier than the cost of extending an existing gap. However, the weighting that should be used for gap penalties often depends on the context and type of sequences.

There is still no perfect system for scoring biological sequence alignments. However, most of the systems that have been developed have attempted, in one way or another, to take into account the different types of mutation and the probability that they may occur, the inferred mechanisms involved, and their evolutionary consequences. Each scoring scheme will be appropriate for each varying situation. It is important to use the appropriate gap penalties as well as choosing an appropriate substitution matrix for the sequences you are aligning. In this chapter, we will review the different events that may lead to insertions and deletions in biological sequences and how these events help us to better understand sequence alignments and their scoring.

MUTATIONS

Before we can begin to explain the mechanisms of indel events, we must first review the different types of mutation that may occur at the DNA level and the resulting effects that these different types have on the translated protein sequence. Novel protein sequences will normally evolve through the mutation of existing sequences via indel events or point mutations. In order to become accepted, the mutations occurring within the sequence must be either neutral or advantageous.

If a mutation leads to an alteration in the protein structure that is sufficient to inactivate the protein, then it is unlikely that the mutation will become accepted. The vast majority of mutations in protein sequences are deleterious and are therefore eliminated through natural selection. However, in some cases, mutations can be neutral; for example, an amino acid substitution that does not alter the structure or function of the resulting protein and is neither damaging nor beneficial. In rare cases, a mutation may be advantageous and as a result may propagate throughout the population (Alberts et al. 1994).

Point Mutations

Point mutations occur when one nucleotide is exchanged for another. When these mutations occur within coding regions of DNA, they can be classified as one of three types: *silent, nonsense,* or *missense* (Weaver and Hedrick 1992).

Silent Mutations When a silent mutation occurs, the codon is altered such that it results in the translation of the same amino acid. Therefore, no effect is seen at the protein sequence level and such mutations can be seen as neutral.

Nonsense Mutations A nonsense mutation occurs when the base change results in the production of a stop codon. This leads to the translation of truncated proteins and is often deleterious. However, a nonsense mutation toward the beginning of the sequence will have a potentially more dramatic effect than one that occurs near the end of the sequence.

Missense Mutations In this case, the codon is altered such that it results in the translation of a different amino acid. These mutations can be deleterious, effectively neutral if they lead to no change in biological activity, or, in rare cases, advantageous.

Point accepted missense mutations are accounted for in protein sequence alignments by using mutation matrices. These matrices are essentially tables of scores that weigh changes in amino acids according to the differences in their physical, chemical, or structural properties.

Certain mutations that appear to be point mutations could be considered to have occurred through the deletion of one amino acid followed by the replacement of another. Intuitively, it would seem that these events will be far less likely than a true point mutation. True indel events have a greater effect on the gene products and can significantly alter the amino acid sequences. Often, in the case where a frameshift has occurred, indels may lead to a deleterious nonsense mutation.

Insertions and Deletions (Indels)

For the purpose of a sequence alignment, we do not differentiate between insertion and deletion events; however, there are subtle differences as to how they may occur and potentially major differences in their effects.

Simply, an insertion occurs when one or more nucleotides are added to the DNA sequence. A deletion occurs when one or more nucleotides are removed from the DNA sequence. Insertions and deletions can occur through errors during DNA replication followed by failure of repair, unequal crossover during recombination, or the introduction of transposable elements. Insertions and deletions observed in protein sequence alignments may also have occurred through alternative splicing, where different combinations of exons from the same gene are translated into different protein sequences.

Mutations caused by insertions of DNA subsequences can be reverted, or back-mutated, for example, through correctly functioning DNA repair mechanisms or through the excision of the transposable element. Similarly, a deletion may be reverted through the reintroduction of the lost subsequence. However, such reversions are highly unlikely to occur naturally (Weaver and Hedrick 1992). The mechanisms of indel events are discussed in more detail later in the chapter.

Frameshift Mutations Insertions and deletions of bases may cause alterations to the reading frame of the gene. Such alterations are known as *frameshift mutations,* and they may have a drastic effect on the coding regions. The insertion or deletion of one or two nucleotides will alter every codon downstream of the mutation, which can lead to translation of a different amino acid sequence. This may also lead to a nonsense mutation occurring downstream, which causes the premature termination of translation (Streisinger et al. 1966). Figure 2.2a illustrates how the insertion of a single base leads to a frameshift mutation.

Triplet indels may have milder consequences because the reading frame of the gene downstream of the mutation will be unaffected. Figure 2.2b illustrates how the insertion of a triplet maintains the reading frame downstream of the mutation. Multiple triplet indels may also occur, which may lead to severe consequences. For example, variable numbers of trinucleotide repeats occur in human genetic disorders such as Huntington's disease and Fragile X (Weaver and Hedrick 1992).

Splice Site Mutations In eukaryotic organisms the coding regions of DNA (*exons*) are interrupted by noncoding regions (*introns*). Prior to translation, the introns are cut out and exons are stitched back together in a process known as *RNA splicing.*

Indels can lead to the disruption of the specific sequences denoting the sites at which splicing takes place. These mutations can result in one or more introns remaining in the mature messenger RNA or lead to one or more exons being spliced out. This has severe implications for the resulting protein sequence.

Molecular Mechanisms of Indels

Original sequence

AAG	AAT	ATC	GAG	AAC	CCG	TAT	AGA	...
LYS	ASN	ILE	GLU	ASN	PRO	TYR	ARG	...

A) ↓

AAG	AAT	ATC	GAG	AAC	CCC	GTA	TAG	A..
LYS	ASN	ILE	GLU	ASN	PRO	VAL	Stop	...

B) ↓↓↓

AAG	AAT	ATC	GAG	AAC	CCC	CCG	TAT	AGA	...
LYS	ASN	ILE	GLU	ASN	PRO	PRO	TYR	ARG	...

Figure 2.2. Insertions leading to frameshift mutations. Original sequence: A hypothetical DNA sequence is divided into codons with the amino acids translated below. (A) Insertion of a single cytosine nucleotide (in bold) leading to a frameshift and nonsense mutation. (B) Insertion of three cytosine nucleotides leading to additional proline inserted into the protein sequence. The reading frame downstream of the mutation is preserved.

Alternative splicing can also be a mechanism explaining the occurrence of insertions and deletions in eukaryotic proteins. The mechanisms of alternative splicing are discussed in later in the chapter.

PROPOSED MECHANISMS OF INDEL FORMATION

There are many different molecular events that can lead to the formation of indels in biological sequences. Errors can occur during DNA replication; these may be missed during proofreading by polymerases and DNA repair mechanisms. Alternatively, insertions can occur in DNA due to transposable genetic elements such as simple insertion sequences and more complex transposons and retroviruses. At the RNA level, alternative splicing of genes can give rise to many different protein sequences encoded from various combinations of exons. Larger insertions and deletions may also be explained by DNA recombination mechanisms occurring during meiosis.

Errors in DNA Replication

The main molecular mechanism proposed that causes frameshift mutations at the DNA level involves errors occurring within the replication machinery (Levinson and Gutman 1987). When a gene is inaccurately copied, one or many base pairs may be introduced or omitted from the original sequence.

Streisinger and Owen (1985) developed a model for the mechanisms of frameshift mutations; the idea for the model arose from their studies on the lysozyme gene sequences in bacteriophage T4. The mutations were found to occur most frequently in regions of repeated sequences, and their model provided an explanation for this observation.

In the model, the insertion or deletion of bases is explained by the slippage of one strand occurring during DNA synthesis, which creates a loop of one or many bases. The loop is stabilized by pairing that occurs at an alternative position within the repetitive sequence. An insertion occurs when the loop is in the newly synthesized strand, and a deletion occurs when the loop is in the template strand. Figure 2.3 shows a simplified view of Streisinger and Owen's model and illustrates how the DNA replication machinery can "slip a cog" occasionally.

If the proofreading stages and the DNA repair mechanisms fail to correct the insertion or deletion of a subsequence, then it will remain in the sequence. In some cases, after slippage of the replication machinery has caused a frameshift, the DNA repair mechanism may attempt to revert the frameshift by inserting or deleting several extra bases, thereby incorporating new insertions or deletions (Levinson and Gutman 1987).

Transposons

Transposons are mobile genetic elements that move around the genome of an organism, causing both insertion and deletion events in DNA sequences. The discovery of transposable elements by Barbara McClintock, for which she won a Nobel Prize in the early 1980s, has had a major impact on our understanding of genetics. Transposable elements are considered to be a major contributor to natural genetic indel variation among humans (Mills, Bennett, et al. 2006). In fact, it is estimated that up to approximately 44% of the human genome is made

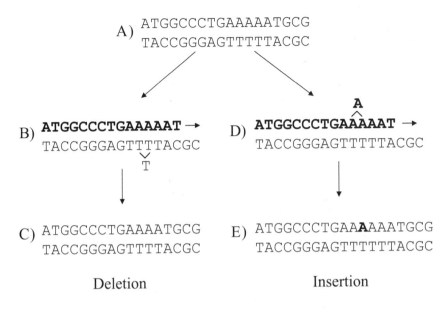

Figure 2.3. Model for frameshift mutation proposed by Streisinger and Owen (1985). (A) Original DNA duplex with adenine repeat. (B) Slippage occurs during replication causing the template strand to loop out. (C) Single-base deletion occurs after another round of replication. (D) Newly synthesized strand loops out during replication. (E) Single-base insertion occurs after another round of replication. Figure adapted from Weaver and Hedrick (1992).

up of transposons or remnants of transposons (International Human Genome Sequencing Consortium 2001). Transposons are a major cause of mutations and variation in the amount of DNA within a genome.

Transposons are generally classified into two categories based on the mechanisms of their transposition. First, conservative or simple transposition occurs when both strands of DNA are conserved as they move from place to place; that is, the DNA is "cut" from one location and "pasted" into another. The second mechanism is replicative transposition, where the DNA is "copied" from one location and "pasted" to another. All transposons generate direct repeats within the DNA sequence at their point of insertion (Weaver and Hedrick 1992).

Class I Transposons Class I transposons, which are also known as retrotransposons, generally use a replicative "copy and paste" mechanism. An RNA copy of the transposon is made, which is then reverse transcribed into DNA and inserted back into the genome. Retrotransposons use a mechanism similar to that of retroviruses (such as HIV) and are very probably their evolutionary ancestors. The retrotransposon often carries the gene encoding the reverse transcriptase enzyme required to facilitate transposition. Long interspersed nuclear elements (LINEs) and short interspersed nuclear elements (SINEs) are highly abundant types of retrotransposons found in mammalian genomes (Weiner 2000).

Class II Transposons Class II transposons mainly transpose via a "cut and paste" mechanism. This requires an enzyme known as a *transposase*, which is often encoded within the transposon itself. In general, the transposase acts by binding to DNA and making a staggered cut at the target site to produce sticky ends. The transposon is then cut out and ligated into the target site, leaving gaps at each end, which are then filled by a DNA polymerase. Some Class II transposons may also transpose using a "copy and paste" mechanism. The two proposed mechanisms for transposition of Class II transposons are shown in Figure 2.4.

Alternative Splicing

In eukaryotes, alternative splicing is an important mechanism that allows for the production of many different protein sequences from the same gene. This is achieved through splicing of different combinations of exons (Breitbart et al. 1987). Using this process, eukaryotic organisms can achieve more efficient data storage at the DNA level, and theoretically faster evolution of new proteins.

Alternative splicing either involves the substitution of one subsequence encoded by an exon for another, which is known as substitution alternative splicing, or results in the insertion or deletion of a subsequence, which is known as length-dependent alternative splicing (Kondrashov and Koonin 2003). Figure 2.5 illustrates how length-dependent alternative splicing may lead to the insertion or deletion of a new subsequence within a protein.

Computational studies have been carried out to investigate the structural and functional influences of alternative splicing leading to a potentially vast repertoire of proteins. Alternative splicing is thought to be a key process in modulating gene function and influencing protein networks through the generation of structures with many varying conformations (Yura et al. 2006).

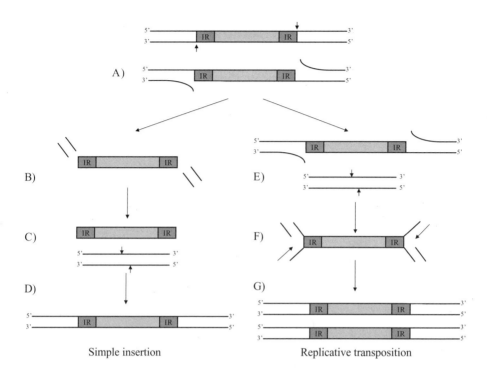

Figure 2.4. Mechanisms of transposition adapted from Tavakoli and Derbyshire (2001). (A) Transposable element with inverted repeat regions at either end. The first stage involves nicking occurring at the 3′ ends. (B) With simple insertion, the transposon is cut from the original location and pasted into the new location; this requires cleavage at the 5′ ends. (C) The transposon interacts with the target sequence which has a staggered cut. (D) The transposon is inserted into the new location. (E) With replicative transposition, the transposon is copied and pasted, which occurs via strand transfer. The 5′ ends of the transposon remain intact, and the free ends interact with the nicked target sequence. (F) The free ends in the target sequence serve as primers for DNA replication. (G) The result is a cointegrate structure and replication of the transposon.

Unequal Crossover and Chromosomal Translocation

Recombination events can create indels at the chromosomal level. If sister chromatids misalign during meiosis, for example, due to repeated regions within the sequence, unequal crossover may occur during recombination. This can lead to the duplication or deletion of thousands of base pairs within chromosomes. Figure 2.6 illustrates unequal crossover leading to an insertion in one chromosome and a deletion in the other.

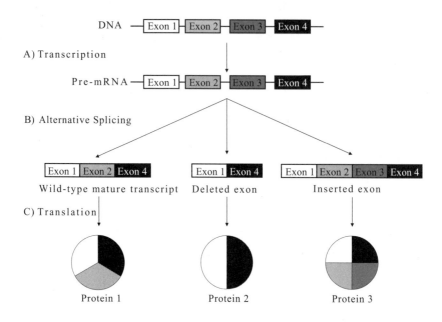

Figure 2.5. Length-dependent alternative splicing. (A) Pre-mRNA is transcribed from the DNA template including introns and several possible exons. (B) Alternative splicing leads to alternative mature mRNA molecules made up of different combinations of exons. (C) In this hypothetical example, the wild-type protein 1 is translated from exons 1, 2, and 4. Deletions and insertions may occur due to alternative splicing, which will lead to different protein sequences, such as protein 2, where exon 2 has been deleted, and protein 3, where exon 3 has been inserted.

Unequal crossover during meiosis leading to chromosomal insertions and deletions has been shown to be an important factor in disease. For example, neurofibromatosis is caused by microdeletions, a result of a homologous recombination between misaligned repeat regions (Lopez-Correa et al. 2000). In addition, the CATCH-22 group of syndromes and Williams–Beuren syndrome are correlated with high levels of unequal meiotic crossover occurring at the site of chromosomal deletions (Baumer et al. 1998).

Indels caused by unequal crossover lead to variation in microsatellite and minisatellite repeat regions. Variable number of tandem repeats (VNTRs) are useful for gauging genetic variation among populations and are widely exploited for DNA fingerprinting (Harding et al. 1992).

Molecular Mechanisms of Indels   35

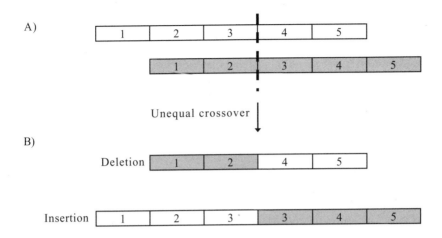

Figure 2.6. Unequal crossover during recombination leading to insertions and deletions, adapted from Alkan et al. (2002). (A) Sister chromatids misalign during meiosis due to repeat units. The site of crossover is indicated by the dashed line. (B) Strand breaks on sister chromatids lead to different numbers of repeat units. A deletion occurs in one chromatid and an insertion occurs in the other.

Indels at the chromosome level may also occur due to chromosomal breaks followed by translocation events. Several widely studied diseases in humans are known to be caused by chromosomal translocations. For example, translocation of the long arm of human chromosome 21 to chromosome 14 can lead to Down syndrome (Weaver and Hedrick 1992).

IMPACT OF INDELS ON SEQUENCE ALIGNMENTS

Whatever their scale and underlying mechanisms, insertions and deletions are indicated by gaps in sequence alignments. These gaps pose a significant problem and greatly affect the accuracy of the alignment. The way we treat the scoring of these gaps is therefore crucial to effective sequence alignment and is dependent on the context of the insertion or deletions. The penalty incurred by a gap should reflect the expected frequency of the indel event. Therefore, it is crucial to know the type of sequence in which the gap is occurring and the likelihood that a gap will occur in that context. For example, a large gap in an important protein that causes a major conformational change may be less likely to occur than indel events leading to VNTRs.

Global and Local Alignments

Perhaps the most widely used algorithm for alignment of two sequences is dynamic programming. Needleman and Wunsch (1970) were the first to describe the use of dynamic programming to perform a global alignment between pairs of biological sequences. Smith and Waterman (1981b) later refined the algorithm to allow for optimal local alignments; the refined algorithm proved to be more accurate for aligning repetitive regions in sequences. (See Chapter 3 for further discussion about global and local alignments.)

Dynamic programming is a computationally efficient way of optimally aligning sequences and is dependent on a scoring scheme that is based on both the substitution of bases or amino acids and the creation and extension of gaps. Methods to improve sequence alignments have often focused on the optimization of substitution matrices and gap penalties.

Gap Penalties

Most programs will implement a cost for introducing a gap when aligning two biological sequences. This cost contributes to the overall score of the alignment and is often closely dependent on the substitution matrix used. For example, when determining the cost of introducing a gap, we must also consider the costs associated with aligning dissimilar bases or amino acids. The placement of gaps in a sequence alignment is a difficult task, and a number of solutions to the problem have been proposed.

Affine Versus Linear Gap Penalties The most basic gap penalty is the linear gap penalty, where the cost of introducing a new gap in a sequence is the same as the cost of extending an existing gap. For example, the introduction of three gaps each of one amino acid in length is seen as equally likely as the introduction of one large gap of three amino acids in length and therefore incurs the same cost.

The affine gap penalty, however, differentiates between the opening of a gap and extension of a gap. The opening of a new gap is considered to be of greater cost than extending the gap. This is based on the conjecture that one long gap cause by a single indel event is more likely to occur than several shorter gaps caused by several different indel events.

A) GATCGCGCGCGCGCATGC
    GATC--G--C--G--CATGC

B) GATCGCGCGCGCGCATGC
    GATCGCG--------CATGC

Figure 2.7. Affine gap penalties versus linear gap penalties. (A) Alignment of two hypothetical DNA sequences. Using a linear penalty, all gaps are scored equally, which could result in a "gappy" alignment. (B) The affine gap penalty scheme involves two scores: one for opening a gap and another for extending the gap. The cost of opening a new gap is higher than that for extending a gap, which results in a more accurate alignment. A single indel event indicated by a single long gap is assumed to be more likely than a number of separate events indicated by a series of smaller gaps.

Thus, according to the affine gap penalty scheme, gaps are scored using the formula $o + e \times l$, where o is the cost for opening a gap, e is the cost for extending the gap, and l is the length of the gap. Figure 2.7 shows a hypothetical example highlighting the advantage of the affine gap penalty scheme.

The parameters representing the costs of affine gap penalties were rigorously optimized by Barton and Sternberg (1987a). Since then several groups have attempted to further improve or generalize the scoring of the affine gap penalty (Altschul 1998; Mott 1999). Despite being an obvious improvement over a linear gap penalty, the affine gap penalty is still an imprecise treatment of real gaps. It has been criticized for having no sound theoretical basis and no real supporting experimental evidence (Goonesekere and Lee 2004).

Improving Gap Penalties by Understanding the Context of Indel Events

The context and mechanisms of indel events have been recognized as important factors for determining optimal gap penalty functions. There have been many attempts to deduce gap penalties empirically based on observation of patterns of indel in aligned sequences, such as the studies carried out by Benner et al. (1993), Reese and Pearson (2002), and Chang and Benner (2004).

The gap penalties may be manually adjusted when using common sequence alignment tools, such as ClustalW (Thompson et al. 1994), to account for rules determined by protein structure, for example. Indels may have a greater impact on the protein structure, whereas point mutations usually have less effect. It may be preferable to allow gaps only in coil regions, which are often more variable, rather than introduce a gap in a strand or helix.

Recent work on the production of improved gap penalties for homology modeling has focused on the structural context. Goonesekere and Lee (2004) suggest a simple modification of the affine gap penalty based on their observations of the patterns of gaps occurring in a database of structurally aligned proteins. More recently, Madhusudhan et al. (2006) have benchmarked a new gap penalty, which automatically varies depending on the structural context, against the standard affine gap penalty scheme. This novel variable gap penalty is reported to significantly increase the number of correctly aligned residues.

CONCLUSION

It is important to understand the context of insertions and deletions and their mechanisms when carrying out sequence alignments. Some indel events can cause major variations in sequences, which will have a great effect on the inferred relationship. Other events may have a less severe impact on the variation and should be taken into account when the sequence alignment is scored. Thus, gap penalties should be dependent on where the gap occurs, and the scoring should be appropriate to the situation in which the indel has arisen. Gap penalties incurred for one sequence alignment may not be appropriate for another.

We have begun to develop different scoring schemes for sequence alignments, which take into account the context of indels. However, the scoring of sequence alignments can be further improved, perhaps through our improved understanding of the mechanisms causing indel events.

CHAPTER 3

Local versus Global Alignments

BURKHARD MORGENSTERN
University of Göttingen, Germany

Global Pairwise Alignment: The Basic Approach 40
Global Multiple Alignment. 42
Local Pairwise Alignment: The Basic Concept. 43
Optimal Local Pairwise Alignment by Dynamic Programming 45
Local Multiple Alignment. 46
Combining Global and Local Multiple Alignment. 47
Alignment of Genomic Sequences 52

For a given set of input sequences, the overall goal of pairwise and multiple sequence alignment is to identify those parts of the sequences that are related to each other by common structure, function, or evolution. As with other bioinformatics approaches, computational methods for sequence alignment have to make a number of *a priori* assumptions on the data to be analyzed, either explicitly or implicitly. For example, a basic assumption made by almost all alignment methods is that homologies between the input sequences, if there are any at all, appear in the same relative order in all sequences. Obviously, if two regions A and B in one of the input sequences are related to regions A' and B', respectively, in a second input sequence, and A is to the left of B in the first sequence, a simultaneous alignment of both homologies, A,A' and B,B' is possible only if A' appears on the left-hand side of B' in the second sequence. This *consistency* or *colinearity* condition is a fundamental assumption and a limitation of most alignment methods. On the other hand, this restriction can help to find weak homologies. A region of weak similarity between two sequences may not appear to be statistically significant if considered out of context, but it may well be significant if it is consistent with the constraints given by other homologies among the input sequences. Only recently, some alignment programs have been proposed that do not assume that homologies appear in the same order within the input sequences (Brudno, Malder, et al. 2003; Darling et al. 2004; Didier and Guziolowski 2007; Phuong et al. 2006; Raphael et al. 2004). Such algorithms are particularly important if large genomic sequences are aligned, since here rearrangements have to be taken into account.

Another *a priori* assumption made by most alignment programs concerns the existence and extent of detectable similarities among the sequences under study. In the simplest case, pairwise and multiple alignment programs make three *a priori* assumptions on the sequences under study: (*a*) that all input sequences are related by common structure, function, and evolution; (*b*) that their biological relatedness corresponds to detectable similarity at the primary sequence level; and (*c*) that similarity extends over the entire length of the input sequences. Under these three conditions, homologies in a set of input sequences can be properly represented by a global alignment, that is, in an alignment that extends from the beginning to the end of the sequences. In principle, it is possible to calculate a reasonable alignment automatically, solely based on primary sequence information. A basic task in comparative sequence analysis is therefore to calculate a global alignment for a set of two nucleic acid or protein sequences.

GLOBAL PAIRWISE ALIGNMENT: THE BASIC APPROACH

As with many problems in computational biology, the problem to calculate a global alignment of two sequences consists of two parts: an *objective function* assigns a *quality score* to every possible alignment of the two input sequences, and an *optimization algorithm* calculates the best possible alignment according to the given objective function. The standard objective function for global pairwise alignment is defined as follows: each pair (a, b) of nucleotides or amino acids is assigned a similarity score $s(a, b)$, and a penalty $\gamma(l)$ is imposed for a gap of length l in an alignment. The score of a given alignment is then defined as the sum of the similarity scores of the aligned residues minus the corresponding penalty for every gap. The simplest type of gap penalties are linear gap penalties, where a gap of length l gets a penalty of $\gamma(l) = g \times l$, where g is some constant value. This can be interpreted as penalizing every single gap character "-" by g, so the overall score of a pairwise alignment can be expressed as the sum of scores for each column of the alignment. This simplifies the optimization algorithm explained below.

A technique called *dynamic programming* can then be used to calculate an optimal alignment efficiently (Needleman and Wunsch 1970; Sellers 1974); see also Durbin et al. (1998) and Chapter 1 of this book. Dynamic programming is widely used in bioinformatics, for example, for RNA secondary structure prediction (Nussinov and Jacobson 1980; Sankoff 1985), decoding in hidden Markov models (Viterbi 1967), gene prediction (Burge and Karlin 1997), or phylogeny reconstruction (Felsenstein 1981; Fitch 1971; Sankoff and Cedergren 1983). In dynamic programming, a complex optimization problem is broken down into smaller subproblems that can be dealt with one by one in such a way that the results for smaller subproblems are used to deal with larger subproblems. More precisely, for each of the subproblems, the score of an optimal solution is calculated based on the previously calculated scores of smaller subproblems until the score of the original problem is known. In a final step, the so-called *trace-back* procedure, an optimal solution of the original problem itself is calculated.

The dynamic programming concept is applied to pairwise global sequence alignment in the following way: given two input sequences $X = X_1...X_m$ and $Y = Y_1...Y_n$ of length m and n, respectively, the subproblems are optimal alignment of the prefixes $X_1...X_i$ and $Y_1...Y_j$, where i and j have values between 0 and m or n. Generally, a prefix i of a sequence or string X is the sequence that consists of the first i characters. This includes the special cases $i = 0$ (the empty prefix) and the full sequence X. For increasing values of i and j, the score $F(i, j)$ of an optimal alignment of the prefixes of length i and j, respectively, is calculated. If either i or j (or both) are 0, $F(i, j)$ is trivial to calculate, because in this case there is only one possible alignment consisting only of gaps. For larger values of i and j, the following recursion formula can be used, provided that linear gap penalties are used:

$$F(i, j) = \max \begin{cases} F(i-1, j-1) + s(X_i, Y_j) \\ F(i-1, j) - g \\ F(i, j-1) - g \end{cases} \quad (1)$$

The recursion holds, since for linear gap penalties, the score of an alignment can be calculated as the sum of the scores of the alignment columns. For the last column, there are three possibilities: the last residues of the prefixes X_i and Y_j are aligned, there is gap in the first sequence, or there is a gap in the second sequence. Thus, the score of the last column is either $s(X_i, Y_j)$ or $-g$, and the score of the remaining alignment columns is $F(i, j)$, $F(i-1, j)$, or $F(i, j-1)$, depending on which one of the three alternatives is true. Finally, $F(m, n)$ is the score of an optimal alignment of the input sequences X and Y. An optimal alignment of X and Y can be calculated by dynamic programming in $O(m \times n)$ time and memory, that is, the computing time and memory required to calculate an optimal alignment are proportional to $m \times n$. For arbitrary gap penalties, the complexity is $O(m \times n \times \min\{m, n\})$, or $O(L^3)$ if L is the maximum of m and n. Instead of similarity scores $s(a, b)$ for residue pairs (a, b), it is also possible to use distance scores (Sellers 1974). Smith et al. (1981) have shown that these two scoring schemes are equivalent. The dynamic programming algorithm for pairwise global alignment is known as the *Needleman–Wunsch algorithm*. Interestingly, the same algorithm was developed independently by Wagner and Fisher (1974).

GLOBAL MULTIPLE ALIGNMENT

In theory, the above-outlined objective function and the corresponding dynamic programming algorithm for pairwise alignment can be generalized to align $N > 2$ sequences simultaneously. This would require $O(L^N)$ time and memory, however; so calculating an exact multiple alignment is feasible only for very small sets of sequences. Instead, heuristics are used to calculate optimal multiple alignments, for example, progressive alignment algorithms (Corpet 1988; Feng and Doolittle 1987; Higgins and Sharp 1988; Taylor 1988). Here, a so-called *guide tree* is calculated in a first step to cluster the input sequences based on their degree of pairwise similarity. Sequences and clusters of sequences are aligned two by two until all of the input sequences are aligned in one single multiple alignment. The most time-consuming step in this procedure is finding a good guide tree. Recently developed programs for global multiple alignments use efficient heuristics to speed up this step. This way the computing time for multiple sequence alignment can be dramatically improved (Edgar 2004b; Katoh et al. 2005; Katoh et al. 2002). Other methods use iterative refinement strategy to improve the quality of the output alignments (Gotoh 1996). Another recent development in global multiple alignment is the program *ProbCons* (Brudno et al. 2003b; Do et al. 2005). This approach uses pair hidden Markov models (Durbin et al. 1998) to calculate optimal alignments based on a probabilistic sequence model.

LOCAL PAIRWISE ALIGNMENT: THE BASIC CONCEPT

In the previous section, we made the general assumption that two given input sequences are related over their entire length. This assumption is often wrong. A pair of sequences may be related by common structure, function, or evolution, but for various reasons similarity at the primary sequence level can be restricted to some local region—because the remainder of the sequences is not related at all, because homologies do not correspond to similarity at the primary sequence level, or because homologies are not colinear and cannot be represented in one single alignment.

In this situation, it makes no sense to calculate a global alignment as outlined earlier. At best, such a global alignment could contain a proper alignment of some true local homologies, surrounded by a meaningless alignment of the nonrelated parts of the sequences. In this case, the user would have to post-process the produced alignment to discriminate the meaningful from the meaningless sections of the alignment. However, application of a global alignment algorithm to locally related sequences can also result in a completely nonsensical alignment where only nonrelated parts of the sequences are aligned to each other, if an alignment of these nonrelated sections has a higher numerical score than an alternative biologically correct alignment of the related parts of the sequences. In particular, such misalignments are likely to happen in situations where the real homologies are at different relative positions within the two input sequences, for example, at the 5′ end in one DNA sequence and at the 3′ end of another sequence.

If no global similarity at the primary sequence level is given, a meaningful alignment has to be restricted to (possible) local homologies. That is, a meaningful alignment would align these homologies and ignore the nonrelated parts of the sequences. In the simplest case, the assumption is that two input sequences contain one single region of homology and are not related to each other outside this region. This is the situation where local pairwise alignment algorithms can be used. These algorithms align a pair of subregions of the input sequences and ignore the remainder of the sequences. Somewhat more formally, the local counterpart of the above-outlined global alignment problem can be stated as follows: given a pair of input sequences X and Y, find a subsequence X' of X and a subsequence Y' of Y and an optimal global alignment of X' and Y' such that the score of this alignment is maximal over all possible choices of X' and Y'. In this context, a subsequence of a sequence S is considered to be a contiguous substring of S; a subsequence may have any length. In the standard approach, the same scoring scheme is used as mentioned earlier, that is, the terms "optimal," "score,", and so forth, refer to the objective function that was explained earlier for global alignment. Note that X' and Y' are allowed to have length zero, in which case no local alignment at all is returned (or, formally spoken, the local alignment has length zero).

A certain difference between local and global alignment approaches is that for local alignment, only similarity-based scoring schemes are meaningful. An algorithm that tries to minimize a distance score between two sequences cannot be applied to local alignment because, in this case, the algorithm would always return the shortest possible alignment, that is, an alignment of length zero. Note also that with the above-explained scoring scheme, an alignment of sequences of length 0 has a score of 0. This implies that the similarity score of an optimal local alignment of two sequences can never be negative. There are two necessary conditions on the similarity matrix used to assess the similarity of two characters to ensure that the local alignment algorithm returns a biologically meaningful alignment. There should be at least one pair of characters with positive score, and the average similarity score for a pair of characters should be negative. The first condition is necessary to ensure that there can be a local alignment with length greater than zero. If all similarity values $s(a, b)$ were negative, any (local) alignment would have a negative score, and the best possible alignment would be of length zero. The second condition is necessary since, with a positive average similarity score for pairs of characters, any extension of a local alignment would increase the expected alignment score, even for completely unrelated sequences; so the algorithm would tend to align the entire input sequences instead of selecting locally conserved subsequences.

The local version of the alignment problem appears to be far more complex than the global version outlined in the previous section because we are looking for the best-scoring alignment among all possible pairs of subsequences X' and Y' of the input sequences. Thus, we have a double optimization problem, finding a pair of subsequences and finding an optimal alignment of these subsequences, these two problems are interconnected because selection of the subsequences S'_1 and S'_2 depends on their alignment score.

A *naive* or *brute-force* algorithm would solve this problem by evaluating all possible pairs of subsequences by calculating their respective optimal alignment. For a sequence of length L, there are $O(L^2)$ possible subsequences; so, for a pair of sequences of (maximum) length L, there are $O(L^2 \times L^2) = O(L^4)$ pairs of subsequences. Finding an optimal (global) alignment for a given pair of subsequences takes again $O(L^2)$ time, so the overall runtime of such a brute-force algorithm would be $O(L^6)$, which would be unfeasible even for sequences of moderate length.

One of the most important developments in bioinformatics is the algorithm by Smith and Waterman (1981a,b) for local pairwise alignment. This algorithm calculates an optimal local alignment in $O(L^2)$ time, that is, with exactly the same time complexity as the Needleman–Wunsch algorithm for pairwise global alignment. Interestingly, some small modifications of the Needleman–Wunsch algorithm are sufficient to solve the local alignment problem, which, at first glance, is far more complex than its global counterpart.

OPTIMAL LOCAL PAIRWISE ALIGNMENT BY DYNAMIC PROGRAMMING

Like the Needleman–Wunsch algorithm, the algorithm by Smith and Waterman considers prefixes $X_1...X_i$ and $Y_1...Y_j$ of the input sequences for growing values of i and j. For every (i, j), it calculates the score $F(i, j)$ of the best local alignment starting at some position i' and j' of the input sequences and ending with positions i and j. As with global alignment, the score of this optimal local alignment can be calculated using the corresponding values for smaller values of i and j. Again, there are three possibilities if the last column of the alignment is considered: alignment of the characters X_i and Y_j, a gap in sequence X, or a gap in sequence Y. Note, however, that a local alignment can have length zero, in which case its score is zero. So, we have to take a fourth situation into account where the best local alignment has a score of zero. If this is the best possible local alignment ending at positions i and j, we have $F(i, j) = 0$. Thus, in addition to the three possibilities that we have in global alignment, we have to consider this fourth case, and the recursion formula for local alignment becomes

$$F(i, j) = \max \begin{cases} F(i-1, j-1) + s(X_i, Y_j) \\ F(i-1, j) - g \\ F(i, j-1) - g \\ 0 \end{cases} \quad (2)$$

Again, this equation holds only for linear gap penalties; we do not consider other gap penalties in this chapter. As in global pairwise alignment, for two input sequences of length m and n, the score of their optimal local alignment can be calculated in $O(m \times n)$ time by calculating all values $F(i, j)$. However, because a local alignment may end anywhere in the input sequences, we have to keep track of the positions i_{max} and j_{max} where $F(i, j)$ is maximal. The score of the optimal local alignment of X and Y is then given as $F(i_{max}, j_{max})$. For the trace-back procedure that retrieves the optimal local alignment, we have to start at positions i_{max} and j_{max}, rather than at m and n as in the global algorithm, and the trace back will then continue until a position (i', j') with $F(i', j') = 0$ is reached.

The most important application of local pairwise sequence alignment is database searching. If a database is searched for homologues to a query sequence Q, local similarities are usually sufficient to infer homology. The Smith–Waterman algorithm has been shown to perform well when used as a database search tool. However, although this algorithm is fast enough to align two sequences of moderate length, it is too slow to search large databases. Heuristic approaches such as BLAST (Altschul et al. 1990) or FASTA (Pearson and Lipman 1988) are therefore usually employed to search sequence databases. These tools do not attempt to find optimal local alignments in the sense of Smith and Waterman; rather, they make a number of simplifying assumptions to increase computational efficiency. Advanced database tools usually include local multiple sequence alignments (Altschul et al. 1990; Krogh et al. 1994; Park et al. 1998).

LOCAL MULTIPLE ALIGNMENT

As with global alignment, the concept of local pairwise alignment can be generalized to multiple alignment of $N > 2$ sequences. In principle, the standard progressive approach to multiple global alignment could also be applied to multiple local alignment. A certain difficulty of such an approach is the fact that pairwise optimal local alignments from different sequence pairs will not necessarily have the same length; we cannot even expect these local alignments to overlap with each other. Therefore, most methods for local multiple alignment take a different approach. Some algorithms search for a gap-free local multiple alignment of a predefined length L such that a certain alignment score is maximized. The first local alignment method of this type was proposed by Waterman (1986). Here, a local alignment with a certain maximum number of mismatches is searched. Other popular methods for gap-free multiple alignment are the Gibbs sampling approach (Smith and Smith 1990) and the program MEME (Bailey and Elkan 1994). These methods use efficient heuristics to maximize a score based on probabilistic considerations. A straightforward extension of this approach is to search for several local multiple alignments in a given input data set. This is done, for example, by the program MatchBox (Depiereux et al. 1997; Depiereux and Feytmans 1992). If several local multiple alignments are to be returned, one often requires them to form a chain.

COMBINING GLOBAL AND LOCAL MULTIPLE ALIGNMENT

The above-outlined approaches for local multiple alignment have been successfully used to identify biologically functional motifs in nucleic acid and protein sequences. However, their general applicability is rather limited because they make two very restrictive assumptions: (*a*) they assume that local homologies are present (and detectable) in all of the input sequences in a given data set, and (*b*) they assume that these homologies have the same length in all of the sequences; their length has often to be specified by the user. Another limitation of traditional (global or local) alignment approaches is the fact that, for a set of N unknown protein or DNA sequences, we can usually not know in advance if they are globally or locally related. Moreover, in a multiple sequence set, some of the sequences could be globally related to each other while other sequences may share one or several local similarities or no similarity at all. A flexible alignment tool should therefore be able to align those regions of the input sequences that share some statistically significant similarity at the primary sequence level, but it should not try to align the nonrelated remainder of the sequences. Thus, such a tool should return a local or global alignment—or no alignment at all—depending on the extent of detectable similarity among the input sequences.

The program DIALIGN (Morgenstern 2004; Morgenstern et al. 1996) has been designed as a combination of local and global alignment approaches. It constructs pairwise or multiple alignments by using gap-free local pairwise alignments as elementary alignments that are used to assemble a resulting pairwise or multiple alignment. Gap-free pairwise local alignments are called *fragment alignments* or *fragments*; in earlier publications, they were referred to as *diagonals*. Thus, a fragment is a pair of segments of the same length (although the two segments may have arbitrary length). Fragments are allowed to overlap in a resulting alignment as long as they involve different sequences, but not if they are from the same pair of sequences.

The objective function in this approach is defined as follows: each possible fragment, that is, each pair of equal-length segments from two of the input sequences, receives a score that is defined by the probability of finding such a fragment by chance. The lower this probability is, the higher the fragment score (see Morgenstern 1999 for details). The score of an alignment is then defined as the sum of scores of the fragments it is composed of. The corresponding optimization problem is to find a collection of fragments with the highest score that is consistent in the sense that all fragments can be included in one resulting (pairwise or multiple) alignment.

For pairwise alignment, an optimal alignment in this approach is a chain of fragments with maximum total score. Such a chain can be calculated efficiently by standard algorithms; in DIALIGN, a space-efficient version of these algorithms is used (Morgenstern 2000, 2002). As in the standard alignment approach, calculating an optimal alignment of $N > 2$ sequences would be computationally feasible only for very small values of N. Thus, a greedy heuristic is used to find near-optimal alignments. In a first step, optimal pairwise alignments are calculated for all possible pairs of input sequences. All fragments contained in these optimal pairwise alignments are then sorted according to their scores (and according to the degree of overlap among them) and included one by one into a growing set of consistent fragments. This is repeated iteratively: in a second round, optimal pairwise alignments are constructed under the consistency constraints imposed by those fragments that have been accepted in the first round, and so forth, until no additional fragments can be found.

The main advantage of DIALIGN, compared with other local or global alignment methods, is that it tries to align all (local) similarities that are statistically significant and consistent which each other, but it tries not to align those parts of the sequences that share no observable similarity at the primary sequence level. In particular, local homologies are not required to involve all of the input sequences. As a result, the program can produce global alignments where similarity among the input sequences extends over their entire length. However, if only local similarity is detectable, the output alignment will be restricted to those similarities. Finally, the program can produce combinations of local and global alignments where locally aligned homologies may be interrupted by nonaligned parts of the sequences or where some of the input sequences are globally aligned to each other, whereas others are aligned only locally or not aligned at all. A reimplementation of DIALIGN, called *DIALIGN-T*, uses a number of algorithmic improvements in the multiple alignment procedure (Subramanian et al. 2005). To process large data sets, Dialign has been parallelized (Schmollinger et al. 2004). Another recent development is the option to utilize user-defined anchor points where the program can be forced to align regions specified by the user (Morgenstern et al. 2004).

Another program combining local and global alignment features is Notredame's T-Coffee (Notredame et al. 2000). In a first step, this method constructs a primary library of aligned residue pairs. These aligned residue pairs can come from arbitrary input alignments. The default version of the program uses one global alignment for each sequence pair calculated using ClustalW (Thompson et al. 1994) and ten local alignments calculated with SIM (Huang and Miller 1991). In a second step, these aligned residue pairs are compared to each other in order to find pairwise similarities that support each other, that is, similarities involving more than two sequences. Based on such overlapping pairwise similarities, a secondary library of sequence similarities is constructed. In a final step, T-Coffee uses the traditional progressive approach to construct a multiple alignment of the input sequences based on the similarities in the secondary library.

T-Coffee is now one of the most successful methods for multiple sequence alignment. Benchmark studies have shown that it performs very well on globally related sequence sets. Here it is often superior to traditional progressive approaches such as ClustalW (Thompson et al. 1994), but it also works well on locally related data sets, as long as there is sufficiently strong local similarity at the primary sequence level. If weakly conserved local motifs in otherwise unrelated sequences are

CLUSTAL W Alignment
```
seq1  -IYWTDVRIPKCAHNAIETGDNEQLNHHCCVQGWTVRHDRDWVEFYNLWRPDDLALHCEQ
seq2  NIYSDEP----CWQVESLWPSILAPDLECVEYWWMSEHQWQRDSCMQKLN-NDIGFGEPT
seq3  RNFAMDLD---GKTTVDRWKVECYSSHRYGRQLPHKNHIPEYQWCTVPANQDSKFLKMQD

seq1  WIWFVIND--TERHGTRVMTNDSDIILSVVEKAPGCTMNQKKW--
seq2  WPSPNEYSRCTHPHLVVGMVTQYAIELRLVVRVDVDQYPPLPETI
seq3  YYHSDYTNEAHQKTCKLGCGLWPCSEWWPIVHTRNTLWVCQFF--
```

T-COFFEE Alignment
```
seq1  IYWTDVRIPKCAHNAIETGDNEQLNHHCCVQGWTVRHDRDWVEFYNLWRPDDLA--LHCE
seq3  ---RNFAMD--------------LDGKTTV-------DRWKVECYSSHR---YGRQLPHK
seq2  ----------------------NIYSDE-------PCWQVE--SLW-PSILAPDLECV

seq1  ------QWIWFVINDTE------------------------R----HGTRVMTN----
seq3  NHIPEYQW--CTVPANQ-DSKFLKMQD---------YYHSDYTN----EAHQKTCKLGCG
seq2  ------EYWWMSEHQWQRDSCMQKLNNDIGFGEPTWPSPNEYSRCTHPHLVVGMVT----

seq1  -DSDIILSVVEKAPGCTMNQK---KW
seq3  LWPCSEWWPIVHTRNTLWVCQ---FF
seq2  -QYAIELRLVVRVDVDQYPPLPETI-
```

DIALIGN Alignment
```
seq1  --IY------  ----------  ----------  ----------  ----------
seq2  -NIYSDepc-  -----WQVE-  ----------  ----------  ----------
seq3  rNFAMDldgk  ttvdrWKVEc  ysshrygrql  phknhipeyq  wctvpanqds

seq1  ----------  ----------  --------WT  DVRIPkcahn  aietgdneql
seq2  ----------  ----------  ------SLWP  SILAPdle--  ----------
seq3  kflkmqdyyh  sdytneahqk  tcklgcGLWP  ----------  ----------

seq1  nhhcCVQ-GW  ----------  ----------  --------TV  RHDRD--WVe
seq2  ----CVEyWW  msehqwqrds  cmqklnndig  fgeptwpsPN  EYSR------
seq3  ----CSE-WW  ----------  ----------  --------PI  VHTRNtlWV-

seq1  fynlwrpddl  alhCEqwiwf  vindterhgt  rvmtndsdii  lsvvekapgC
seq2  ----------  ----------  ----------  ----------  ---------C
seq3  ----------  ---CQff---  ----------  ----------  ----------

seq1  Tmnqkkw---  ----------  ----------  -----
seq2  Thphlvvgmv  tqyaielrlv  vrvdvdqypp  lpeti
seq3  ----------  ----------  ----------  -----
```

Figure 3.1. Specificity of global and local methods for multiple alignment. A set of three totally unrelated random sequences has been aligned using a method for global alignment (ClustalW), and two methods that combine global and local alignment features, one with a stronger tendency toward global alignment (T-Coffee) and one with a tendency to produce local alignments (DIALIGN-T). ClustalW aligns the sequences over their entire length with only a few small gaps, although no significant similarity can be detected. T-Coffee introduces large gaps, so it becomes clear that no global similarity can be found. DIALIGN-T aligns only small parts of the sequences (in upper-case letters) and leaves most of the sequences unaligned (in lower-case letters).

to be found, however, DIALIGN seems to be superior to T-Coffee. A particularly attractive feature of T-Coffee is that arbitrary user-defined (local or global) alignments can be utilized as input for the primary library. This feature has been used to include structure information for improved alignment of RNA or protein sequences (Siebert and Backofen 2005). A recent version of T-Coffee, called 3DCoffee, includes structural information for improved multiple protein alignment (Poirot et al. 2005).

Alignment methods are usually evaluated and compared to each other based on benchmark sequence sets for which a "true" alignment is known. These benchmark data are either real-world sequences, for example, protein sequences with known 3D structure, or simulated artificial sequences. The most important benchmark database for multiple protein alignment is *BAliBASE* (Thompson, Koehl, et al. 2005; Thompson et al. 1999a); this database consists mainly of globally related sequences. Other databases for global multiple protein alignment are *PREFAB* (Edgar 2004b) and *OXBench* (Raghava et al. 2003). Benchmark databases for local multiple protein alignment are *IRMBASE* (Subramanian et al. 2005) and a database compiled by Lassmann and Sonnhammer (Lassmann and Sonnhammer 2002). Other benchmark are available for alignment of genomic sequences (Pollard et al. 2004) and for RNA sequences (Wilm et al. 2006).

So far, all benchmark studies for sequence alignment have evaluated the sensitivity of the commonly used methods, that is, their ability to correctly align regions of real or simulated sequence homology. Different measures are used to evaluate how accurately these regions are aligned, for example, column scores or sum-of-pairs scores. For global alignment tools, this approach is clearly appropriate. However, for local or "mixed" global-local alignment methods, it is also important to know what these methods do with sequences or parts of the sequences that are not homologous to any other sequences or parts of sequences in the input data set. In other words, for local methods, it would be important to evaluate their specificity in addition to measuring their sensitivity. An ideal alignment program should align those parts of the input sequences that are biologically related but should not align any nonrelated parts of the sequences. An example is a set of totally nonrelated random sequences. In this case, a meaningful alignment would not align these sequences to each other at all; that is, the alignment would consist of large gaps only.

To my knowledge, no systematic study has been carried out so far on the specificity of multiple alignment programs. Figure 3.1 shows the results of three widely used multiple alignment programs on a set of unrelated random sequences. A global method (ClustalW) and two combined local-global methods (T-Coffee and DIALIGN) were applied to a set of 100 nonrelated random sequences. ClustalW aligns the sequences over their entire length with only a small number of tiny gaps, although they share no significant similarity. T-Coffee is much more specific; it inserts larger gaps and leaves a substantial fraction of the sequences unaligned. DIALIGN is the most specific program; it returns only a few short local similarities and does not align the rest of the sequences. MAFFT and ProbCons produced results similar to those of T-Coffee; the outputs of MUSCLE and PRRN were more similar to that of the ClustalW output, although the former programs inserted slightly more and larger gaps than the least specific method, ClustalW. Thus, if one assumes that a set of input sequences contains substantial regions of homology, T-Coffee, MAFFT, and ProbCons may be good methods. They have been shown to produce good global alignments, but at the same time they are reasonably specific and do not attempt to align sequences over their entire length if no global similarity is detectable. For sequences sharing only small conserved motifs surrounded by otherwise unrelated sequences, DIALIGN may be the best method to detect these motifs and to distinguish them from nonrelated parts of the sequences.

ALIGNMENT OF GENOMIC SEQUENCES

Since large-scale genomic sequence data are available, cross-species sequence alignment has been used as a powerful tool to identify functional sites in anonymous genomic sequences. The general idea is that functional parts of the sequences are under evolutionary pressure and are therefore more conserved than nonfunctional regions of the sequences. Thus, functional sites can be revealed by pairwise or multiple alignment of related sequences. This general approach has been used to identify noncoding functional elements, such as regulatory sites (Chapman et al. 2003; Dubchak et al. 2000; Göttgens et al. 2000), and also for gene prediction in eukaryotes (Alexandersson et al. 2003; Bafna and Huson 2000; Carter and Durbin 2006; Flicek et al. 2003; Korf et al. 2001; Meyer and Durbin 2002; Stanke et al. 2006) to identify signature sequences for pathogenic microorganisms (Chain et al. 2003; Fitch et al. 2002) or to detect functional noncoding RNAs (Pedersen et al. 2006; Washietl et al. 2005).

Local versus Global Alignments

Large-scale alignment of genomic sequences involves several challenges, compared with the problems of more traditional alignment. First, the sheer size of the sequences makes it impossible to apply standard alignment programs where the time complexity for pairwise alignment is proportional to the product of the sequence lengths. Second, if syntenic regions from different genomes are compared, conserved regions are usually interrupted by nonconserved parts of the sequences. Thus, neither global nor local alignment methods can be used to represent the homologies within a set of input sequences. Finally, homologies do not necessarily appear in the same relative order in a set of input sequences; it is therefore necessary to take genomic rearrangements into account.

To reduce the computing time for large input sequences, most programs for genomic alignment use some kind of anchoring approach (Brudno et al. 2003a; Morgenstern et al. 2006). In a first step, some fast local alignment method is used to identify a list of strong local similarities. A chain of such similarities is selected in a second step, and, in a third step, the regions between those selected local similarities are aligned using some more traditional alignment method. Delcher et al. (1999) and Höhl et al. (2002) use suffix trees (Ukkonen 1995; Weiner 1973) to find gap-free local exact or near-exact matches between the input sequences (see also Batzoglou et al. 2000). Regions between those matches are aligned with ClustalW. Because the suffix tree for a sequence can be constructed in linear time, these algorithms are extremely fast. However, their application is limited to closely related sequences since standard suffix-tree algorithms can detect only exact matches, and their variants that can deal with mismatches become inefficient if too many mismatches are allowed. Consequently, those alignment methods have been mainly used to align very closely related species, for example, different strains of the same bacterium. Other tools have been developed that are flexible enough to calculate multiple alignments of more distantly related genomes, such as GLASS (Batzoglou et al. 2000), a combination of CHAOS and DIALIGN (Brudno and Morgenstern 2002; Brudno et al. 2004), LAGAN (Brudno et al. 2003b), or AVID (Bray et al. 2003). These tools, however, are slower than the methods that are based on suffix trees.

CHAPTER 4

Computing Multiple Sequence Alignment with Template-Based Methods

CÉDRIC NOTREDAME

Centre for Genomic Regulation, Spain

The Most Common Algorithmic Frameworks
for MSA Computation 57
Consistency-Based MSA Methods..........................58
Meta-Methods as an Alternative to Regular MSA Methods 61
Template-Based MSA Methods............................62
Validation and Benchmarking of MSA Methods................65
    Alignment-Free Benchmarking Methods....................66
    Alignment of Very Large Datasets........................67
Conclusions...68

An ever increasing number of biological modeling methods depend on the assembly of an accurate multiple sequence alignment (MSA). These include phylogenetic tree reconstruction, hidden Markov modeling (profiles; HMM), secondary or tertiary structure prediction, function prediction, and many minor but useful applications, such as PCR primer design and data validation. Assembling an accurate multiple sequence alignment is not, however, a trivial task, and none of the existing methods have yet managed to overcome the biological and computational hurdles preventing the delivery of biologically perfect MSAs. These limitations combined with a growing reliance of biology on the computation of accurate MSAs probably explain the recent surge of research activity in this field. Most of the important methods described during these last few years have been extensively reviewed (Edgar and Batzoglou 2006; Gotoh 1999; Wallace, Blackshields, et al. 2005), and I will focus here on the latest developments, including the development of meta-methods and the emergence of template-based alignment techniques. I will also address some aspects of the validation and benchmarking of these novel methods.

Two major issues surround the computation of an accurate MSA. The first one is computational. Given any sensible biological model, the computation of an exact MSA is NP-complete and therefore impossible for all but unrealistically small datasets—hence the development of a wealth of approximate heuristics (more than 50 over these last 20 years). The second aspect is biological. We still lack an objective measure for precisely quantifying the biological correctness of an alignment. Although it is generally agreed that structural correctness would constitute a good measure (Ginalski et al. 2005), its effective use is hampered by our poor understanding of the sequence/structure relationship, and evaluating the structural correctness of an MSA is possible only when experimental data are available (that is, when the sequences have a known 3D structure). Since, in practice, sequences with a known structure constitute a tiny minority, structure-based evaluations are possible in only a handful of cases.

The simplest alternative to estimate an MSA correctness is sequence similarity, mostly because of its ease of computation. However, it is well known that MSAs optimized for similarity are not necessarily correct from a structural point of view, especially when distantly related sequences are considered (for review, see Blackshields et al. 2006). This gap between similarity and structural correctness is a central issue in a context where MSA packages are routinely evaluated by comparing their output against established collections of structure-based reference alignments, used as gold standards. These gold standards have had a considerable effect on the development of MSA methods, refocusing the entire methodological development toward the production of structurally correct sequence alignments.

Well-standardized reference datasets have also gradually pushed the MSA field toward becoming a fairly well codified discipline, where all contenders try to improve over each other's methods by developing increasingly sophisticated algorithms. Whenever such improvements are reported, it is important to insure that they do not result from some overfitting on one of the reference datasets. This can be checked by considering alternative datasets, as shown recently by Blackshields et al. (2006). Even so, all of the reported improvements are not necessarily of equal importance from a methodological point of view. One may argue that it is important to precisely distinguish between smart variations around established procedures, alternative parameterization, and genuine innovations.

THE MOST COMMON ALGORITHMIC FRAMEWORKS FOR MSA COMPUTATION

For various reasons, innovations are often less competitive in their early development stages. For instance, the first HMM packages were shown to deliver poor alignments, although HMM modeling has now become an important element of the most accurate packages, such as ProbCons or PROMALS. Likewise, a partial order alignment graph (POA) (Lee et al. 2002) is a very innovative data structure, extremely well suited to many problems that are hard to address with the current MSA formalism. The use of POAs, however, has not yet resulted in improved alignments, either because the reference collections are not well suited to reveal the merits of POA, or because the method has not yet been integrated into a framework that would let it show its best qualities.

Yet, POA is only one of the many MSA algorithms developed. The NP-complete nature of the MSA problem is probably at the source of a surprising algorithmic diversity and creativity. Few optimization methods exist that have not been, at one point or another, applied to the MSA problem. One can cite in bulk the branch and bound algorithms; stochastic methods, such as simulated annealing; genetic algorithms or tabu search, divide, and conquer approaches; or agglomerative approaches, such as the one described in DIALIGN. Most of these algorithms are extensively reviewed in Notredame (2002).

Gold standards seem to have had a strong effect on this diversity, and it is a rather striking observation that most methods have converged and are now built around one of the first MSA methods ever described: the *progressive algorithm* (Hogeweg and Hesper 1984). This popular MSA assembly algorithm is a straightforward computational procedure where the sequences are first compared two by two in order to fill up a distance matrix on which a clustering algorithm, such as neighbor joining (Saitou and Nei 1987), is applied to generate a guide tree (rooted binary tree). The algorithm then follows the tree topology to incorporate the sequences one by one into the MSA while proceeding from the leaves (sequences) toward the root. During this process, each node ends up containing an alignment resulting from the pairwise alignment of its two children. The pairwise algorithm associated with this strategy must therefore be able to align two sequences, a sequence to an alignment (often formalized as a profile or HMM), and pairs of alignments.

Most pairwise algorithms described in this context are dynamic programming variations, based on either the Needleman and Wunsch (1970) or the Viterbi algorithm (Durbin et al. 1998). These pairwise alignment procedures always depend on a scoring scheme (substitution matrix or consistency), the purpose of which is to define the optimal matching of the two considered sequences or profiles. The main drawback of the progressive alignment strategy is its greediness. When progressing from the leaves toward the root, the progressive algorithm starts aligning pairs of sequences while ignoring the information contained in the rest of the dataset. These initial alignments can be wrong, especially if the sequences are not very closely related. Nonetheless, these alignments will be kept, and the errors they may contain may then propagate into the rest of the MSA. In an attempt to minimize this effect, ClustalW delays the incorporation of distantly related sequences, and aligns them only when an MSA framework has been assembled. Another simple and effective way to address the greediness issue is to embed the progressive alignment within an iterative loop where the guide tree and the MSA are successively reestimated until reaching some convergence. It was recently shown that these iterations almost always improve the MSA accuracy (Wallace, O'Sullivan, et al. 2005), especially when they are deeply embedded within the assembly algorithm. Iterations have also become a common feature of most MSA procedures (Do et al. 2005; Edgar 2004a; Katoh et al. 2005; Simossis and Heringa 2005).

CONSISTENCY-BASED MSA METHODS

While the framework described above forms the basis of most MSA packages (Armougom, Moretti, Poirot, et al. 2006; Do et al. 2005; Edgar 2004a; Katoh et al. 2005; Lassmann and Sonnhammer 2005b; Notredame et al. 2000; O'Sullivan et al. 2004; Pei and Grishin 2006, 2007; Pei et al. 2003; Simossis and Heringa 2005; Thompson et al. 1994; Wallace et al. 2006), it is well known that the MSA devil lurks in the details. As a consequence, authors have had to adapt the generic canvas more or less extensively to make improvements over existing methods. Listing all reported variations is beyond our current scope, but these have been extensively discussed and analyzed in a series of recent reviews (Edgar and Batzoglou 2006; Gotoh 1999; Wallace, Blackshields, et al. 2005). Table 4.1 points out

TABLE 4.1. SUMMARY OF ALL OF THE METHODS DESCRIBED IN THE REVIEW

Method	Score	Templates	Prefab	HOMSTRAD	Server
ClustalW (Thompson et al. 1994)	Matrix	—	61.80[a]	—	www.ebi.ac.uk/clustalw
Kalign (Lassmann and Sonnhammer 2005b)	Matrix	—	63.00[b]	—	msa.cgb.ki.se
MUSCLE (Edgar 2004a)	Matrix	—	68.00[c]	45.0[d]	www.drive5.com/muslce
T-Coffee (Notredame et al. 2000)	Consistency	—	69.97[a]	44.0[d]	www.tcoffee.org
ProbCons (Do et al. 2005)	Consistency	—	70.54[a]	—	probcons.stanford.edu
MAFFT (Katoh et al. 2005)	Consistency	—	72.20[a]	—	align.genome.jp/MAFFT
M-Coffee (Wallace et al. 2006)	Consistency	—	72.91[a]	—	www.tcoffee.org
MUMMALS (Pei and Grishin 2006)	Consistency	—	73.10[c]	—	prodata.swmed.edu/mummals
EXPRESSO (Armougom et al. 2006b)	Consistency	Structures	—	71.9[c]	www.tcoffee.org
PRALINE (Simossis and Heringa 2005)	Matrix	Profiles	—	50.2[d,f]	zeus.cs.vu.nl/programs/pralinewww
PROMALS (Pei and Grishin 2007)	Consistency	Profiles	79.00[c]	—	prodata.swmed.edu/promals
T-Lara (Bauer et al. 2005)	Consistency	Structures	—	—	www.planet-lisa.net

NOTE: Validation figures were compiled from several sources, and selected for the compatibility. "Prefab" refers to some validation made on Prefab Version 3. The HOMSTRAD validation was made on datasets having less than 30% identity. The source of each figure is indicated below the table.

[a]Wallace et al. (2006).
[b]Lassmann and Sonnhammer (2005b).
[c]Pei and Grishin (2007).
[d]Simossis and Heringa (2005).
[e]O'Sullivan et al. (2004).
[f]The EXPRESSO figure comes from a slightly more demanding subset of HOMSTRAD (HOM39) made of sequences less than 25% identical.

some aspects discussed in this chapter. Readers should also be aware that publications sometimes lag behind the packages, thus making a detailed analysis of the code the most secure way to precisely determine computational details.

Because it defines the mathematically optimal MSA, the pairwise alignment scoring scheme is probably the most central piece of the progressive strategy, and one of its most influential components. Scoring schemes can be divided in two categories: matrix- and consistency-based. Matrix-based algorithms, such as ClustalW (Thompson et al. 1994), MUSCLE (Edgar 2004a), or Kalign (Lassmann and Sonnhammer 2006), use a substitution matrix and a dynamic programming algorithm to align pairs of sequences or profiles. Although profile statistics can be more or less sophisticated, the score for matching two positions depends only on the considered columns or their immediate surroundings.

Consistency-based algorithms are more recent; their purpose is to incorporate a larger share of information into the evaluation. This result is achieved by using a recipe initially developed for T-Coffee (Notredame et al. 2000) and inspired by DIALIGN overlapping weights (Morgenstern et al. 1996). Sequences are first aligned two by two, and each pair of aligned residues thus obtained is added to a stack called the primary library, where it receives a weight proportional to its estimated reliability (that is, the estimated probability that it may be part of the final MSA). This library is then used to compute for every residue pair a match score called the extended weight that quantifies the consistency (compatibility) of the considered pair with all of the pairs contained in the library. The purpose of the extended weight is to concentrate into the match score of every residue some information coming from the entire dataset (library). This means that even when aligning two sequences only, the scoring scheme depends on the entire library and therefore incorporates within this pairwise alignment some information from the entire dataset. Such a pairwise alignment is therefore more likely to be compatible with the entire MSA than if it had been computed merely on the basis of the information contained by the two sequences. Overall, this approach is meant to address the greediness of the progressive algorithm by decreasing the number of early mistakes. The library is also a powerful concept that makes it possible to evaluate and construct an alignment on the basis of any kind of information.

T-Coffee was the first implementation of a consistency-based evaluation coupled with the progressive algorithm. One of its characteristics is to combine local and global alignments when compiling the library. Many recent packages have built up on this initial framework. For instance, PCMA (Pei et al. 2003) decreases T-Coffee computational requirements by prealigning closely related sequences. ProbCons (Do et al. 2005) uses Bayesian consistency and fills the primary library using the posterior decoding of a pair HMM. The substitution costs are estimated from this library using Bayesian statistics. MUMMALS (Pei and Grishin 2006) combines the ProbCons scoring scheme with the PCMA strategy while including secondary structure predictions in its pair HMM. The most accurate flavors of MAFFT (Katoh et al. 2005) (G/L-NS) use a T-Coffee-like evaluation. A majority of studies indicate that consistency-based methods are more accurate than their matrix-based counterparts (Blackshields et al. 2006), although they typically require an amount of CPU time N times greater than simpler methods (N being the number of sequences).

META-METHODS AS AN ALTERNATIVE TO REGULAR MSA METHODS

The wealth of available methods and their increasingly similar accuracies makes it harder than ever to choose one over the others. A common strategy is to select the method shown to perform best, on average, on most reference datasets. Such a strategy, however, is only an educated probabilistic guess, since the best method on average is merely a good candidate for performing best. In practice, alternative methods tend to behave differently on different datasets, even when they have identical average accuracies. It is also important to note that so far no method has been reported to outperform all of the others on every individual reference dataset; to make things worse, the best average method can sometimes be outperformed by the worst one on specific datasets.

An alternative to the "average best" is to align a dataset using several methods, and then select the best resulting MSA. The main difficulty with this strategy is the lack of any reliable quality criterion to objectively select one MSA over the others. Similarity is not a very good criterion, and structural information is rarely available. The last and most recent alternative is the use of a meta-method, such as M-Coffee (Wallace et al. 2006). M-Coffee (Meta-Coffee) is a consensus meta-method. Given a sequence dataset, it computes alternative MSAs using any publicly available methods (ClustalW, MAFFT, MUSCLE . . .) and turns each of them into a primary library. These libraries are then merged by T-Coffee and used to compute an MSA consistent with the original alignments.

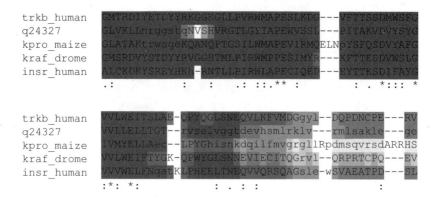

Figure 4.1. Typical output of M-Coffee. This output was obtained on the kinase1_ref5 BaliBase dataset by combining MUSCLE, MAFFT, POA, DIALIGN-T, T-Coffee, ClustalW, PCMA and ProbCons with M-Coffee. Correctly aligned residues (as judged from the reference) are in upper case; noncorrect ones are in lower case. In this output, each residue has a shade that indicates the agreement of the individual MSAs with respect to the alignment of that specific residue (normally this would be indicated by color). Darker shades indicate residues aligned in a similar fashion among all of the individual MSAs; lighter shades indicate a very low agreement. Residues that are medium gray or darker can be considered to be reliably aligned.

When eight of the most accurate and distinct MSA packages are combined, 67% of the time M-Coffee produces a better MSA than ProbCons, the best individual method (Wallace et al. 2006). Aside from its ease of extension, M-Coffee's main advantage is its ability to estimate the local consistency between the final alignment and the combined MSAs. This measure (the CORE index) (Notredame and Abergel 2003) not only estimates the agreement among the various methods (Figure 4.1), but also gives precious indication on the local structural correctness (Lassmann and Sonnhammer 2005a; Notredame and Abergel 2003). M-Coffee and the CORE evaluation are available online (www.tcoffee.org).

TEMPLATE-BASED MSA METHODS

Although it delivers MSAs on average better than the combined methods, M-Coffee is not dramatically more accurate than any of the individual methods. This is not surprising given the relative correlation of the best methods, the accuracy of which has tended to get even and to stagnate these last few years. In 2000, T-Coffee was reported to outperform

ClustalW by about 10 points; ProbCons was then shown to outperform T-Coffee by 1–2 points, and improved HMM modeling has resulted in a couple of extra points in accuracy shared among MAFFT, MUMMALS, and ProbCons. Further improvement while using only sequence information remains an elusive goal. This does not seem unnatural since the delivery of structurally correct sequence alignments of remote homologues would somehow imply dramatic improvements of the sequence/structure relationship.

During the wait for these progresses to materialize, the most realistic option is probably to shift from the regular MSA rules and incorporate within the datasets any information likely to improve the alignments, such as structural and homology data. This approach defines a new category of methods called template-based sequence alignment (Armougom, Moretti, Poirot, et al. 2006) that uses either structural or homology extension of the initial dataset. Structural extension takes advantage of the increasing number of sequences having an experimentally characterized homologue in the Protein Data Bank (PDB) database. Given two sequences with a homologue in PDB, it is straightforward to superpose the PDB structures (templates) and map the resulting alignment onto the original sequences. Such a mapping will result in a structurally accurate sequence alignment as long as each sequence is related enough to its template. Although this approach merely defines a pairwise alignment, the alignments thus produced can be compiled into a library and later used by T-Coffee or a similar method to assemble the final MSA.

Homology extension works along the same lines. Relatives of each sequence are rapidly identified and aligned using PSI-BLAST. The resulting profiles are then aligned to one another, thus inducing a profile-based alignment of the original sequences. This pairwise alignment can then be compiled in a library and used by a consistency-based method.

This homology extended library amounts to using the profile as an evolutionary enriched template. The only difference between homology and structural extension is the template's nature and the associated alignment method. This makes the overall strategy outlined in Figure 4.2 very generic and suitable to be an extended template of various nature, as long as the sequence/template mapping is straightforward and as long as the template/template alignment is possible and leads to improved sequence alignments. For instance, EXPRESSO (Armougom, Moretti, Poirot, et al. 2006) uses SAP (Taylor and Orengo 1989) and FUGUE (Shi et al. 2001) to align structural templates identified by a BLAST against the PDB. PROMALS (Pei and Grishin 2007) and PRALINE

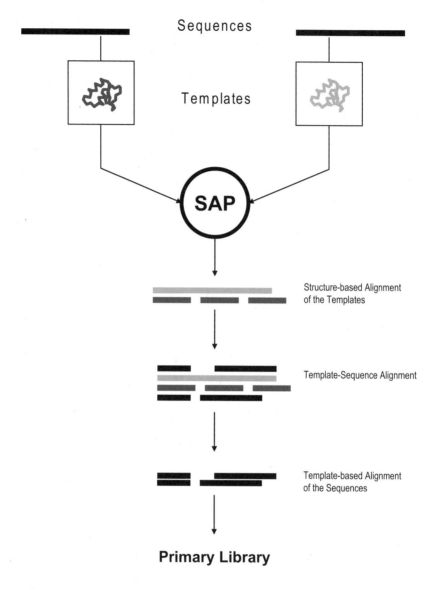

Figure 4.2. Framework of a template-based method. Structural templates are first identified, mapped onto the sequences, and aligned using SAP. The sequence/template mapping is then used to guide the alignment of the original sequences. This alignment is integrated into the library that is used to compute the final MSA.

(Simossis and Heringa 2005) make a profile/profile alignment of templates made of PSI-BLAST profiles. PROMALS uses ProbCons Bayesian consistency to fill its library with the posterior decoding of a pair HMM. T-Lara (Bauer et al. 2005) uses RNA secondary structure predictions as templates and fills a T-Coffee library with the Lara pairwise algorithm, and MARNA works along the same lines but using a different pairwise algorithm. Most of these methods, with the exception of PRALINE, are consistency-based. Some of them, such as T-Lara, EXPRESSO, and MARNA, take advantage of the modular nature of T-Coffee.

Aside from the possibility of mapping new types of information onto the MSA, the main advantage of template-based alignment methods is their increased accuracy. Recent benchmarks on PROMALS (Table 4.1) show that homology extension results in a ten-point improvement over existing methods. Likewise, structure-based methods such as EXPRESSO produce alignments much closer to the structural references than any of their sequence-based counterparts. However, one must be careful not to over-interpret figures like the EXPRESSO one in Table 4.1, since both the reference and the EXPRESSO alignments were computed using the same structural information. This simply means that standard structure-based reference collections may not be the most effective way to evaluate template-based sequence alignment when the template is a structure. In the end, such approaches amount to comparison to alternative structure-based alignment recipes, rather than evaluation of a sequence alignment procedure.

VALIDATION AND BENCHMARKING OF MSA METHODS

This last point raises the important issue of method validation and benchmarking. Quantifying the accuracy of an MSA package is a complicated task that requires being able to precisely define what a biologically correct MSA really is. Such a definition is very hard to give, and in practice it has become widely accepted that structural correctness is probably the best proxy we currently have for biological correctness.

The main justification for using structural information is its evolutionary resilience. During the course of evolution, homologous sequences often retain similar and comparable structures, even when their sequences have diverged beyond recognition (Lesk and Chothia 1980). A structurally correct alignment is therefore more likely to be correct from an evolutionary point of view than its sequence-based counterpart, especially when considering distantly related sequences.

An important consequence of this observation has been the compilation of collections of structure-based sequence alignments meant to be used as standards of truth for sequence alignments. These collections have been set up in order to validate MSA packages, the principle being that a package is as accurate as its management to reproduce a correct structure-based MSA while using only sequence information. Precise quantification of package accuracy is obtained by counting positions identically aligned in the reference and in the sequence alignment. Technical details regarding the setting up of these collections are discussed in Chapter 8. In the context of this chapter, we will merely consider the effect these reference benchmarks have had on the recent development of MSA packages and their potential short comings.

Alignment-Free Benchmarking Methods

As discussed earlier in this chapter, reference collections have had a strong influence on the most recent developments in the files of MSA method developments, and it has become customary to use the benchmark figures obtained on reference datasets as a magic number, summarizing all the merits of a method. Table 4.1 is a typical example of such a collection of figures, making it possible to pinpoint at a glance the current champion. Yet, this approach can be reductive, and some care should be taken when analyzing such figures.

The first issue is a sampling problem. Only a fraction of known proteins have had their structures experimentally determined. As a consequence, the benchmarks carried out on the references are merely an estimation of how well a given package may fare on real datasets made of proteins whose structure is unknown. This would not necessarily be a problem if the sampling was even among all protein types. This, however, is not the case and databases of proteins with known structures have several strong biases. They tend to contain proteins considered to be of interest, proteins that can easily be cloned, expressed, purified, and crystallized. This means an over-representation of small globular and highly soluble proteins. For these reasons, trans-membrane proteins are strongly under-represented. Ribonucleic acids structures are also poor relatives of these databases. As a consequence, one may need to consider that the current accuracy figures published for MSA packages can be safely extrapolated only to globular proteins.

Another complication has to do with the transformation of a structure superposition into an alignment. This task is not trivial, as suggested by the wealth of scientific literature dedicated to this subject. Many alternative sequence superposition methods exist, based on slightly different principles and delivering alignments that can be fairly different. For instance, DALI (Holm and Sander 1993) and CE (Shindyalov and Bourne 1998), two popular structural alignment methods, deliver alignments that are 40% different, as measured on the Prefab reference collection. Unfortunately, it is difficult to objectively evaluate the respective merits of these alternative methods, and this probably explains why at least six collections of reference alignments have been established during these last few years. These collections, extensively reviewed by Blackshields et al. (2006), are based on slightly different strategies for gathering the structures, aligning them, and annotating the reliable portions of the alignments, on which the benchmark is carried out. An extensive comparison of these datasets indicates that, with the exception of artificial datasets, analyses made with these various collections tend to deliver fairly coherent results. Unsurprisingly, the most accurate methods have no difficulty aligning sequences with more than 25% identity. In this range of identity, the methods have become virtually indistinguishable, and the MSA problem may be considered as good as solved.

The alignment of remote homologues is a different story and probably one of the new frontiers of MSA methods development. Remote homologues are harder to align because on these sequences, sequence similarity is not a good indicator of structural similarity. Furthermore, it is not entirely clear whether using reference alignments is a suitable approach when considering such sequences. It has been well established that translating the 3D superposition of remote homologues can result in ambiguous sequence alignments. As a consequence, several alternative alignments may have similar structural accuracy, thus making it incorrect to let the benchmark depend on a single reference alignment (Lackner et al. 2000). Bypassing the reference alignment stage is a possible alternative. It can be achieved by directly comparing the sequence alignment one wishes to evaluate to some idealized 3D superposition. Such procedures, independent of any reference alignment, have been recently described and are becoming increasingly popular (Armougom, Moretti, Keduas, et al. 2006; O'Sullivan et al. 2003; Pei and Grishin 2006).

Alignment of Very Large Datasets

Another limitation of current MSA method validation is the alignment of very large datasets. Although, for many years, the purpose of an MSA was to make the best of the few available members of most protein families, large-scale sequencing projects have induced dramatic changes. Many applications now exist that require the alignment of several hundreds, if not thousands, of sequences. Most packages have accommodated this need and contain highly efficient algorithms able to align thousands of sequences in an amount of time linearly proportional to the number of sequences (Edgar 2004a; Katoh et al. 2005; Lassmann and Sonnhammer 2005b). It remains unclear, however, to which extent the published accuracy figures can be extrapolated to these large datasets. In this context, the main issue seems to be the over-alignment (alignment of unrelated residues) induced by the progressive approach. By nature, the progressive alignment induces an amount of errors increasing with the number of sequences being aligned.

Assessing the accuracy of large datasets using the existing framework is probably feasible. For instance, the Prefab strategy that involves embedding sequences with known structures within a very large dataset could be used. One can then evaluate the induced alignment of the known structures and extrapolate this figure to the full alignment. Estimating the accuracy of the induced alignment can be done either by comparison with a reference alignment, or by using one of the alignment-independent methods described earlier (Armougom, Moretti, Keduas, et al. 2006; O'Sullivan et al. 2003; Pei and Grishin 2006). Two of these are available on the T-Coffee online server (www.tcoffee.org).

The last issue with structure-based reference collections is much harder to address because it relates to the very nature of these collections. They all share the "one size fits all" assumption that structurally correct alignments are the best possible MSAs for modeling any kind of biological signal (evolution, homology, function). A report on profile construction (Griffiths–Jones and Bateman 2002) has recently challenged this view by showing that structurally correct alignments do not necessarily result in better profiles. Likewise, it may be reasonable to ask whether better alignments always result in better phylogenetic trees, and, more systematically, to question and quantify the relationship between MSAs accuracy and the biological relevance of any model drawn upon them.

CONCLUSIONS

In this chapter, we have seen some of the latest additions to the MSA computation arsenal. The first improvement was the emergence of consistency as a useful tool for increasing MSA accuracy. Another interesting milestone was the development of a meta-method able to seamlessly combine the output of several methods. Aside from easing the user's work, the main advantage of such a consensus method is probably the local estimation of reliability it provides (Figure 4.2). Using this estimation to filter out unreliable regions has already proven useful in homology modeling (Claude et al. 2004) and could probably be used further. The main improvement, however, is probably the notion of template-based alignment. Template-based alignment is more than a trivial extension of the consistency-based methods. Under this new model, the purpose of an MSA is no longer to squeeze a dataset and extract all of the information it may contain, but rather to use this dataset as a starting point for exploring and retrieving all of the related information contained in public databases, so that this information can be mapped onto the final MSA and drive its alignment. This makes template-based methods a real paradigm shift and a major step toward global biological data integration.

CHAPTER 5

Sequence Evolution Models for Simultaneous Alignment and Phylogeny Reconstruction

DIRK METZLER

Ludwig-Maximilians-Universität, Munich, Germany

ROLAND FLEISSNER

Center for Integrative Bioinformatics, Vienna, Austria

Sequence Evolution Models with Indels 74
   Insertions and Deletions of Single Positions 74
   Fragment Insertion-Deletion Models. 76
   Extensions of the Fragment Models 79
Estimation of Multiple Alignments and Phylogenies 80
   Multiple Alignment with Multiple HMMs 80
   Alignment Probabilities Induced by Indel Models. 81
   Multiple HMM Induced by Indel Models. 86
   The Forward Algorithm 88
   Markov-Chain Monte Carlo Methods for Sampling Multiple
   Alignments 90
Discussion and Outlook 92

While it has long been recognized that the problems of multiple sequence alignment and phylogeny reconstruction are interdependent, they traditionally are tackled with different methodologies. The principles of statistical inference, be it Bayesian or based on likelihood maximization, form the sound foundation for the most widely used phylogeny estimation methods; yet, for sequence alignment, heuristic optimization of more or less arbitrary scoring schemes is still common. The main reason for this split may have been the lack of insertion-deletion models which are both realistic and computationally tractable. In this chapter, we give an overview of recent advances in modeling insertions and deletions and explain how these can be used to unify the two fields in a stochastic framework.

Usually, phylogenetic inference is organized in a stepwise fashion. First, we are searching for highly similar sequences using local alignment tools such as BLAST (Altschul et al. 1990) or FASTA (Pearson and Lipman 1988). Then, the retrieved sequences are aligned again with some multiple alignment program, thereby minimizing the distances between the sequences or maximizing their similarity (Gotoh 1999). Finally, this multiple sequence alignment is used to infer the details of the substitution process and to reconstruct the phylogenetic tree of the sequences, again minimizing distances or maximizing the probability to observe the data (Swofford et al. 1996). Because each of these steps results in a more or less detailed hypothesis about the evolution of the sequences, the whole procedure appears to be rather circular. In order to find the sequences and to align them, we already have to make assumptions about the substitution process, and while we are using the multiple alignment to reconstruct the tree, many alignment methods already need a tree to align the sequences. Hence, it is not surprising that this way of proceeding is the cause of some problems. As was demonstrated by Morrison and Ellis (1997), the multiple alignment method may influence the reconstructed tree more severely than the usage of different tree reconstruction methods. Other studies have shown that the accuracy of phylogenetic inference is greatly reduced by alignment errors (Ogden and Rosenberg 2006). This effect can be so strong as to make it completely impossible to reconstruct the correct tree topology or to get unbiased estimates of the substitution rates (Fleißner 2004). In addition, neither the Bayesian sampling of trees (Ronquist and Huelsenbeck 2003) nor bootstrapping the aligned sequences (Felsenstein 1985) will hint at these errors as long as we keep the alignments fixed.

For all of these reasons, it would be desirable to reorganize the process of phylogenetic inference in a way that allows us to judge alignment uncertainties and to reconstruct phylogenies without having to rely on one single optimal or nearly optimal multiple alignment. Methods that produce multiple alignments and phylogenies at the same time are not new. One could argue that this idea is already implemented by alignment programs that use iterative refinements to change their guide trees (e.g., Gotoh 1999). However, it is doubtful that these guide trees, which are computed from pairwise distances alone, are very accurate. Moreover, in order to get rid of alignment artifacts, it is not enough to modify the guide tree, but one would want to consider different scoring schemes and gap penalties, too. Even the method of Vingron and von Haeseler (1997), which minimizes the branch lengths of the phylogeny and the distances within the alignment simultaneously, exhibits this dependency on arbitrary scoring functions. What we need instead is an explicit model of the insertion-deletion process. Only if we have an insertion-deletion model can we draw conclusions about the nature of this process via statistical estimation. Such a model would also provide us with a considerable amount of additional information on the tree (Mitchison 1999). Most important, however, we would then be able to jointly sample multiple alignments, phylogenies, and model parameters. Thus, we would not only get an accurate view of the uncertainties within the reconstructed tree and the parameter estimates, but we would also see which alignment positions are more reliable than others.

Various attempts have been made to model insertions and deletions (*indels*, for short). The approach of McGuire et al. (2001) is a straightforward extension of the usual nucleotide substitution models. It simply is defined by a substitution rate matrix that includes an additional row and column for the gap symbol '-' as the fifth character. As appealing as this may seem, it has several drawbacks. Because this substitution process treats every site independently of the others, it corresponds to alignment algorithms that use linear gap costs and thus does not treat longer indels adequately. Moreover, the model has to allow the substitution of a nucleotide with '-', which later on may be substituted again with another nucleotide. Hence, such a substitution model might consider two positions to be homologous where in fact they are not, since one was deleted and another one was inserted. Another approach to modeling indels are the so-called *tree-HMMs* (Mitchison and Durbin 1995; Mitchison 1999). There, the information about each of the sequences of a multiple alignment is augmented with a path through a series of *match* and *delete* states. The match states correspond to actual residues of the sequence, whereas a sequence's path goes through a delete state whenever this sequence contains a gap in the alignment. Indels can then be modeled using simple Markov rate matrices whose states are the transits between the match and delete states. Although tree-HMMs recognize the principle difference between substitutions and indels, they, too, may keep a memory of deleted subsequences and restore them later (Holmes and Bruno 2001).

There is, however, a whole class of models that offer both realism and tractability. In the rest of this chapter, we will explain these models, which go back to Thorne et al. (1991), in more detail. First, we will focus on the basic assumptions of these models, and then we will see how they can be used for phylogenetic inference.

SEQUENCE EVOLUTION MODELS WITH INDELS

Insertions and Deletions of Single Positions

We begin with a short overview of the TKF91 model (Thorne et al. 1991), since it already exhibits most of the features of the more complicated models. This model consists of two stochastic processes acting side by side: a substitution process and a process that inserts and deletes single residues. While substitutions occur according to one of the usual substitution models for amino acids or nucleotides (Müller and Vingron 2000; Tavaré 1986), insertions and deletions are produced quite differently. Every position may be deleted at rate μ, and between any two positions as well as at the ends of a sequence, single positions are inserted at rate λ, with $\lambda < \mu$. Whenever such an insertion occurs, the corresponding nucleotide or amino acid is drawn from the equilibrium distribution of the substitution process. The length of a sequence under the TKF91 model is determined by a birth and death process (Taylor and Karlin 1998). If n is the current sequence length, the death rate of this process is $n \cdot \mu$ and the birth rate is $(n + 1) \cdot \lambda$. Thus, the stationary distribution of the sequence length will be geometric, with mean $\lambda/(\mu - \lambda)$. This means that a sequence in equilibrium has length n with probability $(\lambda/\mu)^n \cdot (1 - \lambda/\mu)$.

Now consider two sequences r and s, r being the ancestor and s its descendant, after some time t. If the TKF91 model is to be of any use in computing the probability of r evolving into s during time t, it has to fulfill two criteria: first, every alignment of the two sequences must have a clearly defined probability under the model, and second, it must be possible to efficiently sum up the probabilities of all of the individual alignments.

To meet the first criterion, it has to be specified how a realization of the insertion-deletion process is represented in an alignment. Although it is almost canonical how this is to be done, there is one source of ambiguity: When a position is deleted and another position is inserted at its place, then it is not canonically defined in which order they are denoted in the alignment. According to a convention suggested by Thorne et al. (1991), however, insertions are regarded as offspring of the position to their left and are denoted immediately to the right of their ancestors. This clarifies notation and defines the necessary probability distribution on the set of alignments. Note, however, that this convention leads to the alignments having different probabilities depending on which sequence is taken as the ancestral one. Yet, since this effect can be accounted for by the appropriate reordering of adjacent gaps, and since it does not affect the overall probability of observing the two sequences, the TKF91 model is reversible in a certain sense, and one is free to pick the ancestral sequence at will.

In order to see that the TKF91 model also fulfills the second criterion, consider the possible histories of positions of the descendant sequence s and the possible fate of positions of the ancestral sequence r (see Figure 5.1). Every position of s is either a surviving position of r or, due to the notation convention, an offspring of a position in r or of its left end. Every position of r, on the other hand, either will be present in s, or at least have offspring in s, or will have died out without a trace. Following Lunter, Miklós, Song, et al. (2003), one can now recode every *indel history* of the two sequences, that is, their bare alignment without the information about the individual nucleotides or amino acids, by denoting every match and mismatch with an H for "homology", every insertion with a B for "birth", and every deletion with an E for "extinction".

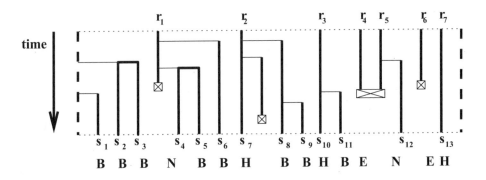

Figure 5.1. Evolution from ancestral sequence $(r_1, \ldots, r_7)$ to offspring sequence $(s_1, \ldots, s_{13})$ according to a fragment insertion-deletion model and alignment in $\{B, H, N, E\}$ notation (Lunter, Miklós, Song, et al. 2003). Events such as the simultaneous deletion of r_4 and r_5 or the simultaneous insertion of s_4 and s_5 are possible in the TKF92 model and the FID model, but not in the TKF91 model.

To make things simpler, *EB* is always abbreviated by *N* for "new". It is easy to see that, apart from its left end, every indel history can be decomposed into blocks of the form (*HB* ... *B*), (*NB* ... *B*), or (*E*), where every *H* and every *N* is followed by, geometrically, many *B*s. Since each of these blocks is independent of the blocks to its left, and due to the geometric number of insertions within the blocks as well as at the indel history's left end, the whole sequence of *B*s, *H*s, *N*s, and *E*s is Markovian (Norris 1997). In other words, a random alignment that is distributed according to the TKF91 model's probability law can be regarded as a sequence of columns read from left to right, which is produced by a Markov chain. Thus, TKF91 induces a *pair hidden Markov model* or *pair HMM* (Durbin et al. 1998) whose transition probabilities can be computed from λ, μ, and t (Hein et al. 2000; Holmes and Bruno 2001). This means that several dynamic programming algorithms become applicable to the TKF91 model. One can easily find the most probable alignment for a given pair of sequences by a variant of the Viterbi algorithm, and the so-called forward algorithm can be used to compute the likelihood of given values for the model parameters. The time complexity of these dynamic programming algorithms is asymptotically proportional to the product of the lengths of the input sequences.

Hence, if we restrict ourselves to the case of pairwise alignments, and if we accept the assumption that only single positions get inserted and deleted, the TKF91 model provides us with everything we need. By optimizing the rate parameters of the model for every sequence pair, one can obtain a neighbor-joining tree for a data set without knowing the multiple alignments (Thorne and Kishino 1992). Furthermore, intermediate results computed in the forward algorithm can be employed for the sampling of pairwise alignments according to their posterior probability distribution (Metzler et al. 2001), yielding images, such as Figure 5.2 and Figure 5.3, that show which parts of the alignment are more reliable.

Fragment Insertion-Deletion Models

Extending the TKF91 model to the case of longer indels is not trivial, since there is no pair HMM that is equivalent to a general insertion-deletion process that may insert more than one position between any pair of sites and that may delete any subsequence at one instant. One way around this problem is the use of a pair HMM, which only approximates such a general indel process (Knudsen and Miyamoto 2003).

Simultaneous Alignment and Phylogeny

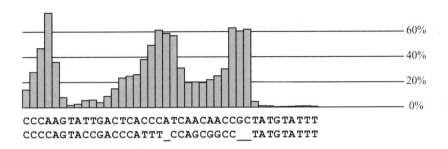

Figure 5.2. Most probable alignment for subsequences of *HVR1* (Handt et al. 1998) from human and orangutan. Bars indicate the uncertainty of the alignment. The height of a bar is the ratio of sampled alignments that do *not* coincide with the most probable alignment at this position.

Another approach is taken by the TKF92 model presented in Thorne et al. (1992), which allows insertions and deletions of indivisible sequence fragments that may contain more than one position. These fragments then behave like the single positions in the TKF91 process. This means, for example, that a newly inserted fragment can be deleted only as a whole and that no other fragment can be inserted between its positions. As the positions within a fragment are inseparable, and as the fragments follow the birth-death dynamics of the TKF91 process, one can come up with a dynamic programming recursion to compute the probability of two related sequences via such a process, given an arbitrary distribution of fragment lengths (Fleißner 2004). In the TKF92 model, things get even easier because the model assumes the fragment lengths to be geometrically distributed. Thus, every fragment has geometrically many sites and every fragment of an ancestral sequence r has geometrically many fragments as offspring in its descendant sequence s. The sum of geometrically many independent random variables that are all geometrically distributed with the same mean, however, follows a geometric distribution itself. Hence, every indel history under the TKF92 process can be decomposed into blocks of the form ($HB \ldots B$), ($NB \ldots B$), or (E) in just the same way as a TKF91 indel history (see Figure 5.1). Therefore, the TKF92 model also has a pair HMM structure, though with different transition probabilities than TKF91, and the usual dynamic programming algorithms are applicable.

The fragment insertion-deletion (FID) model (Metzler 2003) is a slight simplification of the TKF92 model. It sets the rate of fragment insertion equal to the rate of fragment deletion. Consequently, there

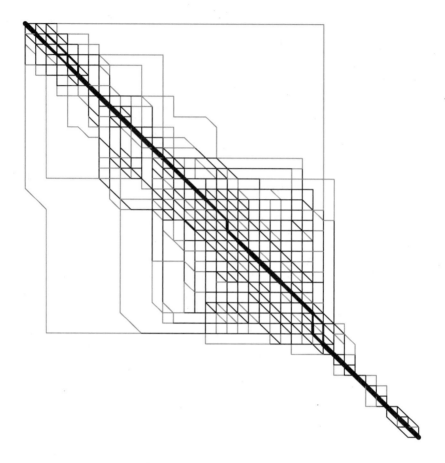

Figure 5.3. Graphical representation of sampled alignment paths for the subsequences of *HVR1* from human and orangutan used in Figure 5.2. Thickness corresponds to frequency in the sample.

is no stationary distribution for the sequence length in the FID model. Instead, it is assumed that the sequence lengths are given and nonrandom, which means that one thinks of the sequences as embedded into longer sequences. We conjecture that neither this assumption of neighboring positions to the left and the right of the actual data nor the distribution of the sequence length have a strong influence on alignments that have been estimated by methods relying on the FID or TKF92 model.

Usually it is impossible to reliably locate the ancestral sequence in a phylogenetic tree from sequence data if the molecular clock assumption is not met. Therefore, a sequence evolution model should be reversible, which implies that the likelihood of the data does not depend on the direction of the evolution. In general, this is ensured by assuming that the ancestral sequence was taken from the reversible equilibrium of the mutation process. However, for the FID model, there is no equilibrium on the set of finite sequences, and Felsenstein (2004) raises the question as to whether the FID model is still reversible. Yet, Metzler et al. (2005) show that the FID model is reversible in the sense that the conditional probability of a homology structure for given sequence data does not depend on the location of the root in their genealogy.

Extensions of the Fragment Models

As already stated, assuming indivisible fragments is necessary to equip the TKF92 and the FID model with a pair HMM structure and to allow for the application of efficient dynamic programming algorithms. Models that come along without this fixed-fragment assumption, such as the general insertion-deletion (GID) model in Metzler (2003) or the long-indel model in Miklós et al. (2004), might appear to be more plausible, but are computationally far more costly to use. Simulation studies in Metzler (2003), however, give evidence that methods for estimating the parameters of the FID model from a pair of unaligned sequences are robust against violations of the fixed-fragment assumption. Miklós et al. (2004), on the other hand, show that falsely applying a fixed-fragment assumption may decrease the precision of alignment methods. A possible explanation for these contradictory observations could be that the gap length distribution observed in real data and also in data simulated according to the long-indel model of Miklós et al. (2004) is more long-tailed than the geometric gap length distribution in the fixed-fragment model. We conjecture that for most biological data sets a mixture of two geometric distributions would be a good approximation for the indel length distribution. Recursions and pair HMMs that use these mixed distributions are straightforward (Fleißner 2004). There simply are two possible parameters for a geometric fragment length distribution. Each fragment randomly picks one of the parameters, with a higher probability of taking the one favoring shorter fragments. Because the type of each fragment is a hidden parameter, we obtain a pair HMM with three additional hidden states.

Another model with different types of fragments is presented in Arribas-Gil et al. (2007). The purpose of this model is the alignment of sequences that contain some slowly evolving region, for example, genes with conserved exons and variable introns, protein sequences with conserved alpha-helices and beta-sheets and variable coils, or RNA sequences with conserved stem regions and variable loop regions. These regions are modeled by fragments of geometrically distributed lengths that can be neither deleted nor inserted. They are affected only by substitutions. The type of the fragment to which a position belongs is a hidden parameter. Between the "slow" fragments, fast-type fragments evolve according to the TKF92 model. At this point, TKF92 is a better choice than FID because the length of segments between the slow fragments really does matter. The FID model would tend to increase the distance between the slow fragments, and this would contradict the requirement of reversibility.

It is straightforward to set up a three-type fragment insertion-deletion model that combines the model of Arribas-Gil et al. (2007) with mixed-geometric indel-length in the fast regions. This model would have at least nine hidden states, three for each fragment type, namely, homologous pair, gap in first sequence, and gap in second sequence.

ESTIMATION OF MULTIPLE ALIGNMENTS AND PHYLOGENIES

Multiple Alignment with Multiple HMMs

Assume that more than two sequences are to be aligned and that these sequences have evolved from a common ancestral sequence along a phylogenetic tree, which might be unknown. One approach to finding efficient methods for multiple alignments is to use models for the sequence evolution along the tree which carry a hidden Markov structure, thus providing a basis for the application of dynamic programming algorithms. The *evolutionary (multiple) hidden Markov models* in Holmes and Bruno (2001) are generalizations of the TKF91 model for more than two sequences at the tips of a tree. Their hidden Markov chain is the indel history of all internal and external nodes, and it emits the observable sequences at the tips. Finding the most probable alignment by dynamic programming is feasible in this model if the number of sequences is low and if their phylogeny is known. Finding the exact solution for a high number of sequences is still intractable, as the computational complexity increases exponentially with the number of sequences. However, the good performance of a program such as TREE-PUZZLE (Schmidt et al. 2002) is quite promising for applying heuristics for phylogeny reconstruction which are based on exact solutions for small subsets of sequences. The

TKF91 model is inappropriate for most data sets because it allows insertions and deletions of solely single positions. The fragment indel models discussed in the previous section may be used to model the indel process along each edge of the tree. Yet, what happens to the fragments at the internal nodes of the tree still has to be specified. The two major possibilities for a fragment that has been inserted are (1) that it is fixed in the whole tree, in the sense that it may be deleted only as a whole, or (2) that it is fixed only on the edge where it has been inserted and the fragmentation structure is forgotten in each internal node. In general, the second possibility is preferable since it is more likely to be a good approximation for models such as GID (Metzler 2003) and the long-indel model of Miklós et al. (2004) that seem more plausible.

In Fleißner et al. (2005), we present a simulated annealing strategy for the joint estimation of multiple alignments and phylogenies for a given set of sequences. Starting with an initial guess for the alignment and the phylogeny, ancestral sequences for all internal nodes are estimated, and the pairwise alignments of sequences belonging to neighboring nodes, the phylogeny, and the sequences for the internal nodes are iteratively re-estimated. The method is based mainly on a TKF92 model with fragmentation structures varying from edge to edge. Exceptions, however, are the subroutines that estimate internal sequences and alignments of sequence pairs belonging to edges which newly appear in the tree after a re-estimation of its topology. In these procedures, we keep the fragmentation structure fixed for all edges touching the new nodes. With this simplification, we obtain a TKF92-adaptation of the Holmes and Bruno (2001) multiple HMM with a relatively low number of possible hidden states and transitions.

In order to obtain a probability distribution on the set of possible multiple alignments from an indel model, it is necessary to abolish some indeterminacies in alignment notation. Therefore, one must determine the order in which insertions occurring in distinct parts of the tree are to be notated in the alignment. This is an essential step in Holmes and Bruno (2001) for transferring the TKF91 model to three-leaved trees and in Fleißner et al. (2005) for the TKF92 model on three- or four-leaved trees with global fixation of the fragments.

Alignment Probabilities Induced by Indel Models

Given a phylogenetic tree, a sequence evolution model such as TKF91, TKF92, or FID defines a probability distribution for the evolutionary history **H** of the sequences. The crucial point is to define a map **A**, which turns each possible realization H of **H** into a suitable alignment $\mathbf{A}(H)$ of the sequences at the tips of the phylogeny. This will equip the set of possible alignments with a probability distribution via $Pr(\mathbf{A}(H) = A) = Pr(\mathbf{H} \in \mathbf{A}^{-1}(A))$, where the inverse image $\mathbf{A}^{-1}(A)$ of A is the set of all evolution histories that lead to the alignment A. In the rest of this section, we define such a map **A** from the realizations of the evolution process to the sequence alignments. We will take care that the induced probability distribution on the set of possible sequence alignments is equipped with the structure of a multiple HMM, which we will discuss in the next subsection.

Lunter et al. (2005, 2006), Lunter, Miklós, and Jensen (2003), and Lunter, Miklós, Song, et al. (2003) define elegant notations and rules for statistical alignment that reveal a multiple HMM structure for the TKF91 model on any n-leaved tree. In the notation of Lunter et al., the information about the indel history between an ancestral and an offspring sequence, which is usually contained in the alignment of the sequences, is summarized in a sequence of letters from $\{H, B, E, N\}$. H stands for a homologous pair of positions, B for a newly inserted position, and E for a deleted position. EB is always replaced by a single N because this simplifies the transition probabilities for the hidden Markov chain. An example of a sequence evolution and the corresponding alignment in this notation is given in Figure 5.1. If the relationship between a sequence position i and a position j in an ancestral position is denoted by H or N, then i is called an heir of the ancestral sequence's position j.

Figure 5.4 shows how the indel history of four extant sequences can be represented with this notation. Labeling each edge of the sequences' phylogeny (right side of Figure 5.4) with a sequence over $\{H, B, E, N\}$, which specifies the alignment of the (perhaps unknown) sequences at its adjacent nodes, determines the multiple alignment of the sequences at the tips of the tree. In fact, it even determines an alignment of all sequences belonging to the nodes of the tree (however, without specifying the nucleotides or amino acids of the sequences at the internal nodes). In the left part of Figure 5.4, the phylogenies of the sequence positions are visible. Such a tree connects sequence positions if and only if they are homologous, and if and only if this is the case, the sequence positions are to occur in the same column of the alignment summarizing their indel history. In order to achieve a multiple HMM structure, which will make our algorithms more efficient, we will have to specify an appropriate order on these sequence position trees, and thus also

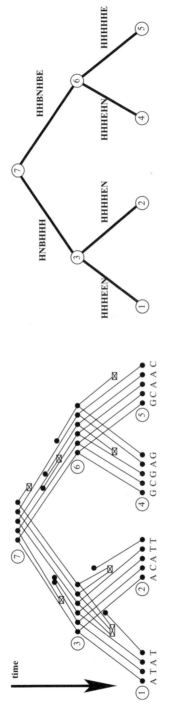

Figure 5.4. Sequence evolution along a tree according to a fragment insertion-deletion model (left) and the corresponding pairwise alignments for all edges in {B, H, N, E} notation.

on the alignment columns. It is clear that an alignment column will be denoted to the left of another one if the first one contains a position i of a sequence and the latter contains some position $j > i$ of the same sequence, but this rule alone does not always specify a complete order. As an intermediate step, we join sequence position trees t and t' if the root of t is an heir of t'. Thus we obtain trees in which position i in a sequence is a child of position j in its ancestral sequence, not only if i and j are homologous but, more general, if i is an heir of j. We label each edge of these trees with H, B, E, or N according to its type. For edges labeled with H or N, both nodes are associated with nodes of the sequences' phylogeny. For B-labeled edges, this is true only for the child node, and for E-labeled edges only for the parental node. The labeled trees defined in this way are called *events* in Lunter, Miklós, Song, et al. (2003). We prefer using an unambiguous term and therefore call them *tree-indexed heirs lines* or *tihls* for short (Metzler et al. 2005). The tihls for the indel history in Figure 5.4 are shown in the top of Figure 5.5. N-labeled edges are suggestively drawn with a disruption. The numbers labeling the nodes associate them with the nodes of the sequences' phylogeny (right side of Figure 5.4).

To fully determine the multiple alignment of the sequences, a complete order must be defined on the tihls that arise from a realization of the indel process along the tree. If a position i in one of the sequences labeled to the nodes is involved in tihl t_1, and some position $j > i$ of the same sequence is involved in tihl t_2, then it is clear that $t_1 < t_2$, which means that t_1 is listed to the left of t_2. This defines a partial order on the tihls but it does not specify an order between, for example, the third and the fourth tihl in the top of Figure 5.5. Lunter, Miklós, and Jensen (2003) and Lunter, Miklós, Song, et al. (2003) propose to complete this partial order according to some order on the nodes of the tree with the property that $n_1 < n_2$ if n_2 is an ancestor of n_1. For example, this is fulfilled for the nodes' order given by the numbering shown in Figure 5.4 and Figure 5.5. If the root of a tihl t corresponds to the root of the sequences' phylogeny, let $r_0(t)$ be this root. Otherwise, the root of t is the parent node of a B-labeled edge, and we define $r_0(t)$ then as its child node. Let $r(t)$ be the node of the sequences' phylogeny that corresponds to $r_0(t)$. If the partial order of the tihls as defined above does not relate tihls t_1 and t_2, then we set $t_1 < t_2$ if $r(t_1) < r(t_2)$.

We still have to specify how the series of tihls defines an alignment (see Figure 5.5). The crucial point is that a tihl may emit more than one alignment column, as, for example, the second, the fifth, and the

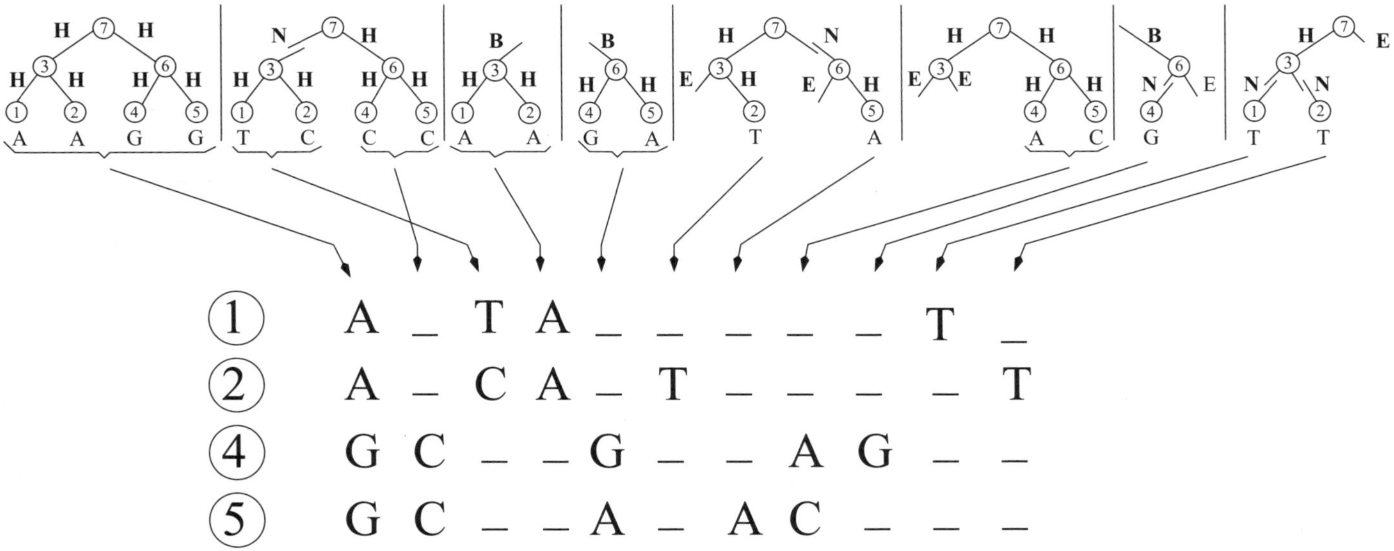

Figure 5.5. Sequence of tihls of the insertion-deletion history of four contemporary sequences and the corresponding alignment of the leave-labeling sequences in standard notation.

rightmost tihl in Figure 5.5, and we have to define an order between the columns belonging to these tihls. We do that by subdividing each tihl into trees of homologous positions, which correspond to alignment columns. If t and t' are such trees and the root of t is the heir of a node from t', then the column corresponding to t' is denoted to the left of the column emitted by t. This applies to the second and the fifth tihls in Figure 5.5. If this rule does not suffice, as, for example, for the rightmost tihl in Figure 5.5, then we proceed analogously as done for tihls and set $t < t'$ if $r(t) < r(t')$. This specifies the order of the last two columns in the alignment shown in Figure 5.5. Even if their order is perhaps biologically inconsequential, it must be specified to complete the definition of the map $\mathbf{A}$ from the evolutionary histories to the alignments, which induces a probability distribution on the set of possible sequence alignments.

So far, we have assumed that the sequences' phylogeny is rooted *a priori*. If this should not be the case, a root is arbitrarily set somewhere in the tree. Like the convention of Thorne et al. (1991) for pairwise alignment notation, the rules for tihl column ordering make the alignments root-dependent. However, these dependencies affect only biologically inconsequential details such as the order of the last two columns of the alignment in Figure 5.5. The location of the root does not influence the initial partial orders on tihls and columns and has no effect on the positional homologies in the multiple alignment (Metzler et al. 2005).

Multiple HMM Induced by Indel Models

The evolution of a sequence along a phylogenetic tree according to TKF91 is a stochastic process. Hence, the resulting sequence of tihls is a random sequence. Unfortunately, it cannot serve as the hidden Markov chain in the multiple HMM because it is not Markovian. Yet there is a Markovian sequence that is closely entangled to the tihl sequence. Assume that the first $n - 1$ tihls of this tihl sequence are given, and that t and t' are two candidates for the nth tihl. Furthermore, assume that t and t' do not intersect and that $r(t) < r(t')$. If t and t' were the next two tihls, then, according the rule of Lunter et al. for the completion of the order on the tihls, t would be the nth tihl and t' the $n + 1$th. This implies that if t' is the nth tihl, then the $n + 1$th cannot be t, and the same holds for any other tihl $\tilde{t}$ with $r(\tilde{t}) = r(t)$. In this sense, t' deactivates the node $r(t)$ (or more precisely its ingoing edge). It may also activate nodes because any tihl $\hat{t}$ will be allowed as a successor of t' if $r(\hat{t})$ is a node of t'. In this sense, t' will activate each currently inactive node v that corresponds to a node of t'. Thus, the tihl sequence induces a sequence of *sets of active nodes* (*soan*, for short, called *states* in Lunter, Miklós, and Jensen 2003, and Lunter, Miklós, Song, et al. 2003), and this sequence turns out to have two nice properties if the underlying indel model is TKF91. First, it is Markovian, and second, given the first $n-1$ tihls, the probability of the nth tihl depends only on the current soan. Therefore, the sequence of soans is an appropriate hidden Markov chain for a multiple HMM based on the TKF91 indel model. The root of the entire tree is always active. Instead of the nodes, one may also consider their ingoing edges to be (de)activated.

For the TKF91 model, Lunter et al. present a dynamic programming algorithm for finding the most probable alignment for a given set of sequences with known phylogeny. It computes step-by-step, for all tuples of sequence positions and each soan, the probability of the most probable alignment of the sequence beginnings up to the corresponding positions ending in the corresponding soan. The forward algorithm computes the probability of a given set of sequences. The use of dynamic programming implicitly sums up the probabilities of all possible alignments. Lunter, Miklós, Song, et al. (2003) have found a clever way of grouping soans in the forward algorithm for the computation of the probability of given sequence data. With this procedure, it suffices to compute, for each tuple of sequence positions, the probability of the sequence beginnings up to these positions, without the need to compute this separately for all soans. This makes the computation more efficient, but for large data sets the runtime grows faster than the product of all sequence lengths and is thus exponential in the number of sequences.

In Metzler et al. (2005) we discuss how the notations and rules of Lunter et al. can be applied to the fragment indel models TKF92 and FID. Indeed, this leads to a multiple HMM, but with a more complex state space for the hidden Markov chain. Since the transition probabilities are more complex in the fragment indel models, it does not suffice to record which nodes are active. What the last tihl that affected the ingoing edge mapped to this edge is also relevant. It is not necessary to distinguish B and N here because they have the same effect on the probability distribution for the subsequent tihl. Furthermore, E is always deactivating. Therefore, the appropriate state space for our hidden Markov chain is the set of all labelings of the edges of the tree with letters from $\{H, B, h, b, e\}$, where capital letters label active edges and small letters inactive edges. Figure 5.6 gives an example of how a tihl t changes the labeling of the edges. Each edge that corresponds to an edge

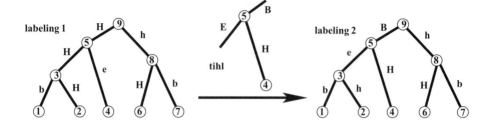

Figure 5.6. Example of a tihl operating on the labelings.

in t takes on its label. Each of the other edges is deactivated if $v < r(t)$ holds for its child node v.

For a tree with n edges, there are 5^n $\{H, B, h, b, e\}$ labelings, as compared to only 2^n soans. However, we assume that the initial state of the Markov chain labels each edge with H. Thus, for a three-leaved tree only 41 and for a four-leaved tree only 437 labelings are relevant as other labelings cannot be reached from the initial one by any combination of tihls (Lokau 2006).

The Forward Algorithm

We will now have a closer look at the forward algorithm for the multiple-alignment FID model and its application for Bayesian sampling of alignments by randomized backtracing.

For a given tree with n leaves $v_1, \ldots, v_n$, let Ψ denote the set of tihls and let Φ be the set of all edge labelings that can be reached by applying a series of tihls to the initial labeling $\eta \in \Phi$ which labels each edge with H. If a tihl $\tau \in \Psi$ transforms a labeling $\varphi \in \Phi$ into a labeling $\vartheta \in \Phi$, we write $\vartheta = \tau(\varphi)$. For $\tau \in \Psi$, let $e(\tau) \in \{0,1\}^n$ be a vector whose ith component $e_i(\tau)$ is 1 if and only if τ emits a position (i.e., labels B, H, or N) to the ith leaf v_i of the tree. For a multi-index $k = (k_1, \ldots, k_n) \in \mathbb{N}^n$ and $\varphi \in \Phi$, let $P_\varphi(k)$ be the probability that

1. the initial labeling η is followed by the sequence of tihls $\tau_1, \tau_2, \ldots, \tau_M$, which emit to each leaf v_i the first k_i nucleotides or amino acids in the sequence labeled to v_i.
2. $\tau_M \circ \ldots \circ \tau_2 \circ \tau_1 (\eta) = \varphi$
3. φ is immediately followed by another η.

(Condition 3 is necessary to avoid complications that may arise from the possibility that φ is followed by tihls which do not emit to the leaves and lead to another φ. In this case, condition 3 clarifies to which φ we refer.)

If we are able to compute all values of $P_\varphi(k)$, then we can also compute the probability of the sequence data (which is also the likelihood of the tree and the mutation parameters), because this value is just $P_\eta(k)$ if we let k be the vector of the lengths of the input sequences. Using the Markov structure of our sequence of labelings, we get a forward equation (Taylor and Karlin 1998) of the form

$$P_\varphi(k) = \sum_{\{(\vartheta,\tau):\tau(\vartheta)=\varphi\}} P_\vartheta(k-e(\tau)) \cdot g(\tau,k) \cdot F(\varphi,\vartheta,\tau),$$

where $g(\tau, k)$ is the probability that τ emits the nucleotides or amino acids given in the input data at the positions specified by k, and $F(\varphi, \vartheta, \tau)$ contains the probability that ϑ is followed by τ and some other terms that handle condition 3. The details are given in Metzler et al. (2005).

In general, the forward equation is not a proper recursion, since for tihls τ that do not emit to the leaves, we have $k - e(\tau) = k$. Separating these silent tihls, we get

$$P_\varphi(k) = \sum_{\{(\vartheta,\tau):\tau(\vartheta)=\varphi, e(\tau)\neq\vec{0}\}} P_\vartheta(k-e(\tau)) \cdot g(\tau,k) \cdot F(\varphi,\vartheta,\tau)$$
$$+ \sum_{\{(\vartheta,\tau):\tau(\vartheta)=\varphi, e(\tau)=\vec{0}\}} P_\vartheta(k) \cdot F(\varphi,\vartheta,\tau).$$

With the vectors $\mathbf{P}(k) = (P_\varphi(k))_{\varphi\in\Phi}$ and $\mathbf{V}(k) = (\sum_{\{(\vartheta,\tau):\tau(\vartheta)=\varphi, e(\tau)\neq\vec{0}\}} P_\vartheta(k-e(\tau)) \cdot g(\tau,k) \cdot F(\varphi,\vartheta,\tau))_{\varphi\in\Phi}$, and the matrix $\mathbf{M} = (\sum_{\{\tau:e(\tau)=\vec{0},\tau(\vartheta)=\varphi\}} F(\varphi,\vartheta,\tau))_{\varphi,\vartheta\in\Phi}$, the forward equation can be written as

$$\mathbf{P}(k) = \mathbf{V}(k) + \mathbf{P}(k) \cdot \mathbf{M}.$$

With the unit matrix $\mathbf{I}$, and given that $(\mathbf{I} - \mathbf{M})$ is invertible, we obtain the solution

$$\mathbf{P}(k) = \mathbf{V}(k) \cdot (\mathbf{I} - \mathbf{M})^{-1},$$

which is a proper recursion because the entries of $\mathbf{V}(k)$ contain only $P_\vartheta(k')$ with $\forall_i : k'_i \leq k_i$ and $\exists_j : k'_j < k_j$. The matrix $(\mathbf{I} - \mathbf{M})$ does not depend on k and therefore needs to be inverted only once. This can be done numerically in advance of the forward algorithm. The matrix is very sparse. In order to make the forward algorithm as efficient as possible, one should carefully take advantage of the sparseness of $(\mathbf{I} - \mathbf{M})^{-1}$.

The runtime of the forward algorithm is at least proportional to the product of the lengths of the input sequences. Therefore, in its original form, it should be applied to only a small number of short sequences.

Markov-Chain Monte Carlo Methods for Sampling Multiple Alignments

In this section we describe a collection of ideas for a method for the joint sampling of multiple alignments and phylogenetic trees according to their posterior distribution for given sequence data. An implementation of this method is currently under development in the group of D. Metzler and will be freely available.

Using the idea of Markov-chain Monte Carlo (MCMC) sampling, we construct a Markov chain that converges against the posterior distribution. The state space S of the Markov chain consists of binary trees with edge lengths whose leaves are labeled with the input sequences, whose internal nodes are labeled with sequence lengths, and whose edges are labeled with pairwise bare alignments of sequences with lengths according to the adjacent nodes. A *bare* alignment does not contain information about the nucleotide or amino acid types. It can be given as a sequence over $\{H,B,N,E\}$ if it is specified which of the sequences is considered as ancestral. For the construction of the Markov chain, we use the idea of Gibbs sampling (Geman and Geman 1984; see also Liu 2001). This means that the random steps of our Markov chain will resample partial aspects of the current state, as, for example, a part of the alignment or the tree, from their posterior distributions conditioned on all other aspects of the current state.

One possibility is to resample a part of the alignment; that is, one has to pick an edge in the current tree and resample the pairwise alignment of the adjacent nodes (Fleißner et al. 2005). However, Jensen and Hein (2005) have shown that Gibbs sampling strategies that resample alignments triplewise instead of pairwise have better mixing properties and converge faster. Therefore, in our alignment resampling step, we pick an inner node v of the tree and resample a part of, say, 20 positions of the multiple alignment between v and its neighbors a, b, and c. We also resample the number of positions of the sequence associated with v in this part of the alignment. Then we consider a, b, and c as roots of the three subtrees that are separated by v. For each position associated with a, b, and c that must be realigned, we compute the posterior nucleotide (or amino acid) distribution, given the sequences at the leaves of the corresponding subtree.

Simultaneous Alignment and Phylogeny

This can be done efficiently by Felsenstein's pruning algorithm (Felsenstein 1981). Then we resample a multiple alignment for these sequence profiles and the new subsequence associated with v, whose length will also result from the sampling procedure. This resampling is done by a forward algorithm with randomized backtracing. During the forward algorithm, we compute for every triple (i, j, k) of sequence positions and every φ of the 41 relevant maps from $\{v, a, b, c\}$ to $\{H, B, h, b, e\}$ the probability $P_\varphi(i, j, k)$ that the beginnings of the sequence profiles up to positions i, j, and k are aligned to each other, and that the corresponding sequence of labelings ends with φ, which then will be followed by a tihl in which all positions are homologous. This is done by dynamic programming using the recursive formula given in Metzler et al. (2005).

For the resampling of the phylogenetic tree, we use Metropolis–Hastings steps (Hastings 1970). Thus, it is straightforward to resample the edge lengths. For resampling the topology of the tree, we randomly pick an internal edge, preferring short ones. Let v and w be its adjacent nodes; let $\{a, v\}$, $\{b, v\}$, $\{c, w\}$, $\{d, w\}$, and $\{v, w\}$ be their edges; and let l_a, l_b, l_c, l_d, and l be the lengths of these edges. We toss a fair coin to decide which change of topology is then proposed: (1) let $\{a, v\}$, $\{b, w\}$, $\{c, v\}$, $\{d, w\}$, and $\{v, w\}$ be the edges of v and w; (2) let $\{a, v\}$, $\{b, w\}$, $\{c, w\}$, $\{d, v\}$, and $\{v, w\}$ be the edges of v and w. In both cases the lengths of the edges are (in order of appearance) the old values l_a, l_b, l_c, l_d, and l. Before we decide about the acceptance of the new topology in a Metropolis–Hastings rejection-or-acceptance step, we have to propose new alignments for the six edges. Because the nodes v and w have changed their positions relative to the leaves, we reject their former sequence lengths and alignments. With a variant of the forward algorithm with randomized backtracking, we sample new sequence lengths for v and w, and resample the multiple alignment of a, b, c, d, v, and w. As in the resampling step for triplewise alignments, we compute sequence profiles for the entire sequences labeled to a, b, c, and d according to the sequences at the leaves of their corresponding subtrees. Since the multiple alignment must be sampled for the entire sequence profiles labeled as a, b, c, and d, it would be very time-consuming to compute (in analogy to the triplewise alignment sampling procedure) the values $P_\varphi(i, j, k, h)$ for all possible combinations of indices i, j, k, and h and each of the 437 relevant labelings φ. Therefore, we keep the homologies between the sequences at a, b, c, and d, and condition the new multiple alignment to respect these homologies. This drastically restricts the numbers of possible alignments and index combinations i, j, k, and h for which $P_\varphi(i, j, k, h)$

must be computed. The acceptance probability of the new combination of topology and alignment depends on the ratio of the posterior probabilities of the newly sampled and the former multiple alignment.

Optionally, it is also possible to resample mutation model parameters such as the transition–transversion ratio and the mean length of insertions and deletions.

DISCUSSION AND OUTLOOK

Bayesian and likelihood-based methods for the estimation of phylogenies from unaligned sequences need a probability distribution on the set of possible alignments. For making efficient dynamic programming algorithms applicable, these probability distributions should induce an HMM structure. Moreover, the probability distributions should be defined by a sequence evolution model that allows for insertions and deletions, since this makes the probability distribution interpretable. Another reason for this requirement is that the sequence evolution model defines how the parameters of the HMM depend on the edge lengths of the tree. The simplest indel process that leads to an HMM structure, the TKF91 model, is not suitable for many data sets as it allows only indels of single positions. The models TKF92 and FID permit longer indels and are still HMM-compatible. Extending these models without violating the HMM structure, however, increases the number of hidden states and thus also the computational cost.

The software MrBayes (Ronquist and Huelsenbeck 2003) has become an indispensable tool in modern molecular evolution biology as it makes the uncertainty in phylogeny reconstruction explicit, under the assumption that there are no doubts about the alignment. The MCMC procedure for the Bayesian sampling of phylogenies and alignments would also take the uncertainty in sequence alignment into account. Redelings and Suchard (2005) propose a method for this joint Bayesian sampling and also provide an implementation of their method (Suchard and Redelings 2006). However, they define the probability of alignments by an HMM model that is not exactly compatible with a model of the insertion-deletion process. In particular, the probability of the occurrence of gaps along the edges of the phylogeny does not take the edge lengths into account. This raises the question as to whether the method can be improved by using a multiple HMM model that is induced by the FID or TKF92 model.

Existing programs for joint estimation of phylogenies and alignments are very time-consuming (Fleißner et al. 2005). Therefore, we need to investigate the convergence behavior of different variants of the MCMC procedure and look for possible speed-ups. For example, the following questions are to be assessed: What is a good length or length distribution of the parts of triplewise alignments that are resampled? Is it always better to resample triplewise subalignments or could it be better to resample quadruple subalignments? Is it really necessary to take care of the silent states in the forward equation, or is their contribution so small that they can be neglected? This would turn the forward equation directly into a proper recursion, which would save a significant factor of computational time compared to the numerical solution proposed in this chapter. Since theoretically provable answers to questions such as these are usually out of reach, it is more promising to tackle these questions by simulation studies, which, of course, require an implementation of the algorithms.

Implementing a Gibbs sampling scheme for phylogenies and multiple alignments, such as the method sketched in the discussion of the forward equation, is not trivial. The data structures involved become very complex and, due to the nature of Gibbs sampling, everything must be efficiently changeable. This requires a very clean and foresighted design. Even early versions of the software should provide a high flexibility and extensibility, because both are needed to explore the various methods with respect to applicability, efficiency, speed of convergence, limitations, and possible improvements.

A future challenge for statisticians and computer scientists interested in visualization is to develop graphical methods and interactive software tools for exploring large Bayesian samples of multiple alignments and phylogenies.

Acknowledgments

The authors of this chapter thank Istvan Miklós, Gerton Lunter, Anton Wakolbinger, and Arndt von Haeseler for stimulating discussion.

CHAPTER 6

Phylogenetic Hypotheses and the Utility of Multiple Sequence Alignment

WARD C. WHEELER
American Museum of Natural History
GONZALO GIRIBET
Harvard University

Phylogenetic Hypotheses..................................96
    Observations and Data................................96
The Tree Alignment Problem..............................97
Criteria to Evaluate Hypotheses 98
Heuristic Techniques....................................98
    Multiple Sequence Alignment and the Tree
    Alignment Problem...................................99
Evaluation of Heuristic Techniques.......................99
    Real Data and Heuristics 101
Conclusions..102

Multiple sequence alignment (MSA) is not a necessary, but rather a potentially useful, technique in phylogenetic analysis. By this we mean that we can construct and evaluate phylogenetic hypotheses without MSA, and it may be productive in terms of time or optimality to do so. In order to evaluate this statement, we must first define phylogenetic hypothesis, define the problem, define the criteria we will use to assay the relative merits of hypotheses, define what we mean by the utility of a technique, and then finally compare the results of alternate techniques.

In the following sections, each of these terms and operations is defined. The final section will compare the results of a "one-step" optimization heuristic (direct optimization, per Wheeler 1996) embodied in POY4 (Varón et al. 2007) with the MSA + Search approach ("two-step" phylogenetics *sensu* Giribet 2005) embodied by Clustal (Higgins and Sharp 1988), using a large number of small data set simulations run under a variety of conditions (Ogden and Rosenberg 2007a) and a few larger (hundreds to over 1000 taxa) real data sets.

PHYLOGENETIC HYPOTHESES

A phylogenetic hypothesis is a topology (T), a tree linking terminal taxa [leaves or operational taxonomic units (OTUs)] through internal vertices [or hypothetical taxonomic units (HTUs)] without cycles. More formally, $T = (V, E)$, where V is the vertices (both terminal leaves and internal), and E the edges or branches that link V. Furthermore, there must be an assignment χ of observed data D to V, and a cost function σ that specifies the transformation costs between sequence elements (referred to as A, C, G, T, and GAP or "-" for this discussion). The phylogenetic hypothesis (H) then can be expressed as $H = (T, \chi_D, \sigma)$. For simplicity, the discussion and examples here (keeping the terminology used in Ogden and Rosenberg 2007a) will use the homogenous $\sigma = 1$. In addition, χ_D will always be a function of D; hence, we can rewrite as $H = (T, D)$ or just $T(D)$.

This topology may represent historical relationships or be simple summaries of hierarchical variation, but their form is the same. All hypotheses explain all potential data, just not to the same extent. Hence, those hypotheses that explain the data "best" are favored over others (summarized by Giribet and Wheeler 2007).

Observations and Data

In order to test hypotheses, we require data. Data are the observations an investigator makes in nature. DNA sequence data are gathered as contiguous strings of nucleotides from individual taxa. These data are observed without reference to the sequences of other creatures. Entire genomes can be sequenced without knowledge of any other entity. Nucleotides are observed only in reference to those that are collinear in the same taxon. MSAs are highly structured, inferential objects constructed by scientists either automatically or manually. They do not exist in nature, they cannot be observed, and they are not data.

THE TREE ALIGNMENT PROBLEM

The problem of assigning vertex sequences such that the overall tree cost is minimized when sequences may vary in length is known as the tree alignment problem (TAP) (Sankoff 1975; Sankoff et al. 1976). A phylogenetic search seeks to minimize the TAP cost over the universe of possible trees (Figure 6.1). Such an approach embodies the notion of "dynamic homology" (Wheeler 2001) as opposed to "static homology," where predetermined correspondences and putative homologies are established prior to analysis and applied uniformly throughout tree search.

Unfortunately, this problem is NP-Hard (Wang and Jiang 1994), meaning that no polynomial time solution exists (unless P = NP). In other words, the search for vertex median sequences is as complex as the phylogeny search problem over tree space for static homology characters. As with tree searches, other than explicit or implicit exhaustive searches for trivial cases, we will always be limited to heuristic solutions (see Slowinski 1998 for numbers of homology scenarios).

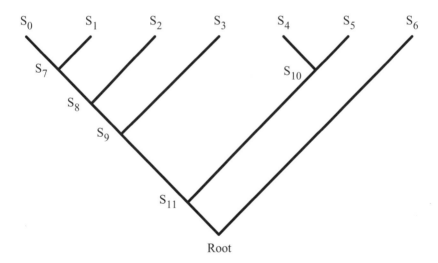

Figure 6.1. The tree alignment minimization assigns medians $\{S_7, \ldots, S_{11}\}$, given observed leaf sequences $\{S_0, \ldots, S_6\}$, such that the overall tree cost summed over all edges E is minimized. The root sequence can be any sequence between S_{11} and S_6.

CRITERIA TO EVALUATE HYPOTHESES

In order to compare hypotheses, there must be an explicit objective criterion. At its most fundamental level, a distance function d is specified to determine the pairwise cost of transforming each ancestor vertex into its descendant along an edge. This distance may be minimization-based (such as parsimony) or statistical in nature (such as likelihood); the TAP itself is agnostic. The distance must, at minimum, accommodate substitutions, insertions, and deletions. (Although substitutions are not strictly necessary, we will proceed as if they are).

Implementations of these criteria under dynamic homology have been proposed for parsimony (Wheeler 1996), likelihood (Steel and Hein 2001; Wheeler 2006), and posterior probability (Redelings and Suchard 2005). Each of these criteria will allow comparison of topological hypotheses (and their associated vertex sequence assignments). The central idea is that there is an explicit value that can be calculated and compared. The remainder of the discussion here will use equally weighted parsimony as the criterion of choice. Hence, results and conclusions were specific to this flavor of this criterion.

HEURISTIC TECHNIQUES

In principle, we could solve the TAP by examining all possible vertex median sequences. This has been proposed by Sankoff and Cedergren (1983) through n-dimensional alignment, using T to determine cell costs. Another method using dynamic programming over all possible sequences (Wheeler 2003c) would yield the same result. Only trivial data sets are amenable to such analysis.

Discussing heuristic solutions, Wheeler (2005) categorized these heuristic approaches into two groups: those that attempt to estimate vertex median sequences directly, and those that examine candidate medians from a predefined set. Estimation methods calculate medians based on the sequences of the vertices adjacent to them. Direct optimization (DO) (Wheeler 1996) uses the two descendant vertices for a (length n sequences) $O(n^2)$. Iterative pass (Sankoff et al. 1976; Wheeler 2003b) uses all three connected vertices and revisits vertices for improved medians, but at a time complexity of $O(n^3)$. Search methods such as "lifted" alignments (Gusfield 1997), fixed states (Wheeler 1999), and search-based (Wheeler 2003c) employ predefined candidate sequence medians in increasing number (for m

taxa, lifted uses $m/2$, fixed states m, and search-based $> m$). Time complexity of lifted alignments is linear for m sequences and quadratic for both fixed states and search-based (after an $O(m^2n^2)$ edit cost matrix setup). A polynomial time approximation scheme (PTAS) exists for TAP (Wang and Gusfield 1997), but the time complexity is too great to be of any practical use.

Multiple Sequence Alignment and the Tree Alignment Problem

As mentioned earlier, the determination of the cost (for any objective distance function d) of each individual candidate tree is NP-Hard. This is then compounded by the complexity of the tree search. The simultaneous optimization of both of these problems can be extremely time-consuming. The motivation behind MSA is to separate these problems, performing the homology step (MSA) only once. The determination of tree cost for these now static characters is linear with the length of the aligned sequences, and tree search can proceed with alacrity. This is a reasonable heuristic procedure whose behavior will depend on the appropriateness of using that single MSA for all tree evaluations. Obviously, for trivial cases, this will be as effective as more exhaustive approaches. The method can be further refined by linking MSA more closely with the tree search by generating new MSAs based on a TAP Implied Alignment (Wheeler 2003a), and iteratively alternating between static and dynamic searches until a local minimum is found ("Static Approximation, Wheeler 2003a).

Each of these techniques can be evaluated on two bases, the quality of the solution in terms of optimality value, and execution time. Here, we concern ourselves with the optimality of the solution, although it is clear that a good solution (such as PTAS or exact) may be "better" by optimality, but of little use due to its time complexity.

EVALUATION OF HEURISTIC TECHNIQUES

In order to examine the relative effectiveness of MSA, we will use equally weighted parsimony as our optimality criterion. Equal weighting is not used because of some innate superiority of this form of analysis (see Giribet and Wheeler 2007; Grant and Kluge 2005, for some acrimony), but because it offers a clear and simple test (similar reasoning motivated Ogden and Rosenberg 2007a). Other indices could be used, and other results found; hence, our conclusions are restricted.

There has been some discussion in the literature about "true" alignments and their place in evaluating phylogenetic methods (Kjer et al. 2007; Ogden and Rosenberg 2007a), more specifically MSA and DO. These authors, and others, have set up the test of these methods as the recovery of "true" alignment (known by simulation or other inferred qualities). The position taken here is that this is incorrect. Alignments are not an attribute of nature. They cannot be observed, only created by automated or manual means. Whether or not a method can create a MSA directly or as a tree adjuvant that matches a notion based in simulation or imagination is irrelevant to its quality as a solution to the TAP.

Ogden and Rosenberg (2007a) performed an admirably thorough set of simulations (15,400) on a small set of taxa (16) for realistically sized sequences (2000 nucleotides) under a variety of tree topology types and evolutionary conditions/models. To summarize, 100 replicate simulations were performed on seven tree topologies (balanced, pectinate, and five "random" topologies) under ultrametric, clocklike, and non-clocklike evolution, with two rates of change for 154 combinations. They then stripped out evolved (= true) gaps and reanalyzed the sequences in two ways. The first was the traditional "two-step" phylogenetics of alignment and subsequent analysis of static data. This was accomplished with Clustal (Thompson et al. 1994) under default conditions and PAUP* (Swofford 1998). The second was a "one-step" analysis using POY3 (Wheeler et al. 1996–2006). Ogden and Rosenberg (2007a) compared the implied alignments (Wheeler 2003a) generated by POY and offered by ClustalW with the simulated "true" alignments. The POY implied alignments were found to be more dissimilar to the simulated alignments than were those of Clustal.

Although the specifics of Ogden and Rosenberg's (2007a) use of POY could be a subject of discussion (Lehtonen 2008), the objective here is not to take issue with the details of their analysis, but to discuss their general approach. Although they did look at topologies in a secondary comparison, the authors never examined the optimality effectiveness of their competing approaches. No tree costs were reported.

We reanalyzed all 15,400 simulations performed and generously provided to us by Ogden and Rosenberg (2007a). Three analyses were performed in each case, yielding 46,200 runs. In the first, we used the Clustal alignments of Ogden and Rosenberg, running them as static nonadditive characters (all transformations equal), in the second the "true" alignments were used again as static nonadditive characters, and in the third the unaligned data were analyzed under equal transformation costs (indels = 1) using DO. All analyses were performed using POY4 beta 2398 (Varón et al. 2007) with ten random addition sequences and tree bisection-reconnection (TBR) branch swapping on several Mac Intel machines. POY4 replaced PAUP* in the static analyses for consistency of heuristic approach. For such small data sets, large differences are unlikely to appear. Given the settings and the problem here, POY4 differs from POY3 mainly in the efficiency of implementation; the core algorithms for DO are the same. The POY4 runs on unaligned data took approximately 10 to 20 times as long as those of the prealigned data.

In every one of the 15,400 comparison cases (Table 6.1), the unaligned data analyzed with POY4 yielded a lower cost than Clustal+POY4. The average tree cost differences for the 154 experimental combinations (over the 100 replicate simulations) were as low as 2% and as high as 20%. The higher rates of evolution (maximum distance = 2) had greater deficits compared to POY4 than did the lower rates. Interestingly, analysis of the simulated and Clustal analyses showed tree costs that were similar, with neither obviously producing lower cost trees.

As far as these simulations are concerned, the one-step heuristic approach of POY is unequivocally superior to the two-step alignment + tree search approach advocated by Ogden and Rosenberg (2007a).

Real Data and Heuristics

As a reality check, we performed the same pairs of analyses on four larger, real data sets (Table 6.2) used in Wheeler (2007). These data sets were all ribosomal RNA and varied in size from 62 taxa to 1040. They were analyzed with Clustal under the same default conditions as in Ogden and Rosenberg (2007a) (gap opening = 15, extension = 6.66, delay divergent = 30%, transition weight = 0.50, DNA weight matrix = IUB), and using the same ten replicates + TBR for aligned and unaligned data. In each of these four cases, the single-step POY4 tree costs were from 6% to 17% lower.

TABLE 6.1. TREE COST COMPARISONS OF SIMULATED
DATA OF OGDEN AND ROSENBERG (2007A)

Model	Balanced			Pectinate			Random A		
	C/P	T/P	T/C	C/P	T/P	T/C	C/P	T/P	T/C
a	1.0299	1.0266	1.0000	1.0726	1.0604	0.9890	1.0433	1.0335	0.9906
b	1.0993	1.1172	1.0164	1.1795	1.2068	1.0200	1.1276	1.1267	0.9992
RBL-1a	1.0525	1.0372	0.9855	1.0749	1.0562	0.9826	1.0522	1.0526	1.0004
RBL-1b	1.1680	1.1690	1.0009	1.1880	1.2068	1.0161	1.1200	1.1513	1.0268
RBL-2a	1.0502	1.0340	0.9843	1.0763	1.0592	0.9843	1.0565	1.0386	0.9830
RBL-2b	1.1811	1.1664	0.9876	1.1728	1.1961	1.0198	1.2100	1.1884	0.9821
RBL-3a	1.0447	1.0330	0.9900	1.0765	1.0634	0.9879	1.0440	1.0340	0.9904
RBL-3b	1.1453	1.1420	0.9971	1.1836	1.2143	1.0260	1.1271	1.1260	0.9990
RBL-4a	1.0476	1.0356	0.9886	1.0732	1.0578	0.9858	1.0696	1.0507	0.9825
RBL-4b	1.1464	1.1435	0.9975	1.1664	1.1877	1.0183	1.2042	1.2162	1.0100
RBL-5a	1.0423	1.0303	0.9885	1.0892	1.0645	0.9774	1.0511	1.0441	0.9934
RBL-5b	1.1356	1.1391	1.0031	1.2104	1.2390	1.0238	1.1296	1.1480	1.0160
RBLNoC-1a	1.0368	1.0186	0.9825	1.0517	1.0480	0.9966	1.0351	1.0262	0.9915
RBLNoC-1b	1.0827	1.0716	0.9898	1.0942	1.1116	1.0159	1.0984	1.0898	0.9922
RBLNoC-2a	1.0340	1.0188	0.9853	1.0563	1.0366	0.9814	1.0470	1.0325	0.9862
RBLNoC-2b	1.0815	1.0676	0.9872	1.1406	1.1310	0.9916	1.1090	1.1110	1.0018
RBLNoC-3a	1.0436	1.0209	0.9780	1.0805	1.0530	0.9747	1.0396	1.0285	0.9893
RBLNoC-3b	1.0934	1.0807	0.9884	1.1468	1.1771	1.0265	1.1013	1.1013	1.0001
RBLNoC-4a	1.0390	1.0195	0.9812	1.0560	1.0449	0.9898	1.0380	1.0274	0.9898
RBLNoC-4b	1.0860	1.0757	0.9910	1.1274	1.1349	1.0066	1.0860	1.0801	0.9946
RBLNoC-5a	1.0389	1.0236	0.9854	1.0656	1.0518	0.9871	1.0345	1.0260	0.9918
RBLNoC-5b	1.0967	1.0930	0.9966	1.1258	1.1520	1.0232	1.0897	1.0872	0.9980

NOTE: The Model column specifies the evolutionary model and rate of the simulation. Balanced, Pectinate, and Random A–E are the tree topologies. C/P denotes the average (over 100) trials of the cost ratio of the Clustal + Search trees and POY trees, T/P the average ratios of "true" alignment to POY costs, and T/C the ratios of "true" alignment to Clustal + Search trees.

CONCLUSIONS

Given that alignments are not "real" in any natural sense, their role can only be as a heuristic tool in the solution of phylogenetic problems. The core hypothesis of phylogenetic analysis is that topology which optimizes some measure of merit. The simulations of Ogden and Rosenberg (2007a) and the analyses presented here clearly show that, for this type of analysis, MSA is an inferior heuristic as far as generating low cost solutions to the TAP. MSA may be a useful tool in accelerating search heuristics (such as in static approximation), but on its own it falls short. In fact, the only case where MSA could be self-consistent would be if a complete set of optimal implied alignments (there are likely many for any given tree) were to be generated for the optimal set of trees (again there may be many). In this case, static (= two-step) analysis of this complete set would return the same set of optimal source

Random B			Random C			Random D			Random E		
C/P	T/P	T/C	C/P	T/P	T/C	C/P	T/P	T/C	C/P	T/P	T/C
1.0527	1.0423	0.9902	1.0621	1.0540	0.9924	1.0480	1.0351	0.9880	1.0478	1.0380	0.9904
1.1327	1.1437	1.0100	1.1440	1.1736	1.0259	1.1397	1.1412	1.0014	1.1296	1.1344	1.0040
1.0548	1.0416	0.9880	1.0570	1.0494	0.9929	1.0613	1.0445	0.9842	1.0479	1.0358	0.9885
1.1643	1.1680	1.0032	1.1478	1.1650	1.0150	1.1934	1.1939	1.0004	1.1400	1.1327	0.9939
1.0600	1.0430	0.9840	1.0630	1.0576	0.9949	1.0618	1.0573	0.9958	1.0485	1.0430	0.9900
1.1700	1.1713	1.0011	1.1624	1.1870	1.0212	1.1540	1.1860	1.0278	1.1193	1.1330	1.0123
1.0612	1.0520	0.9918	1.0600	1.0439	0.9872	1.0507	1.0378	0.9877	1.0638	1.0471	0.9843
1.1449	1.1652	1.0178	1.1729	1.1710	0.9984	1.1451	1.1497	1.0040	1.2059	1.2019	0.9967
1.0654	1.0476	0.9833	1.0621	1.0474	0.9861	1.0594	1.0425	0.9841	1.0585	1.0418	0.9843
1.2017	1.2013	0.9997	1.1716	1.1837	1.0104	1.1909	1.1876	0.9973	1.1936	1.1848	0.9926
1.0602	1.0432	0.9839	1.0574	1.0503	0.9933	1.0476	1.0373	0.9902	1.0622	1.0558	0.9940
1.1990	1.1888	0.9916	1.1442	1.1596	1.0135	1.1442	1.1444	1.0002	1.1680	1.1958	1.0242
1.0745	1.0536	0.9808	1.0600	1.0396	0.9808	1.0449	1.0280	0.9839	1.0466	1.0326	0.9867
1.1155	1.1467	1.0279	1.1330	1.1425	1.0084	1.1103	1.1108	1.0005	1.1028	1.1050	1.0022
1.0616	1.0450	0.9845	1.0500	1.0403	0.9910	1.0430	1.0307	0.9883	1.0457	1.0301	0.9900
1.1074	1.1315	1.0217	1.1097	1.1131	1.0030	1.0970	1.0981	1.0008	1.1164	1.1120	0.9958
1.0480	1.0340	0.9868	1.0519	1.0411	0.9897	1.0358	1.0243	0.9889	1.0351	1.0230	0.9882
1.1066	1.1065	0.9999	1.1256	1.1300	1.0039	1.0970	1.0783	0.9826	1.0966	1.0872	0.9915
1.0513	1.0405	0.9898	1.0682	1.0545	0.9872	1.0420	1.0246	0.9834	1.0323	1.0196	0.9876
1.0987	1.1105	1.0107	1.1197	1.1490	1.0262	1.1120	1.0830	0.9740	1.0937	1.0740	0.9819
1.0353	1.0190	0.9843	1.0615	1.0484	0.9877	1.0431	1.0237	0.9814	1.0369	1.0265	0.9900
1.0950	1.0741	0.9805	1.1248	1.1413	1.0147	1.1050	1.0909	0.9873	1.1005	1.0943	0.9946

trees. Only then would there be a solid relationship between optimal alignments and optimal trees. Given that both of these sets of objects are unlikely to be found and recognized for any nontrivial data set, this situation exists only in theory.

Multiple sequence alignments are neither real, nor particularly useful. So what keeps them around other than tradition, inertia, and Luddism?

TABLE 6.2. ONE- AND TWO-STEP ANALYSES OF FOUR LARGER REAL DATA SETS

Data Set	Taxa	POY	Clustal+POY	Cost Ratio
Mantid 18S rRNA	62	956	1052	1.1004
Metazoa 18S rRNA	208	26697	30983	1.1605
Archaea SSU	585	37003	39193	1.0592
Mitochondrial SSU	1040	77753	90685	1.1663

NOTE: All transformations were set to unity (indels = 1). The Mantid data are from Svenson and Whiting (2004); the other data are from Wheeler (2007).

Acknowledgments

The authors thank the National Science Foundation for financial support, Heath Ogden and Michael Rosenberg for supplying their simulation files, and Andrés Varón and Louise Crowley for discussion of this manuscript.

CHAPTER 7

Structural and Evolutionary Considerations for Multiple Sequence Alignment of RNA, and the Challenges for Algorithms That Ignore Them

KARL M. KJER
Rutgers University
USMAN ROSHAN
New Jersey Institute of Technology
JOSEPH J. GILLESPIE
University of Maryland, Baltimore County; Virginia Bioinformatics Institute, Virginia Tech

Identification of Goals. 106
Alignment and Its Relation to Data Exclusion. 108
Differentiation of Molecules 110
    rRNA Sequences Evolve under Structural Constraints 111
Challenges to Existing Programs. 114
    Compositional Bias Presents a Severe Challenge 114
    Gaps Are Not Uniformly Distributed .116
    Nonindependence of Indels. 121
    Long Inserts/Deletions 122
    Lack of Recognition of Covarying Sites (A Well-Known,
    Seldom-Adopted Strategy) 123
Are Structural Inferences Justified? 126
Why Align Manually?. 127
    Perceived Advantages of Algorithms. .127

An Example of Accuracy and Repeatability 129
Comparison to Protein Alignment—Programs and Benchmarks ... 136
Conclusion 137
Terminology 139
Appendix: Instructions on Performing a Structural Alignment 141

IDENTIFICATION OF GOALS

What Is It You are Trying to Accomplish with an Alignment? Some of the disagreement over alignment approaches comes from differences in objectives among investigators. Are the data merely meant to distinguish target DNA from contaminants in a BLAST search? Or is there a specific node on a cladogram you wish to test? Are you aligning genomes or genes? Are the data protein-coding, structural RNAs or noncoding sequences? Do you consider phylogenetics to be a process of inference or estimation? Would you rather be more consistent or more accurate? Are you studying the performance of your selected programs or the relationships among your taxa? Different answers to each of these questions could likely lead to legitimate alternate alignment approaches. Morrison (2006) reviewed the many uses of alignment programs, and distinguished phylogenetic alignments as a special subset that requires attention to biological processes. Hypša (2006) reached a similar conclusion and emphasized the importance of adding complexity to multiple sequence alignments and phylogeny estimation. This chapter is devoted to discussing the alignment of structural RNAs, or ribosomal RNA (rRNA) and transfer RNA (tRNA) sequences, for phylogenetic analysis (although our thesis applies to other smaller RNAs, such as tmRNAs, RNase Ps, and group I and group II introns). We seek to have our phylogenetic hypotheses be predictive and accurate, even if accuracy is difficult (or impossible) to demonstrate. By *accurate* we mean that a hypothesis coincides with the true history of branching events.

If You Knew You could Improve Your Alignment, Would You Do It? Figure 7.1 shows an example of a fragment of a computer-generated alignment of the 12S rRNA from a variety of primates with that of murine rodent outgroups. It follows an optimality criterion based on minimizing costs from a Needleman–Wunsch (1970) algorithm and a guide tree.

We hope that this example, almost like the first couplet of a dichotomous key, will indicate where you stand on the issue of adjustment.

Structural Considerations for RNA MSA

```
Mouse       GCTACATTTTCTTA--TAAAAGAACAT-TACTATACCCTTTATGA
Rat         GCTACATTTTCTTTTCCCAGAGAACAT-TACGAAACCCTTTATGA
Gibbon      GCTACATTTTCTA--TGCC-AGAAAAC-CACGATAACCCTCATGA
Baboon      GCTACATTTTCTA--CTTCAGAAAACCCCACGATAGCTCTTATGA
Orangutan   GCTACATTTTCTA---CTTCAGAAAAC-TACGATAGCCCTCATGA
Human       GCTACATTTTCTA---CCCCAGAAAAC-TACGATAGCCCTTATGA
Bonobo      GCTACATTTTCTA--CCCC-AGAAAAT-TACGATAACCCTTATGA
Chimp       GCTACATTTTCTA--CCCC-AGAAAAT-TACGATAACCCTTATGA
```

Figure 7.1. A fragment of an alignment of complete 12S rRNA, generated by ClustalX (Jeanmougin et al. 1998; Thompson et al. 1997).

Notice the "CTTCAGAAAA" in the middle of the figure for both the baboon and the orangutan. If these were your data, would you adjust the sequences to *correct* the *errors* made by the program, or would you leave it alone and let the program make all of the decisions about homology, even when it appears to have erred? Would you adjust the nearly identical sequences between the human and the chimps? Would you make decisions about which nucleotides to exclude? Would it bother you if the baboon grouped inside the rodents? There are no correct answers, and the choices you make have implications that relate to what you wish to discover from your data, and tell a lot about both your background and your objectives. If you adjust the alignment, even in this one instance, you have converted to a manual alignment, with all its strengths and limitations. Adjusting the alignment is an attempt to improve the accuracy of an alignment. Here we define "accurate" as representing true (unknowable) homology, and also propose that accurate homology estimations will probably improve the accuracy of the phylogenetic hypothesis. But how do you know what "accurate" is, and where do you draw the line? Is manual alignment an art form subject to the whims and biases of the aligner, or can we identify a repeatable methodology? Similarly, if we criticize manual alignments as subjective and inconsistent, might these same criticisms apply to computer-generated alignments? If accuracy is a concern, where do current algorithms fail?

Many workers could legitimately state that we cannot objectively define errors made by the computer, and, in fact, the whole concept would be counter to an optimality-based study. If you are looking for the shortest tree, you should favor an alignment that reduces the number of steps. Others might assume that a few errors, even if they

```
Mouse      GCTACATT(TTCT TATA--AA AGAA)CAT--TACTATACCCTTTATGA
Rat        GCTACATT(TTCT TTTCCCAG AGAA)CAT--TACGAAACCCTTTATGA
Gibbon     GCTACATT(TTCT -ATGCC-- AGAA)AAC--CACGATAACCCTCATGA
Baboon     GCTACATT(TTCT -ACTTC-- AGAA)AACCCCACGATAGCTCTTATGA
Orangutan  GCTACATT(TTCT -ACTTC-- AGAA)AAC--TACGATAGCCCTCATGA
Human      GCTACATT(TTCT -ACCCC-- AGAA)AAC--TACGATAGCCCTTATGA
Bonobo     GCTACATT(TTCT -ACCCC-- AGAA)AAT--TACGATAACCCTTATGA
Chimp      GCTACATT(TTCT -ACCCC-- AGAA)AAT--TACGATAACCCTTATGA
```

Figure 7.2. A structurally adjusted alignment of the same data as shown in Figure 7.1. Parentheses indicate the bounds of a hairpin-stem loop, with hydrogen-bonded nucleotides indicated with underlines (Kjer et al. 1994). An unaligned region (the "loop" portion of the hairpin-stem) is delimited with spaces.

could be defined, would be better left alone, because the mass of the data should counterbalance a few random errors. But what if the errors are not random, creating a potential for linking together unrelated groups that share the same systematic biases? What if there were some higher order of conservation that we could examine in making decisions about homology that does not necessarily result in shorter trees? Figure 7.2 shows the same region of rRNA as in Figure 7.1, but has been adjusted to minimize secondary structural changes predicted for this region of the molecule. Minimizing structural change is also an optimality criterion for homology assessment that we support and will explore in this chapter. Structural homology is based on position and connection, and assumes that the same structure existed in a common ancestor. Strict adherence to nucleotide homology may require that the same nucleotide state exists in a common ancestor as in its descendants, and, by this definition, structural homology and nucleotide homology may support different alignments.

ALIGNMENT AND ITS RELATION TO DATA EXCLUSION

One of the things that is not clear from the above comparisons (Figures 7.1 and 7.2) is what we should do with the nucleotides in the "loop" portion of the hairpin-stem loop, between the "TTCT" and the "AGAA." This is an extremely important issue, but somewhat outside the debate about alignment. Frequently, these regions are excluded from the analysis on the grounds that they are too variable to align. Some systematists find any form of data exclusion to be unacceptable, and those are

Structural Considerations for RNA MSA								109

	A	B		C					
				REC		RAA		REC'	
Mouse	?????	[TATA--AA]	1	[T-]	1	[ATAA-]	1	[-A]	1
Rat	?????	[TTTCCCAG]	2	[TT]	2	[TCCC-]	2	[AG]	2
Gibbon	ATGCC	[-ATGCC--]	3	[--]	3	[ATGCC]	3	[--]	3
Baboon	ACTTC	[-ACTTC--]	4	[--]	3	[ACTTC]	4	[--]	3
Orangutan	ACTTC	[-ACTTC--]	4	[--]	3	[ACTTC]	4	[--]	3
Human	ACCCC	[-ACCCC--]	5	[--]	3	[ACCCC]	5	[--]	3
Bonobo	ACCCC	[-ACCCC--]	5	[--]	3	[ACCCC]	5	[--]	3
Chimp	ACCCC	[-ACCCC--]	5	[--]	3	[ACCCC]	5	[--]	3

Figure 7.3. Three suggestions for dealing with the unaligned loop from Figure 7.2. (A) If the in-group is alignable, but the out-group cannot be aligned to the in-group, consider the out-group data as missing. (B) Eliminate the entire region, but recode it as multistate characters. Taxa sharing identical sequences are coded with the same state. (C) Gillespie's (2004) method of finding the unpaired middle, to break up the region into the ambiguously aligned loop, and flanking regions of slippage.

typically the same researchers who would not adjust the alignment given in Figure 7.1. It is possible that these regions contain some potentially informative characters, yet in our observations the ability to objectively retrieve signal from these regions quickly becomes confounded with increased sequence divergence across an alignment. There is a wide variety of treatments for these data. One option would be to transform the out-group states to "missing data" (Kjer et al. 2001), with the idea that if you do not have any reasonable confidence of homology between the rodents and the primates, the data really are "missing," and since you are interested in the relationships among the in-group taxa, and in-group motifs are of uniform length, they could be treated as in Figure 7.3A. Alternatively, you could exclude the unaligned nucleotides, and recode them as multistate characters (Figure 7.3B), or as fixed state characters as in Giribet and Wheeler (2001). Taken a step further, a step matrix that calculates the minimum number of steps to transform one state to another could be applied to these multistate characters (Lutzoni et al. 2000). For example, the fewest number of steps it would take to transform ACCCC (state 5) into ACTTC (state 4) is two, just as ACCCC is two steps from ATGCC, but ACTTC is three steps from ATGCC (Lutzoni et al. 2000; Wheeler 1999). Gillespie (2004) suggested that these regions are difficult to align due to the expansion and contraction of the more variable hairpin-stem loops of rRNA. He proposed a method that defines regions of ambiguous alignment, slippage, and regions of expansion and contraction (called RAA/RSC/REC coding), which subdivides these ambiguous regions based on their structural properties and is directly applicable to the methods of Kjer et al. (2001) and Lutzoni et al. (2000). A demonstration of this alternative treatment is shown in Figure 7.3C.

DIFFERENTIATION OF MOLECULES

It is obvious that the selective processes involved in the effects of insertions or deletions (indels) on the function of a gene (and thus the probability of observing such a change in a living organism) are completely different for structural RNAs and protein-coding genes. An indel of one or two nucleotides in a protein-coding gene results in a reading frame shift, whereas an indel of even three nucleotides adds or subtracts a codon. Thus, a single indel will most likely have a major effect on protein structure and function. Indels in structural RNA genes are very different. The effect an indel has on structural RNA is variable across sites of the gene. For instance, some regions of rRNA are highly conserved in length across phylogenetic domains whose common ancestors stretch back for billions of years (Gutell 1996), implying that there is little or no tolerance for length variation in these regions of a functional ribosome. Other regions freely tolerate insertions and deletions, as observed among the most recently divergent species (Schnare et al. 1996). So the location of indels, their frequency, and their length in ribosomal RNAs are determined by the affects they have on rRNA structure and hence function. Indels in rRNA are not randomly distributed, but typically highly clustered into regions called expansion segments (or variable regions) that are much reduced, or nonexistent, in prokaryotes and lower eukaryotes. These expansion segments are located on the surface of the ribosome in regions not considered critical for ribosome function (Ban et al. 2000; Cate et al. 1999; Schluenzen et al. 2000; Spahn et al. 2001; Wimberly et al. 2000; Yusupov et al. 2001); thus their evolution can be considered less constrained than that of the core rRNA.

There are other differences in alignment protocols that are dependent on the kinds of questions an investigator is attempting to answer. Researchers who study the evolution of genes need to look at structural variation. Information about how a protein evolves across kingdoms includes major rearrangements, missing amino acids, and large insertions, which may make alignment more difficult. At the level where we observe major substitution of codons, there is often a coincident saturation of nucleotides, even at the first and second codon positions. On the other end of the spectrum, if you are a population geneticist looking for patterns among recently diverged populations, you may encounter variation in noncoding regions of the genome. So investigators at both the deepest and the shallowest levels of divergence may confront serious alignment problems that we do not address in this chapter. An alignment of a protein-coding gene with so many indels that render homology assignment ambiguous is probably not an ideal marker for phylogenetic studies (note, for example, how many phylogenetic papers state that their protein-coding genes were length invariant, or that alignment was trivial). Similarly, noncoding regions, such as introns, are relatively rare sources for estimating phylogenies. So, for phylogenetic systematists, alignment problems are most frequently encountered with rRNAs or tRNAs. Those who design alignment algorithms are often interested in serving all investigators, however, and many programs are specifically designed to align proteins, with the default parameters set for protein-coding genes. All of these statements seem intuitively obvious. Yet, how many times in the literature have we seen phylogenetic studies state in their methods sections that rRNAs were aligned with *default parameters*, that is, using a program whose defaults were set to align proteins? There seems to be a basic misunderstanding, or at least a lack of concern, about differentiating alignment processes according to the effect that indels have on the kinds of genes that are being aligned (but see Benavides et al. 2007). We find a disconnection between alignment philosophy and biological and evolutionary constraints. Does constructing an alignment based on maximizing nucleotide identity make sense for rRNA?

rRNA Sequences Evolve under Structural Constraints

That nucleotides in rRNA do not evolve parsimoniously can be unambiguously demonstrated. Put another way, rRNA structures change more slowly than do the nucleotides that they comprise. The Gutell laboratory (http://www.rna.icmb.utexas.edu/) maintains a database, the Comparative RNA Web Site (Cannone et al. 2002), from which secondary structural diagrams can be downloaded. To demonstrate the nonconservation of nucleotides, relative to structural features, we suggest you download and print any two structural diagrams

A)

```
Austroagrion    UUAAUUAAUUUAAUUUGGUUAGUGU--UACAUAACUAUCAAU-AAUAUUUAAUUAG
Platycypha      UUAAU-AAUUUAAUUUGUUUGUUGU--GAUAUAAUUGUCAAU-AAUAUUUA-UUAG
Neoneura        UUAAU-AAUUUAAUUUAUAUAUUGU--UAUAUAAAUAUUAAU-AAUAAUUA-UUAG
Megaloprepus    CUAAU-AUUUUAUUUUAUUCAGUGU--AUCAUAAAUUGUUAAU-AAUAUUAAUUAG-
Dysphaea        CUGGU-AUUUUAAUUUAUUUAUUGU--AGCAUAAAUAUUAAU-AAUAAUUAUCUG-
Uropetalura     ---CUAACAUUAUAAUUUAUUUGAUGUUACAUAAUCAUUAAA-AAUAUAAGUUAG-
Tanypteryx      ---CUAAUUAAAUAAUUUAAUUGGUGUUAUAUAACCAUUAAU-AAUAUAAAUUAG-
Oxygastra       ---CUAAAUUAAUAUUUUAUUUAUUGUAUAUAAAUAUUAAA-AAUAUAAUUUAG-
Libellula       ---UUAAAUUAUAUUUUAGGUUAAUGGGA-AUAAUUAUUAAU-AAUAUAAUUUAG-
Chorismagrion   -CUAUCUAUUUAUUUUAUUGGUUGU--UGCAUAAACGUUAAU-AAUUUAUUGGUAG
Gomphus         ---CUAAAUUUGAAUUGGUGGUGGUGGUAUAUAAUCAUAAUU-AAUUUAAUUUUAG
Argia           -CUAAAUUUUUAAUUUAAUUUAUUGU--AAUAAUAAUUAUUAAU-AAUA-AAUUUUGG
Hypopetalia     ---CUAGUUUAAUAAUUUAUUUAAUGUUUUAUAAUUAUUAGA-AAUAAAACUAG-
Macromidia      ---CUAGAUUAUAGAUUUAUUUAAUGUGAUAAGAUUAUUAAA-GAUUUUAUUUAG-
Neogomphus      ---CUAAAUUUAUAAUUCUUUAAUGUUUUAUAAUUAUGGAA-AAUUUAUUUUAG
Amphiagrion     ---CUAAAACUUUAAUCUGUUUAUUGUUACAUAAAUAUCUGA-AAUAUUUUUUAG-
Progomphus      ---CUAAAACUAUAAUUUUUUAAUGUUUCAUAAUUAUAUAU-AAUAUAGUUUUAG
Hagenius        ---CUAAAACCA-GUUAAAAUUAAUGUGGCAUAAUUAUGUUUAACUGGGUUUUAG
Macromia        ---CUAUGUUAGUAUUUAUUUAAUGUGGAAUAAUUAUUGAU-AAUACAUCAUAG-
Macromia        ---CUAUGUUAG-AAUUUAUUUAAUGUGGAAUAAUUAUUAAU-AAUACAUCAUAG-
Calaphaea       CUGAUUUGUUUG--AUUUGGUUAAUGUGUUAUAAUUAUCUUA-AAUAC-UCAUCUG
```

Figure 7.4. An example of how nucleotide changes should not be used to assess alignment quality. This is an example of a hairpin-stem loop structure, with hydrogen-bonded nucleotides underlined. The first five nucleotides bind with the last five. Then there is a large bulge, followed by another four nucleotide interaction (UAGU/ACUA in the top sequence). (A) Aligned with ClustalX. (B) Structurally adjusted.

of the same rRNA sequence from distantly related taxa, and then superimpose one upon the other; hold them up to the light (or make transparencies of each, and superimpose them on a white piece of paper). If the structures between organisms are conserved, but the nucleotides within these structures are relatively less conserved, then you have proved to yourself that minimizing change among nucleotides does not make biological sense. It is not only the number of nucleotide changes that computer programs should seek to minimize, but, rather, they should seek to minimize change among structural features as a higher level of conservation, and then consider minimizing nucleotide changes after structural conservation has been optimized. Figure 7.4 shows an example of a highly conserved stem of five nucleotides. The first five nucleotides in Figure 7.4 are hydrogen-bonded to the last five nucleotides in each taxon. But that is not how ClustalX, with default parameters, aligned them. If we adjust the final five nucleotides in the two *Macromia* sequences (**CAUAG**) by

B)

```
Austroagrion   UUAAUUAAUUUAAUUUGGUUAGUGUUACAUAACUAUCAAU-AAUAUUUAAUUAG
Platycypha     UUAAU-AAUUUAAUUUGUUUGUUGUGAUAUAAUUGUCAAU-AAUAUUU-AUUAG
Neoneura       UUAAU-AAUUUAAUUUAUAUAUUGUUAUAUAAAUAUUAAU-AAUAAUU-AUUAG
Megaloprepus   CUAAU-AUUUUAUUUUAUUCAGUGUAUCAUAAUUGUUAAU-AAUA-UUAAUUAG
Dysphaea       CUGGU-AAUUUAAUUUAUUUAUUGUAGCAUAAAUAUUAAU-AAUAAUU-AUCUG
Uropetalura    CUAAC-AUUAUAAUUUAUUUGAUGUUACAUAAUCAUUAAA-AAUAUAA-GUUAG
Tanypteryx     CUAAU-UAAAUAAUUUAAUUGGUGUUAUAUAACCAUUAAU-AAUAUAA-AUUAG
Oxygastra      CUAAA-UUAAUAUUUUAUUUAUUGUUAUAUAAAAUUAAA-AAUAUAA-UUUAG
Libellula      UUAAA-UUAUAUUUUAGGUUAAUGGGA-AUAAUUAUUAAU-AAUAUAA-UUUAG
Chorismagrion  CUAUC-UAUUUAUUUUAUUGGUUGUUGCAUAAACGUUAAU-AAUUUAUUGGUAG
Gomphus        CUAAA-UUUGAAUUGGUGGUGGUGGUAUAUAAUCAUAAUU-AAUUUAAUUUUAG
Argia          CUAAA-UUUUUAAUUUAUUUAUUGUAAUAUAAAUAUUAAU-AAUA-AAUUUUGG
Hypopetalia    CUAGU-UUAAUAAUUUAUUUAAUGUUUUAUAAUUAUUAGA-AAUAUAA-ACUAG
Macromidia     CUAGA-UUAUAGAUUUAUUUAAUGUGAUAAGAUUAUUAAA-GAUUUUA-UUUAG
Neogomphus     CUAAA-UUUAUAAUUUCUUUAAUGUUUUUAAAUAUUGGAA-AAUUUAUUUUUAG
Amphiagrion    CUAAA-ACUUUAAUCUGUUUAUUGUUACAUAAAAUAUCUGA-AAUAUUU-UUUAG
Progomphus     CUAAA-ACUAUAAUUUUUUUAAUGUUUCAUAAUUAUAUAU-AAUAUAGUUUUAG
Hagenius       CUAAA-ACCA-GUUAAAAUUAAUGUGGCAUAAUUAUAGUUUAACUGGGUUUUAG
Macromia       CUAUG-UUAGUAAUUUAUUUAAUGUGGAAUAAUUAUUGAU-AAUACAU-CAUAG
Macromia       CUAUG-UUAG-AAUUUAUUUAAUGUGGAAUAAUUAUUAAU-AAUACAU-CAUAG
Calaphaea      CUGAUUUGUUUGAUUUGGUUAAUGUGUUAUAAUUAUCUUA-AAUAC-UCAUCUG
```

Figure 7.4. *(continued)*

inserting gaps at the arrow, the tree length increases. Whether this increase in tree length is justified is dependent on what you are trying to minimize in your algorithm: change among nucleotides or change among structures. We argue that in a structural molecule such as rRNA, secondary structure is more conserved than primary structure (nucleotides) (as we suggested above that you could prove to yourself). It is therefore unambiguous to favor the structurally aligned panel (Figure 7.4B) over the panel that was optimized to minimize change among nucleotides (Figure 7.4A). Similar conclusions have been reached even through the structural and phylogenetic analysis of internal transcribed spacer regions that interrupt subunits of rRNA (Denduangboripant and Cronk 2001; Hung et al. 1999; Hypša et al. 2005; Morgan and Blair 1998).

Gillespie, Yoder, et al. (2005, their Fig. 9A) illustrated a similar empirical example of how automated alignment failed to align nucleotides based on secondary structure in one of the most difficult-to-align regions of arthropod nuclear large subunit (LSU or 28S) rRNA. Gillespie, McKenna, et al. (2005) demonstrated the importance of structural alignments in proofreading the data. Hallmark features of rRNA, which must be present in functional rDNA genes, can be utilized as a means of checking the accuracy of generated sequences in a fashion that is no different than using translated amino acid sequences to validate the correct reading frame within protein-coding genes (Gillespie, McKenna, et al. 2005).

CHALLENGES TO EXISTING PROGRAMS

Compositional Bias Presents a Severe Challenge

One of the appealing things about DNA data is that all of the character states are discrete. With morphological characters, it often seems that as you continue to study more representatives of a taxon, your formerly "good" or "discrete" characters dissolve into a grade of continuous variation. Nucleotide characters are what they are, without intermediates. Even though this property of four discrete character states, evolving under a common mechanism, enhances the justification for models and algorithmic alignments, it also presents some new problems with homoplasy due to limited character-state space (Brooks and McLennan 1994; Lanyon 1988; Mishler et al. 1988). If a nucleotide is free to flicker back and forth among these four states, and if there is some nonrandom bias in the data among independent lineages, then there is the possibility for systematic error in our hypotheses of phylogeny. If life on some other planet had five nucleotides instead of four, then this problem would not be as serious as it is here on Earth. If we had only two nucleotide states, this problem would be much worse. Unfortunately, there are biological systems in which there are effectively only two states. For example, arthropod mitochondrial genomes are notoriously AT rich, but this bias ranges from 65.6% in *Reticulitermes* (Isoptera) to 86.7% in *Melipona* (Hymenoptera) (Cameron and Whiting 2007). It is easy to predict that with A's and T's constituting nearly 87% of the genome, a particular site that can be an A or T (such as silent third and first codon sites), will be. So, taxa that have independently evolved similar compositional biases may be drawn together by rapidly evolving, meaningless sites, and this convergence is more likely with two states than it is with four (Meyer 1994). Simmons et al. (2004) discuss at length the problem of limited character-state space.

Nucleotide compositional bias is particularly problematic in the hypervariable regions of rRNA. The conserved core (the length-invariant, alignment trivial regions) may possess the four nucleotides in nearly equal proportions, whereas the hypervariable regions (which contain many if not most of the parsimony informative characters, and

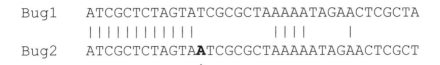

Figure 7.5. A hypothetical alignment from Kjer et al. (2007), showing that if gap costs are too high, Needleman–Wunsch algorithms may favor phenetic solutions in regions of nucleotide compositional bias.

wherein different alignment methods produce different hypotheses) can possess extreme nucleotide compositional bias. This bias can vary a great deal among taxa. For example, analyzing the structural properties of nuclear 18S rRNA across the major lineages of insects, Gillespie, McKenna, et al. (2005) demonstrated that base compositional bias within nearly all variable regions was severe, and that the patterns of these biases were inconsistent with phylogenetic expectations. Interestingly, in instances where pairwise comparisons of base composition were not significantly different, length heterogeneity was significantly different. This suggests that variable sequence length alone is not the only problem encountered in the alignment of rRNA sequences. Base compositional bias is another confounding factor.

Homoplasy presents problems not only in phylogenetic analysis but also in the assessment of homology in alignment programs. Figure 7.5 (from Kjer et al. 2007) shows a pairwise alignment of a hypothetical region in which the top and bottom sequences are identical to one another, except for a single indel, indicated in bold. Computer alignment programs using Needleman–Wunsch (1970) algorithms function by penalizing change through setting up a ratio of costs in inserting gaps, relative to the cost of a substitution (the gap cost-to-change ratio, or "gap cost" for short). If the gap cost used in the alignment is excessively high, the program will not insert a gap in the top sequence where it "belongs," as indicated by the arrow in Figure 7.5. Rather, the algorithm will continue racking up the relatively low mismatch penalties until it reaches a region of biased nucleotide composition, where the program happily lines up A's together, despite their being offset by one base. It is important to note that even in random sequences we would expect every fourth site to "match." In regions of nucleotide compositional bias, the expectations of nonhomologous but identical states is much greater than every fourth site, and approaches 50%.

Biased nucleotide regions are reminiscent of the *bland uniformity* that frustrates morphologists. But by throwing everything into a computer without looking at it, you would miss the fact that, under conditions of high compositional bias combined with rapid evolution and length variation, Needleman–Wunsch algorithms (and their subsequent derivations) can imitate phenetics, wherein taxa are grouped together according to the overall percentages of A's and T's, rather than by synapomorphies. The effects of compositional bias can be amplified if nonhomologous nucleotides are first aligned together with parsimony (with Needleman–Wunsch, minimizing nucleotide change), and then subject to long-branch attraction under a parsimony search. This is why we reject the assertion that alignments and analyses should logically be conducted under the same optimality criterion (Phillips et al. 2000). We believe that the goals of each endeavor (alignment and analysis), while not independent of one another, are different enough to require a separate approach, with each step favoring the best option. Simmons (2004) provides a detailed discussion on the separation of homology and analysis. The regions of rRNA that most commonly accumulate extreme compositional bias are the same regions that are most length heterogeneous, and hard to align. Compositional bias presents a severe challenge to Needleman–Wunsch-based alignment algorithms.

Compositional bias is particularly problematic for the direct optimization program POY (Gladstein and Wheeler 1999) because it depends on accurate reconstruction of ancestral sequences. Collins et al. (1994) showed that under conditions of nucleotide compositional bias, or accelerated substitution rates, parsimony severely underrepresents the rare states in ancestral reconstructions. The Collins et al. (1994) study employed a series of empirical and simulation studies to show this, and the mathematical proof by Eyre-Walker (1998) confirmed their findings. Reconstructing ancestral nodes is what POY *does*, and these studies indicate that results from a POY analysis should be interpreted with caution, and with an understanding of these limitations under conditions that are characteristic of the hard-to-align regions of rRNA.

Gaps Are Not Uniformly Distributed

Not only are substitution rates elevated in the hypervariable regions of rRNA, but also these regions accumulate insertions and deletions at a much more rapid pace than does the "conserved core" (Clark et al. 1984; Hadjiolov et al. 1984; Hogan et al. 1984; Michot et al. 1984).

Structural Considerations for RNA MSA

Anyone who has ever attempted to align rRNA data soon recognizes that gaps are clustered in regions. Kjer et al. (2007) measured the clustering of gaps by simply converting all of the nucleotides in a mammalian rRNA dataset into A's, and all of the gaps into C's. The among-site rate variation of indels from our so-altered NEXUS file was measured on the expected tree using PAUP version 4.0b10 (1998) by estimating the shape of the gamma distribution (Yang 1994). The gamma distribution is best indicated by a value called "alpha," whose values below 1 indicate serious among-site rate variation. The alpha value was 0.45, confirming that variation among sites with respect to the frequency of insertions and deletions is indeed highly variable. This clustering of indels has several important ramifications with respect to alignment. Most importantly, it means that ideal gap costs should vary among sites (Kjer 1995). Typically, computer alignments are performed with fixed gap costs. If biological gap costs vary among sites, then all analyses using fixed gap costs will underrepresent appropriate gap costs at some sites, and overrepresent gap costs at others. The "ideal" average gap cost, even if it were algorithmically and objectively defined, would be inappropriate for most sites. Kjer et al. (2007) demonstrate this with a figure, reproduced here as Figure 7.6. In Figure 7.6, structure is indicated with Kjer et al.'s (1994) notation on the nucleotides, and Gillespie's (2004) structural mask above them. The top panel contains a commonly sequenced region of mitochondrial 12S rRNA from a series of murine rodents. The lower panel contains sequences from a considerably more diverse group: a whale, an ape, an ostrich, a lizard, and a snail. Variation in length among the rodents indicates that the gap cost in those regions should be relatively low, just as invariant lengths among the different phyla should indicate the need for a high gap cost. In this region, we can see that the loop portion of stem 42 (variable region V7) should have a low gap cost, allowing for the easy introduction of gaps. Directly downstream from this hypervariable region is a region of extremely high conservation in length. Apparently, indels in strand 38' are not permitted. Even if we cannot quantify gap costs (which we cannot do because they are arbitrary, and they are arbitrary because we do not have realistic models for indels), you can scan across Figure 7.6 and apply flexible gap costs. For example, the loop between the strands of stem 40 should receive a low gap cost, and the loop between the strands of stem 42, an even lower gap cost. Contrast those low gap costs to the near infinite gap cost within strand 38' and all the undefined gap costs in between for other sites. If you wish to check your own 12S rRNA data for

```
                        H1068   H1074  H1074'                H1068'              H1113                      H1113'     H1047'
                        39      40     40'                   39'                 42                  V7     42'        38'
                        (((.((  (((    )))                   )))..))                              (((                  ))))
                                                                                                                       ))).))...))))))
Pachyuromys duprasi     |GCCAAC| |CCT| [AA-GA] |AGG) [AATAAAAA] |GTAAGC| GAGAT (AAT [AAA--TTAAC--] ATT) AAAA |CGTTAGGTCAAGGTGTAAC|
Tatera afra             |GTAAAC| |CCT| [AA-AA] |AGG) [-ATTCAAA] |GTAAAC| AAAAG (AAT [CA-----AACA-] ATG) AAGA |CGTTAGGTCAAGGTGTAGC|
Tatera leucogaster      |GTAAAC| |CCT| [AA-AA] |AGG) [-ATTCAAA] |GTAAAC| AAAAG (AAT [CA-----AAC--] ATG) AAGA |CGTTAGGTCAAGGTGTAGC|
Tatera indica           |GTAAAC| |CCT| [AA-AA] |AGG) [ACGTAAA] |GTGAGC| AAAAT (AAT [CA-----AAC--] ATG) AAGA |CGTTAGGTCAAGGTGTAGC|
Mus musculus            |GCAAAC| |CCT| [AA-AA] |AGG) [TATT-AAA] |GTAAGC| AAAAG (AAT [CA-----AAC--] ATA) AAAA |CGTTAGGTCAAGGTGTAGC|
Mus musculoides         |GCAAAC| |CCT| [AA-AA] |AGG) [-AGGAACA] |GTAAGC| ACAAG (AAT [AT-----CC--] ATA) AAAA |CGTTAGGTCAAGGTGTAGC|
Myospalax sp.           |GCAAAC| |CTT| [AA-AA] |AAG) [-AACAAAA] |GTAAGC| AAGAT (CAT [C------CC--] ATA) AAAA |CGTTAGGTCAAGGTGTAGC|

Balaenoptera physalus   |GCAAAC| |CCT| [AA--A] |GGG) [-AGCAAAA] |GTAAGC| ATAAC (CAT [CC-----TAC--] ATA) AAAA |CGTTAGGTCAAGGTAAC|
Pan troglodytes         |GCAAAC| |CCT| [GATGA] |AGG) [-TTACAAA] |GTAAGC| ACAAG (TAC [CC-----AC--] GTA) AAGA |CGTTAGGTCAAGGTGTAGC|
Struthio camelus        |GCCCGC| |CTC| [AT-GA] |GAG) [---AATA] |GCGAGC| ACAAT (AGC [CC-----ACCC] GCT) AACA |AGACAGGTCAAGGTATAGC|
Squalus acanthias       |GCTCAC| |CCT| [GT-GA] |AGG) [-ATAAGAA] |GTAAGC| AAAAA (GAA [CT-----AAC--] TCC) CATA |CGTCAGGTCGAGGTGTAGC|
Ceilana tramoserica     |GTTAAC| |CTT| [AT--A] |AAG) [-AAAAAAA] |GTTAAC| AATAA (AGA [ATATAAAAACTT] TCT) CATA |GGTCAGATCAAGGTGCAGC|
```

Figure 7.6. A structural alignment of a 12S rRNA fragment from murine rodents (top panel) and other diverse taxa (bottom panel) from Kjer et al. (2007). Structural notation follows Kjer et al. (1994), with underlines indicating hydrogen bonds, parentheses indicating hairpin-stem loops, and vertical lines delimiting longer-range interactions. Unaligned regions are placed in brackets (and ignored in NEXUS format). Notation above the sequences follows the structural mask of Gillespie (2004). This figure shows the contrast and close proximity of relatively length-heterogeneous regions to length-invariant regions.

alignment errors, you will probably find them by lining up stem 39. Using fixed integers as gap costs and applying them across a molecule that is demonstrably length-heterogeneous as a result of regional-specific clustering of indels is a commonly used, but biologically unrealistic, approach to the "alignment problem."

Hence, the practice of exploring parameter space with *sensitivity analyses*, that is, the testing different gap costs, in an effort to select among an infinite pool of gap costs and other parameters is problematic (but supporters of sensitivity analyses would be correct in noting that this is no more problematic than performing no such tests, if you are tied to fixed gap costs). Sampling around a series of parameters, as a means of parameter selection, implies that some parameters are "good," whereas others are "bad." This is a futile endeavor. There is no ideal, single, fixed gap cost for an alignment such as this because we are dealing with a heterogeneous assortment of regions. Sensitivity analyses require that at least some of the analyses are appropriate. But when we look at rRNA sequence data, where the "gappy" regions are clustered, we can see that one-gap costs will work well for one region, and poorly for another. By changing the gap cost, other regions may be well aligned, whereas the regions previously well aligned become worse. Different gap costs may shift the appropriately aligned regions from one region to another without necessarily expanding the proportion of well-aligned sites. Of course, if homology is completely ambiguous and unknowable, one may find it useful to present alternative alignments in assessing alignment uncertainty. However, we find it unreasonable to assume that history happened in multiple ways when structural homology favors a single solution.

One proposal for selecting among parameters is to perform a sensitivity analysis on a variety of parameters, and measure each resultant tree against some external criterion: a tree based on morphological characters, for example, or minimizing incongruence length differences (ILD; Farris et al. 1994) between partitions. However, there are an infinite number of parameters to explore. Wheeler (2005) discussed the problem, and also how this infinite space might be realistically explored. Wheeler (2005) explored gap costs and transversion weights (as did Terry and Whiting 2005; Whiting et al. 1997; and others). These explorations result in a three-dimensional plot of the parameter landscape. What you would want in such a landscape is a single hump containing a distinct peak, because with such a simple distribution, if you are anywhere near the peak, then you can be assured that the combination of

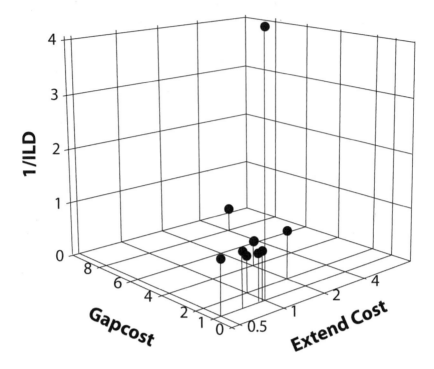

Figure 7.7. Sensitivity analysis, simultaneously exploring the cost of inserting a gap (gap cost), and the cost of extending an existing gap (extend cost), and how combinations of these costs influence ILD scores. The higher the inverse of the ILD scores, the less disagreement there is between the partitioned analyses (in this case, 12S vs. 16S topology). This figure is taken from Kjer et al. (2007).

gap cost and some other metric (such as transversion weights) is near optimal. This is not the shape of the peak found by Kjer et al. (2007), who used an ILD test to maximize congruence between 12S rRNA and 16S rRNA datasets in optimizing gap costs with gap extension costs (the cost of inserting additional gaps, once an initial gap has been inserted). Figure 7.7 shows the shape of this "peak." The problem here is that there was a relatively flat plane of inverse ILD scores, and a single sharp spike, where gap costs = extendcosts = 1. Unlike the "single hump distribution," this kind of distribution does not instill confidence that the best gap cost, relative to the indel extension cost, has been discovered, because an even higher spike may exist among the infinite combinations of parameters that were not explored. Other studies have also found it difficult to select parameters with sensitivity analyses (Terry and Whiting 2005; Wheeler 2005), although the generality of this problem has not been explored. One thing is certain, though; it is arbitrary to perform sensitivity analyses that compare only two analytical parameters (such as gap cost vs. transversion weights, or gap costs vs. extend costs) when there are a multitude of interacting parameters that simultaneously influence phylogenetic hypotheses. Many of these parameters may not be best treated with integers, or with fixed values, and many of them, such as the gap cost, are arbitrary (Doyle and Davis 1998; Hickson et al. 2000; Kjer 1995; Phillips et al. 2000; Vingron and Waterman 1994; Wheeler 1996).

Nonindependence of Indels

One of the reasons that gap costs are arbitrary is that we really do not have a reasonable model for insertions and deletions. One of the most unreasonable assumptions behind many of the existing algorithms is that multiple adjacent indel positions are all independent of one another, when they may have resulted from a single event. Simmons and Ochoterena (2000) discuss at length the problems with nonindependence of gaps and the problems with treating individual gaps as 5th state characters. They convincingly argue that contiguous gap positions are most parsimoniously interpreted as the result of a single event, and propose a system for coding them. Figure 7.8 offers an example of how we do not know the history of events that led to the present condition. However, we do know a number of things with certainty; first, this is a region of compositional bias (82% AT), and second, this is a region in which gaps of multiple lengths have accumulated among these and

```
                                     (((( 				                              ))))
Ptilocolepus          GUCAUUGAG[AAC--------------------CGA-UAAA]CUCAGAGGC
Palaeagapetus         GUCAUUGGG[AAU---------------------CACUAAA]CCCAGAGGC
Anchitrichia          GUCAUUGGG[AAUUUUUCAAACAUA--CAAU---CAUAACUAAA]CCCAUAGGC
Brysopteryx           GUCAUUGGG[AAUAUAUGGAUAAUAAACAAUGAAUCUAACAAAA]CCCAUAGGC
Matrioptila           GUCACUGGG[AG---------------------CGAUUAAA]CCCACGGGC
Rhyacophila fuscula   GUCAUUGGG[AUUUUUUUU----------------ACACUAAA]CCCAGAGGC
Rhyacophila brunnea   GUCAUUGGG[AUUUUUUU-----------------ACACUAAA]CCCAGAGGC
```

Figure 7.8. A structurally aligned region of caddisfly rRNA, with ambiguously aligned nucleotides in brackets. Underlines indicate hydrogen-bonded nucleotides, as do the parentheses above the sequences. Note the compositional bias, and the extreme length variation in the unaligned region.

many other taxa. We also know that this same region is hypervariable and hard to align in a wide range of taxa. In an example taken from caddisfly (Insecta: Trichoptera) rRNA, we can see that the sequences from *Anchitrichia* and *Brysopteryx* (Hydroptilidae, Hydroptilinae) are much longer than those from *Paleagepetus* and *Ptilocolepus* (Hydroptilidae, Ptilocolepinae). If we treat each of the gaps as 5th state characters, that is, as if they were independent of one another, then that would mean that ptilocolepines share 24 independent deletions, relative to *Matrioptila*. Such a treatment of the data is nonparsimonious. It would be more parsimonious to assume that there was a large insertion of multiple nucleotides in Hydroptilinae, followed by subsequent modifications. Although the alignment of this region is ambiguous, it gives strong hints about relationships. The nucleotides between the brackets give strong support to the monophyly of *Rhyacophila*, and the large insert present in the Hydroptilines is likely homologous, since it is long and complex, and there are conserved motifs that indicate a common origin. Yet, for one to infer dozens of synapomorphies from this rapidly changing, unalignable mess, one would have to disregard common sense and assume that all characters are equally informative and independent of one another. Some studies find that 5th state coding for deletions outperform methods that treat deletions as missing data (Ogden and Rosenberg 2007b). While we agree that gaps provide important signals (Freudenstein and Chase 2001), treating them as 5th states falls short to the probability that, while many long inserts *do* indicate phylogenetic relatedness, treating all the gaps as independent characters inflates the support for these nodes (Simmons and Ochoterena 2000), whether they are due to common history, convergence, or alignment artifacts.

Long Inserts/Deletions

Whereas simultaneous deletions of five or six nucleotides at a time, occurring in independent lineages, may draw unrelated taxa together if they are considered to be five or six independent events, much larger indels, sometimes hundreds of nucleotides long, are known to occur (e.g., Giribet and Wheeler 2001 provided a list of atypically long 18S rRNA sequences of metazoans). These long insertions have the potential to wreak havoc on a computer alignment (Benavides et al. 2007). The epitome of this problem is perhaps the bizarre insertions that interrupt both the 28S (Gillespie, unpublished) and 18S (Gillespie, McKenna, et al. 2005) rRNA sequences of the strepsipterans. Not only do inserts occur in the variable regions and expansion segments of the rRNA of these odd insects, but extraordinarily (up to 366 nucleotides) long insertions are known to occur within the hairpin-stem loop of the highly conserved pseudoknot 13/14 in the V4 region of 18S rRNA (Gillespie, Mcenna, et al. 2005). Thus, despite the earlier statement that conserved regions are less tolerant of indels, the Strepsiptera data suggest that large introns can occur in conserved regions. In fact, virtually all of the introns that interrupt rRNAs occur in the most conserved regions of the tertiary structure (Jackson et al. 2002; Wuyts et al. 2001), particularly at the subunit interface or in conserved sites with known tRNA–rRNA interactions (Jackson et al. 2002). While introns are relatively rare in rRNA sequences, only a manual evaluation using a structural model would detect their presence. Still, the majority of large inserts in rRNAs are likely part of the mature molecules and are localized to the surface of the ribosome. However, structural models for the expansion segments and variable regions exposed to the surface of the ribosome are becoming more and more refined with the addition of new taxa and sequences and through refinements in the ribosome crystal structures (e.g., Alkemar and Nygård 2003; Alkemar and Nygård 2004; Buckley et al. 2000; Gillespie et al. 2004; Gillespie et al. 2006; Gillespie, McKenna, et al. 2005; Gillespie, Munro, et al. 2005; Gillespie, Yoder, et al. 2005; Hickson et al. 1996; Kjer 1997, 2004; Mears et al. 2006; Misof and Fleck 2003; Ouvrard et al. 2000; Page 2000; Schnare et al. 1996; Wuyts et al. 2000). Thus, conserved structures within even variable regions and expansion segments will be necessary to guide the assignment of nucleotide homology when high levels of length heterogeneity exist across alignments.

Lack of Recognition of Covarying Sites (A Well-Known, Seldom-Adopted Strategy)

Wheeler and Honeycutt (1988) identified a directed substitution rate within helices of the 5S rRNA of animals and plants that deviates from the neutral theory of molecular evolution explaining rRNA evolution (Kimura 1983; Ohta 1973). This slightly deleterious mode of sequence evolution in rRNA, in which noncanonical base pairings, or bulges, are replaced by compensatory base changes or reversals to the original state, has been identified in subsequent studies (e.g., Douzery and Catzeflis 1995; Gatesy et al. 1994; Kraus et al. 1992; Rousset et al. 1991; Springer and Douzery 1996; Springer et al. 1995; Vawter and

Brown 1993), and appears to be the mechanism orchestrating secondary structural conservation in rRNAs. Paramount to the findings of Wheeler and Honeycutt (1988) was not only the identification of two different selective constraints within the same molecule (pairing versus nonpairing regions), but also the realization that *nucleotides within pairing regions in rRNA datasets are not independent characters, because a change at one site influences the probability of a substitution at another site*. This poses an added difficulty when treating helices in phylogenetic analysis, as opposed to unpaired nucleotides, wherein interdependence with other positions is not easily demonstrated (although the variety of tertiary and stacking interactions could in theory be modeled). Regarding parsimony, analysis of pairing (stems) or nonpairing (loops) regions has been suggested, but not both in simultaneous analysis (Wheeler and Honeycutt 1988). Some workers have implemented a stem-loop weighting approach to accommodate the nonindependence of pairing regions (Dixon and Hillis 1993; Smith 1989; Wheeler and Honeycutt 1988). Although seemingly intuitive, down-weighting stems on the basis of their nonindependence will also relatively up-weight positions that are hypervariable, and often nonpairing and perhaps misaligned, thus inaccurately representing the information contained within pairing regions. Up-weighting compensatory mutations within pairing regions has justification (Ouvrard et al. 2000), particularly if rare substitutions define major clades; however, discerning which characters to weight within an alignment can be puzzling if the ancestral pairing cannot be immediately identified (i.e., before analysis). Simon (1991) warns against stem-loop weighting, and van de Peer et al. (1993) illustrate that stems and loops are both highly heterogeneous in terms of substitution rates. These added difficulties, coupled with the fact that assumptions of certain branch support measures such as the bootstrap (Felsenstein 1985) and Bremer support indices (Bremer 1988; Donoghue et al. 1992) are violated by the nonindependence of rRNA pairing regions, suggests that a parsimony approach to analyzing rRNA alignments may not adequately accommodate these data. There may be interacting operational vs. philosophical factors involved; if compensatory changes in paired stem sites are also relatively more conservative, parsimony may appropriately (although inadvertently) up-weight the slower-evolving characters (Kjer 2004). Similarly, standard likelihood models of DNA substitution, which are all based on a 4×4 rate matrix, are also insufficient for phylogeny estimation using rRNA, because of their failure to account for correlated bases forming helices.

Studies modeling the evolution of pairing regions in rRNA molecules have grown in the last decade. Although most have focused on modeling base pair evolution under likelihood, similar approaches are under development for parsimony (Yoder 2007, Yoder and Gillespie, unpublished). These studies have all centered on implementing a substitution matrix that accommodates the nonindependence of helical regions. Unlike the typical 4 × 4 substitution matrix used for modeling DNA evolution, a matrix modeling rRNA evolution consists of all possible substitutions within a pairing region. Hence, a 16 × 16 matrix is used to model pairing regions, with the most general time reversible (GTR) model allowing for 134 free parameters. A detailed explanation of the simplified families of RNA substitution models was recently provided by Gillespie (2005). Given the attention being addressed to modeling RNA evolution, two software packages have incorporated some of the above-mentioned models into their programs. MrBayes version 3.1 (and earlier versions) (Ronquist and Huelsenbeck 2003) includes model 16B (Schöniger and von Haeseler 1994) and allows for helices to be modeled independently as pairs along with other models for nonpaired sites (i.e., loops, codons, amino acids). Importantly, model 16B should be considered an F81-like model for pairing sites, and when the covarion model in MrBayes is set to REV or HKY85, model 16B becomes different for each case (Jow et al. 2005). The program PHASE version 1.1 (Jow et al. 2002) also provides a means to simultaneously model multiple partitions with different models of evolution. In addition, PHASE contains a suite of RNA models that allow for the evaluation of the performance of different RNA models on a given dataset. Most likely as a result of the study of Savill et al. (2001), those models that allow for base pair asymmetry and a nonzero rate of double substitutions, namely, models 16A, 7A, 7D, 6A, and 6B, are all included in the PHASE program. Thus, PHASE has an appeal over MrBayes 3.1 in that the user can determine the best model of evolution for an RNA dataset, rather than settle for only one RNA model (perhaps with slight modifications). The soon-to-be-released MrBayes 4.0 will contain additional rRNA models.

Doublet models are thus directly related to the alignment issue, because alignments performed within a structural context provide a template that allows for a more realistic modeling of the evolution of these complex biological molecules. Intuitively, they are more desirable to the evolutionary biologist. However, it is not our intention here to criticize the algorithmic approach to alignment just because more biologically sound methods exist. On the contrary, we fully support and prefer algorithmic methods, as long as the algorithm that is applied has some grounding in biological reality. Current methods that ignore the properties of rRNA are not biologically grounded. We hope that in pointing out the challenges faced by current methods, we will accelerate the implementation of algorithmic methods, and eventually eliminate the difficult, tedious, and nonrepeatable manual alignments. Before this can happen, however, if you favor phylogenetic hypotheses that are meaningful and predictive, then manual approaches should not be eliminated until algorithmic methods can be shown to outperform them. Replacement should not occur just because a new method remedies some of the problems and is "cool," new, and computationally expensive.

ARE STRUCTURAL INFERENCES JUSTIFIED?

One of the criticisms of rRNA structural inferences is that they are inferences, not direct observations. Despite great efforts in cryo-electron microscopy (e.g., Frank and Agrawal 2000; Frank and Agrawal 2001; Frank et al. 2000), complete ribosomal RNA secondary structures can be directly observed only through x-ray crystallography. While several atomic structures of ribosomal subunits now exist for yeast (Spahn et al. 2001), the archaean *Haloarcula marismortui* (Ban et al. 2000), and the bacteria *Thermus thermophilus* (Brodersen et al. 2002; Schluenzen et al. 2000; Wimberly et al. 2000; Yusupov et al. 2001) and *Deinococcus radiodurans* (Harms et al. 2001), most rRNA secondary structures are inferred through comparative evidence. Structural alignments identify with very high accuracy (>90%, Gutell et al. 2002) those regions involved in base pairing. Comparative evidence works under the assumption that if multiple sequences can fold into the same conserved structure, and if there is a substitution in one part of the putative stem, it is usually followed by its complementary partner. Inferential, yes, but what are the odds that structures are not real, and are you willing to take that chance? The odds that structurally superimposable structures could arise by chance are easily calculable. For example, if there is an A at one site, what is the probability that there will be a T at the position of its putative partner? Answer: 0.25 according to Jukes and Cantor (1969). So if you align two taxa together, and find 35 compensatory mutations between them, the probability of this happening by chance is 0.25^{35} (that is, a zero, followed by a decimal point, 21 zeros, and then an 8). Adding the thousands of observed compensatory substitutions among all taxa, one arrives at a number so small that the human mind (even among mathematicians who are experienced in thinking about really small numbers) cannot come close to even imagining these numbers in a meaningful way. When one considers how tenuous the whole process of phylogenetic inference is (where we never *know* anything, and the best we can do is come up with a reasonable *guess*, where our data are consistent with our hypotheses, given our assumptions), it seems absurd to argue over whether it is safe to assume whether structural constraints that have a virtually zero (but not *technically* zero) probability of being random should be abandoned on philosophical (or other) grounds. We also note that *translated* amino acids are routinely used to check DNA alignments of protein-coding sequences, even though few (if any) of these studies bother to experimentally demonstrate that the genetic code for the taxon of interest is the same as the model taxon.

WHY ALIGN MANUALLY?

As we were considering our observations about the objectives of phylogenetic alignment, and beginning to write them down for this paper, Morrison (2006) presented a review of procedures and philosophies. This excellent review thoroughly explores the differences among us, and, in fact, much of what we had thought to be intuitive but unproven could now be explained in a series of logical arguments. Morrison (2006) lays out a series of problems with current algorithms that were designed for one purpose, and then used for phylogenetics. He argues that many of the problems we face in alignments stem from a failure to recognize that the program is neither designed nor suited for phylogenetic inference. Whereas we had noticed these problems, we had assumed that some smart person out there must have some reasonable solution to phylogenetic alignment; we just had not read about it yet. Morrison (2006) presents a radical new view, stating, "Our objective should be biological plausibility rather than mathematical optimality." With respect to alignments, we are in complete agreement with this statement. Algorithms that currently align sequences with the goal of reaching a mathematical optimum may fail for phylogenetics if they do not simulate biological reality.

Perceived Advantages of Algorithms

Much about alignment has simply been assumed, without question. One's preferences, alluded to in the introduction, seem more a matter of culture and tradition than experimentally justified or even thoughtfully considered criteria. It seems intuitively obvious that computers are more objective at making alignment decisions than manual alignments. Are they? No, not if the computer requires arbitrary input parameters. The following comparison should be made. Consider a thoughtful systematist, thinking about homology under a series of structural and evolutionary constraints. Contrast this to another reasonable systematist, who believes that homology is best decided objectively with a repeatable optimality criterion implemented by a computer. The former may fail through carelessness. The latter may fail when the computer program is actually an irrational black box. Input parameters, such as gap costs, assigned by the investigator determine phylogenetic hypotheses. If these input parameters are arbitrary, then justifying algorithmic approaches over manual ones under a criterion of "objectivity" is almost impossible to argue. One must justify each of the parameters that influence the analysis. Yet the argument continues. We believe that if input parameters are arbitrary and unpredictable, then alignment methods that use them are also arbitrary and unpredictable. To submit one's data to an algorithm, with no regard for the implications of such an action, is to transfer subjective (and thoughtful) decisions about homology from the human investigator to subjective (and careless) decisions about gap cost determination. Algorithmic methods are not objective if input parameters are subjectively determined.

Another perceived advantage of algorithmic methods is that they are easier than structural alignments. In our experience, many investigators accept that structural alignments make sense, but they do not make the effort to perform them because they assume that their Clustal alignment is "good enough" and that a few alignment errors generated by the algorithm will be overridden by the mass of signal in their data. We find this cavalier attitude toward homology to be surprising, when we consider the effort and expense that goes into collecting the sequences. In our opinion, it is always worth the effort to align the data with care. As systematists, we are often are more interested in resolving controversial nodes and not so interested in re-corroborating well-established relationships. Controversial internodes are often characteristically short, and may be difficult to recover by any means with a variety of datasets. It may be that the characters we discard, because the easy method is applied, are the only ones that are informative. Or more likely, the few characters that inform us about a short internode are overwhelmed by a mass of poorly aligned noise. We could never know how a careful alignment would influence our results without the effort. Those who support sensitivity analyses to optimize parameters with POY would probably agree with us on this point, as they perform months of analysis time on parallel processors or super clusters. Careful alignment, whether performed by hand or computer, takes time, effort, and expertise. We reject the argument that carefully performed algorithmic methods are "easier," and let the reader decide whether "fast and careless" alignments are defendable.

An Example of Accuracy and Repeatability

If algorithmic methods could be shown to be more accurate than manual alignments, then we might be able to overlook the possibility that arbitrary parameter selection may sometimes lead to unpredictable hypotheses. This is not the case, however. Many empirical comparisons have shown that manual alignments tend to recover more reasonable phylogenies (Ellis and Morrison 1995; Gillespie, Yoder, et al. 2005; Hickson et al. 1996; Hickson et al. 2000; Kjer 1995; Kjer 2004; Lutzoni et al. 2000; Morrison and Ellis 1997; Mugridge et al. 2000; Schnare et al. 1996; Titus and Frost 1996; Xia et al. 2003). Phylogenies are hypotheses to be tested, accepted, or refuted by subsequent hypotheses. We never "know the truth." Such hypotheses may be accepted on the grounds that they generally equate to the recovery of expected or corroborated relationships with phylogenetic accuracy. A compelling case can be made for phylogenies generated from manually aligned datasets. Time after time, we recover "more reasonable" phylogenetic hypotheses from carefully aligned data, (while at the same time, analyses justified only on epistemological consistency continue to produce "unexpected" hypotheses). Admittedly, these empirical studies can provide only points for discussion. To demonstrate accuracy, we need either known phylogenies from experimentally manipulated systems (such as sampling evolving viruses, Hillis and Bull 1993) or simulation studies where we know the history of insertions and deletions in a simulated dataset. However, there are problems with both of these approaches, and these problems stem from the nature of rRNA. Viruses do not possess rRNA, so problems specific to rRNA alignment cannot be addressed with manipulated viral sequences. Simulation studies are only as good as the model used to simulate the data. Currently, our ability to model insertions and deletions is limited and unrealistic. Although it is possible to insert gaps into a simulated sequence, any model that assumes that gaps are independent of one another and randomly distributed is not capturing the essence of what is happening in rRNA, where insertions and deletions are frequently multiple nucleotides in lengths, and strongly clustered in variable regions. An accurate model of rRNA evolution would require a proportion of the sites to be covarying, gaps to be nonindependent, and substitution rates and length heterogeneity to be regionally variable. Without these characteristic features of rRNA built into the simulation, any generalizations drawn from these studies must be understood to be only crude approximations of biological reality.

We suggest a reasonable empirical solution to the assessment of accuracy in Kjer et al. (2006, 2007). Although accuracy cannot be fully explored with empirical data, we see at least one example where an "expected tree" is justified. For taxa whose entire mitochondrial genomes are sequenced, it can be expected that partitions of the data share the same history. We suggest that relationships that are corroborated with both nuclear genes and morphology are candidates for identifying sets of phylogenetic expectations from the mitochondrial data. If these independently corroborated nodes are also supported by the combined mitochondrial genome data, then these relationships could be used to assess alignment strategies of any partitions of the data, such as the 12S and 16S mitochondrial rRNAs. Kjer et al. (2007) used the relationships shown in Figure 7.9 to compare phylogenetic accuracy and repeatability of manual and direct optimization methods. These taxa possess complete mitochondrial genome data, and each of the nodes is supported by morphological characters (McKenna and Bell 1997; Novacek 1992; Novacek et al. 1988; Simpson 1945) and nuclear genes (Amrine-Madsen et al. 2003; Delsuc et al. 2002; Waddell and Shelley 2003). The phylogeny shown in Figure 7.9 is recovered from parsimony, Bayesian, and Likelihood analyses of the complete mitochondrial genomes (Gibson et al. 2005; Kjer and Honeycutt 2007; Reyes et al. 2004). One need not accept this as the "true tree," but merely a tree that is recovered by the entire dataset, and corroborated by multiple independent sources. By definition, partitions of the same linked dataset contain less data. It is therefore reasonable to use a tree derived from ten times the number of linked nucleotides in order to test alignment accuracy. There is a risk to judging an alignment method according to this sort of phylogenetic expectation. Namely, the risk would be that the "expected tree" was later shown to be inconsistent with a tree derived by some future superior method. As such, the results from

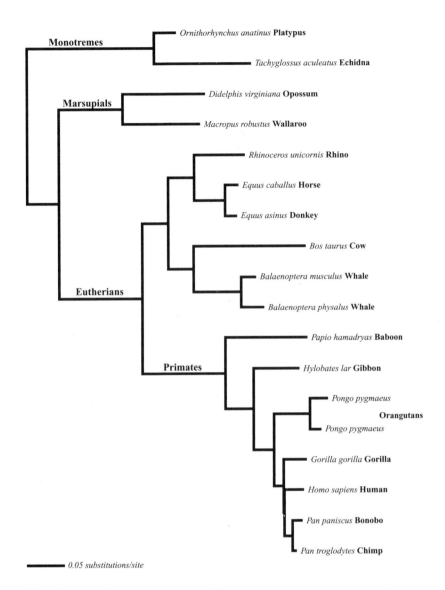

Figure 7.9. Expected tree, generated from analysis of the entire mitochondrial genome. Branch lengths calculated as likelihood, with a GTR + I + G model (Kjer et al. 2007).

the experiment could be modified by the new phylogenetic expectations. In other words, if one is transparent about how phylogenetic expectations are used to assess alignment performance, then conclusions can easily be overturned with future illumination. Science is about laying out one's assumptions, and testing hypotheses according to whether or not the data fit those assumptions.

The experimental design of Kjer et al. (2007) was simple. The 16S rRNA sequences from the taxa shown in Figure 7.9 were assembled, the taxon names were disguised, and the taxon order was shuffled within the matrix. The masked data were then sent to three investigators with simple instructions: "Align these data with secondary structure, and also with POY." It was predicted that if secondary structure could provide a reasonable means of homology assessment, then different investigators would come to similar decisions about structurally influenced homology. More simply stated, if the structures were real, we would all find them, and structurally aligned data would lead to similar phylogenetic conclusions among investigators, because they would be using a nonarbitrary means of homology assessment, even if the alignments themselves were not identical. The second prediction was that if parameter decisions were arbitrary (Doyle and Davis 1998; Hickson et al. 2000; Kjer 1995; Phillips et al. 2000; Vingron and Waterman 1994; Wheeler 1996), and these parameters had a strong influence on phylogenetic conclusions, then different investigators using an algorithmic approach to a phylogenetic problem would arrive at different phylogenetic hypotheses, given the same data.

Figure 7.10 shows the results of the experiment. All three of the structural alignments yielded nearly identical results, with the only difference being on the Chimp/Human/Gorilla branch (which is a reasonable reflection of reality, because this branch has no perceivable length). The structurally aligned data also recovered the expected tree. The hypotheses generated from the independent POY analyses (not shown; see Kjer et al. 2007, Fig. 5) resulted in each investigator proposing a different phylogenetic hypothesis, none of which were the expected tree. Each of the POY analyses resulted in different opinions on how to present the confusing array of trees that were generated from the many explorations of alternative parameters. We have already discussed the ambiguity of sensitivity analyses when each of the explorations of parameter space is biologically unrealistic. Alarmingly, even when the same parameters (but not necessarily the same search heuristics) were used, each of the investigators recovered different trees with POY. In all probability, this

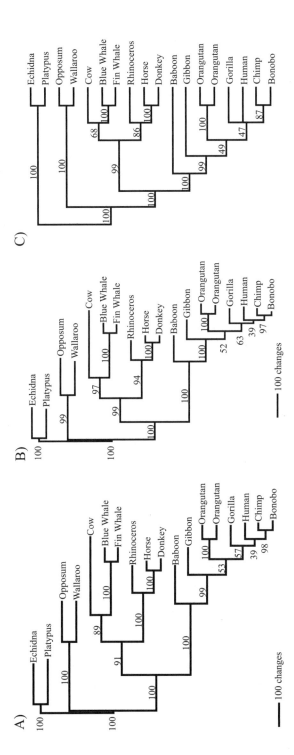

Figure 7.10. Trees generated from a manual alignment of the 16S rRNA data. Branch lengths calculated with parsimony. (A), (B), and (C) generated by Kjer, Ober, and Gillespie, respectively. (C) is a consensus of two trees. Numerals near the nodes are bootstrap proportions.

was the result of an insufficient search strategy. All heuristic exercises, including routine tree searches, can suffer from this problem, and if you start from the same random seed, you will get the same tree. However, direct optimization is more complex than simple tree searches, and figuring out how long to run the analysis is another decision that needs to be made. In this example, structurally aligned data resulted in phylogenetically identical hypotheses that conformed to the corroborated tree. Is that not what we *want* in terms of repeatability? Consider the scenario of baking a cake. We follow a recipe: 2 eggs, a cup of flour, a cup of sugar, and so forth, . . . mix well, and then bake in the oven at 375 °F for 30 minutes. Perhaps one person uses 5% more flour than another, or one person stirs with greater vigor. Regardless, the end product is a cake. With manual alignments guided by secondary structure, we all get cake in the end (Figure 7.10). Repeatability in science has always been defined in this way. We describe the methods, and then see if others can repeat it. Taken even further, if the alignment is presented, the analyses can be precisely repeated, and the decisions that went into it can be assessed and changed. Kjer et al. (2007) show that when you follow the cake recipe, and pop the data into POY, you do not know what will come out; it might be a loaf of bread. Of course, our scenario is one of exaggeration, but it serves to make a point. When you are cooking (or applying mathematics), you can tell the difference between the end results of a cake and a loaf of bread. In phylogenetics, the end products cannot be so easily distinguished from one another with respect to which is correct.

Not everyone agrees with the generalizations we reported in Kjer et al. (2007), and as with any work, there are, no doubt, legitimate criticisms that we would like people to consider. This work (Kjer et al. 2007) was presented as an opinion piece to foster some discussions about the ambiguities of alignments, both manual and computer generated. We asked one of our critics, Gonzalo Giribet, to summarize the basic weakness of our work here.

> I think that we both agree that secondary structure information is valuable for refining homology hypotheses but we differ in the way we incorporate such ancillary information into our homology-assignment techniques, being those multiple alignments or simply putative synapomorphies in "direct optimization" techniques (what I have called "single-step phylogenetics"). We have different understanding of what reproducibility may mean, and although I see an alignment as a pure topology-dependant hypothesis you may view it as something that is fundamentally knowable, i.e., that there is "one" alignment. This is what causes that you may search for

"the" alignment while I am more interested in exploring what alternative parameter sets may have to do with my homology hypothesis, i.e., assessing the stability of my results to alternative parameter sets.

Naturally we do not agree with everything Giribet has to say about sensitivity analysis. There seem to be two purposes for sensitivity analyses; one would be to find the most justified set of parameters, and therefore select a favored hypotheses derived from the preferred input parameters. The other purpose would be to explore the stability of the data to a variety of parameters, and present a phylogenetic hypothesis that includes measures of alignment uncertainty. Giribet supports the second of these in his statement above, and there may have been instances where we have confused these two justifications for sensitivity analyses. We appreciate his effort to clarify this difference, but still find both uses of sensitivity analyses problematic.

We think that phylogenetics is a near impossible enterprise, and the best we can do is to do our best. We should not be ashamed of pursuing accuracy, even if it is impossible to assess. We agree with Giribet that sensitivity to alignment uncertainty should be explored, and that uncertainty in an alignment should eventually be part of our estimates of support. We think that this would probably require a Bayesian method, similar to that proposed by Redelings and Suchard (2005).

If one is interested in a "best estimate" of phylogeny, this estimate will likely come from an analysis in which homology (alignment) is optimized. Another area of ambiguity is amplified when multiple alignment parameters offer many different trees; it becomes difficult to informatively select among them. The utility of a phylogenetic hypothesis is drastically reduced when no hypothesis can be considered better than another, because all of the trees are devoted to explorations of the data under different alignment input parameters. For example, what can we take from Wheeler et al. (2001) as a phylogenetic hypothesis? Surely not the "discussion tree," but if not that, should we favor the myriad of other trees that collapse to a near meaningless polytomy? We believe that it is the responsibility on an investigator to clearly present his or her hypothesis. The best way to do so would be to state "the best estimate of phylogeny we could make comes from analysis 'X,' shown in Fig. 'Y.'" It is the responsibility of the reader to then evaluate the results and either agree or disagree with the findings. For this dynamic process to occur, one must present the data, present the alignment, and justify the decisions that were made. The current feasibility of justifying one's decisions with an algorithmic approach is not as straightforward as is the case with a manual approach.

COMPARISON TO PROTEIN ALIGNMENT— PROGRAMS AND BENCHMARKS

Protein alignment programs have seen much development in recent years, beginning with ClustalW (Thompson et al. 1994) to current state-of-the-art approaches, such as Probalign (Roshan and Livesay 2006), Probcons (Do et al. 2005), and MAFFT (Katoh et al. 2005). These programs use a variety of techniques, such as hidden Markov models (Durbin et al. 1998), maximal expected accuracy (Durbin et al. 1998), fast Fourier transforms (Katoh et al. 2005), profile alignment (Edgar 2004a), and consistency alignment (Do et al. 2005). Most protein alignment programs aim to align parts of proteins conserved in sequence or structure. This is facilitated by amino acid substitution-scoring matrices estimated from real data (such as PAM, Dayhoff and Eck 1968; and BLOSUM, Henikoff and Henikoff 1992) and manually created protein alignment benchmarks, also based on real data (such as HOMSTRAD, Mizuguchi et al. 1998b; and BAliBASE, Thompson et al. 2005b). These benchmarks, which are primarily structure-based alignments, not only allow for comparison of different programs, but also enable optimization of gap penalty parameters on real data. As a result, protein alignment programs have shown a steady increase in accuracy over the years. The most accurate programs use a combination of techniques, such as PSI-BLAST profiles, and predicted secondary structures as found in the PROMALS program (Pei and Grishin 2007).

Most protein sequence alignment programs can be used for RNA alignment in principle. However, substitution scoring matrices and alignment benchmarks (analogous to BLOSUM and BAliBASE, for example) were not developed for RNA until recently. The BRaliBASE RNA alignment benchmark (Gardner et al. 2005), similar to BAliBASE for proteins, is the first RNA alignment benchmark produced by aligning sequence while taking into consideration secondary structure. Subsequent efforts have expanded BRaliBASE (see Wilm et al. 2006). Yet, BRaliBASE still lacks the size and diversity of its protein counterparts. The RIBOSUM scoring matrices (Klein and Eddy 2003) for RNA are comparable to the BLOSUM matrices for proteins; however, recent studies show that they perform more poorly than simpler scoring matrices when used in the ClustalW program on enhanced BRaliBASE benchmarks (Wilm et al. 2006).

Further development of RNA alignment benchmarks, better substitution scoring matrices, and adaptation of techniques used in state-of-the-art protein alignment programs will eventually lead to better algorithms for aligning RNA. Alignment of conserved regions (in sequence or structure) is accepted as a measure of correctness in the protein domain. In light of our discussion in the preceding sections, the same criteria should apply when aligning RNA.

CONCLUSION

When initially asked to contribute to this book, we thought that we would provide a chapter on the problems with algorithmic methods. However, we find this to be an overly negative approach. We all have different backgrounds—and experience. Sometimes we see things in different ways, and our experience differs greatly from that of many of the other contributors to this book. These differences are a good thing, as differing points of view should be openly discussed and debated. Thus, our science progresses. It is all too easy in science to take an adversarial approach to those who disagree with us. It is not our intention in this chapter to be overly critical. We do support properly invoked algorithmic methods, and clearly stated optimality criteria. It is our hope that, by our pointing out some of the problems we have experienced in the alignment of rRNA data, program developers can incorporate solutions to these issues in their algorithms. Biologically realistic algorithms could make manual alignment less and less relevant. Here are some of the issues we see as most important toward the improvement of alignment programs.

In molecules whose function is dependent on structure, the conservation of the structure should be part of the optimality criterion. Minimizing structural change is as justified as minimizing change among nucleotides. Perhaps a program could be developed that could locate covarying sites in a multiple alignment. Some multistepped combination of calculating minimum free energy structures that are shared among multiple taxa, and then confirming those hypothesized structural interactions based on compensatory base changes, seems possible. Sites containing such compensatory base changes should be aligned together. A research group led by P. Stadler at the University of Leipzig and a research group led by B. Misof at the University of Bonn have developed a promising program called RNAsalsa that promises to do these things (B. Misof, personal communication), but we have not had a chance to evaluate it.

Gap costs should vary among regions. Manual alignments contain flexible gap costs, in that when a person comes to a hypervariable region with a lot of variation in length among sequences, gaps are more freely inserted. With an algorithm, there should be a way to locate conserved regions by some criteria, and then measure the range of length variation among taxa in the regions between them. Gap costs could then be regionally assigned based on how much length variation was observed, giving the lowest gap costs to regions that contain the most length heterogeneity. The standard deviation of the lengths could also play a role in gap cost determination, and in data exclusion criteria.

The iterative process of moving from guide trees to multiple alignments should be improved upon. Perhaps the initial guide tree should be developed from unambiguously aligned regions; we could then iteratively move through more difficult regions with guide trees developed only from data whose homology reaches some confidence threshold. It is fairly easy to see by eye when one lineage cannot be aligned with another. Reconstructing ancestral states to the root of a particular lineage may yield sequences so different from other lineages that it would be foolish to attempt to homologize them. If it is that easy to see for the human eye, there should be a way for a computer to measure this incompatibility as well, and reach objective criteria for data exclusion. Gblocks (Castresana 2000) provides a conservative means of data exclusion. Another interesting program called "Aliscore," based on the identification of randomness in sequence alignments using a Monte Carlo approach, is being developed by Misof and Misof (in press). We need more information about how gaps accumulate and evolve in rRNA to model these characters. The greatest challenge is that gaps are not independent of one another, and are not randomly distributed across sites. Alignment programs must recognize both of these properties.

An ideal program would have a means to assess alignment uncertainty (as in Redelings and Suchard 2005). But alignment uncertainty is linked with the model, so it is important to remember that if the model for gaps is biologically unrealistic, then the "uncertainty" cannot be disentangled from those limitations. It is our impression that the differences among trees that are attributed to alignment methods are more often associated with different data exclusion criteria and philosophies. It should be possible to produce a program that incorporates some data exclusion criterion with alignment uncertainty.

Alignments involving moderate to extreme length heterogeneity across sampled sequences will undoubtedly invoke some degree of subjectivity from the investigator, regardless of the methodological approach (Kjer et al. 2007). The legitimate disagreements about the kinds of subjectivity that are justified will likely continue. Here we state our beliefs. Phylogenies are hypotheses only. We think that even though we can never prove a phylogeny to be true, phylogenies that are wrong are worse than worthless because they promote further inaccurate predictions. For phylogenies to be predictive, they must be accurate, and even if we cannot prove accuracy, none of us should be embarrassed to pursue it. Given that, we think it is imperative to intervene when it can be demonstrated that existing methods are failing. Reasoned subjectivity, with all assumptions defined and the alignment made public, is far more accessible than black box analysis justified under some philosophical principle. We suggest that you should do the best you can today with what you have, because if something better comes along later, the data are still available in GenBank for reanalysis. And the addition of new sequences to existing alignment templates is likely the better approach, not only for the re-estimation of phylogenies, but also for the evaluation of structural and functional predictions derived from said alignments (see Morrison 2006). Thus, we disagree with favoring purely algorithmic approaches, such as POY and others based only on the perceived future of direct optimization, simply because no algorithms to date match the level of empiricism ingrained within the biological (manual) method.

TERMINOLOGY

Comparative evidence for secondary structure base pairing comes predominantly from the observation of covarying Watson–Crick pairs (see the early works of Gutell, Noller, and Woese, reviewed in Noller 2005). Typically, contradiction of a covarying position is as follows: AA, CC, UU, GG; AC, AG, and CU (and their symmetrical equivalents) cause disruptive bulges. Gutell and others have observed that some of these pairs actually do covary, mostly within highly conserved regions of rRNA (e.g., see Lee and Gutell 2004), wherein selection favors noncanonical base pairs to foster a variety of tertiary interactions (reviewed in Noller 2005); however, for the alignment of variable regions of rRNA, consider them forbidden. Remember also that G↔U is a permitted hydrogen bonding pair in RNA. C↔A pairs do not appear to be as disruptive as the other noncanonical pairs listed above, and therefore, if there is comparative evidence of a site, C↔A pairs should not contradict the site. Contradiction of core helices and other strongly supported features of rRNA cannot come from a single taxon because sequencing errors can happen, just as evolution can happen. Some deleterious substitutions do not result in the immediate extinction of the lineage, so a single bulge or even a few bulges across an alignment cannot contradict a stem. However, if you see multiple bulges, treat the site as suspect and evaluate the support for the helix by a variety of means. Base pair frequency tables provide a means to measure the degree of covariation for a given base pair. Other statistics, such as mutual information and Kramer's statistic, further illustrate the manner of dependence within sites of a base pair. The entire stem need not be rejected, however, particularly if there are covarying sites at other positions. Ideally, an objective approach would be to accept only base pairs that have less than a certain percentage of noncanonical base pairs within proposed helices.

- **Helix (stem):** A right-handed double helix composed of a succession of complementary hydrogen-bonded nucleotides between paired strands.
- **Single strand (loop):** Unpaired nucleotides separating helices.
- **Hairpin-stem loop:** Helix closed distally by a loop of unpaired nucleotides (terminal bulge).
- **Terminal bulge:** Succession of unpaired nucleotides at the distal end of a hairpin-stem.
- **Lateral bulge:** Succession of unpaired nucleotides on one strand of a helix.
- **Internal bulge:** Group of nucleotides from two antiparallel strands unable to form canonical pairs.
- **Compensatory base change:** Subsequent mutation on one strand of a helix to maintain structure following initial mutation of a complementary base (aka CBC).
- **Insertion:** A single insertion of a nucleotide relative to the rest of the multiple sequence alignment (dependent on frequency and determination of direction of event relative to out-group).
- **Deletion:** A single deletion of a nucleotide relative to the rest of the multiple sequence alignment (dependent on frequency and determination of direction of event relative to out-group).
- **Indel:** An ambiguous position (column) within a multiple sequence alignment that cannot be described as an insertion or deletion.
- **Region of ambiguous alignment (RAA):** Two or more adjacent, nonpairing positions within a sequence wherein positional homology cannot be confidently assigned due to the high occurrence of indels in other sequences.
- **Region of slipped-strand compensation (RSC):** Region involved in base pairing wherein positional homology cannot be defended

across a multiple sequence alignment; inconsistency in pairing likely due to slipped-strand mispairing.
- **Region of expansion and contraction (REC):** Variable helical region flanked by conserved base pairs at the 5' and 3' ends, and an unpaired terminal bulge of at least three nucleotides; characteristic of RNA hairpin-stem loops.

Acknowledgments

JJG acknowledges support from NIAID contract HHSN266200400035C awarded to Bruno Sobral (VBI).

APPENDIX: INSTRUCTIONS ON PERFORMING A STRUCTURAL ALIGNMENT

(NOTE: *A conceptual approach for structural alignment of rRNA sequences and further preparation of the data for RNA maximum likelihood models of evolution is available at http://hymenoptera.tamu.edu/rna/methods.php. Below we provide a didactic example that will help the reader begin manually adjusting his or her own data*).

It is most often the mechanics of manual alignment that trip people up. It is not hard to convince people that doing structural alignments is a good idea. It is just that it becomes "too hard," and if people can get "close enough" with Clustal, they call it "good enough." We understand their pain, but disagree. The following suggestions should make the whole process a little easier:

The goal of a manual structural alignment is to make objective and repeatable decisions, using minimizing structural changes without being arbitrary. An example of being arbitrary would be to say: "retain all nucleotides that are identical among all taxa for ten nucleotides or more." Although it makes good sense, and results in a repeatable criterion, the selection of ten is arbitrary. In our opinion, it is better to admit to ambiguity than to hide from it and pretend you are being objective, in the hope that nobody will notice.

1. Align the sequences with Clustal or any other computer alignment program as a starting point. It works best to avoid a "gappy"-looking alignment, because you will need to manually adjust the gaps. The computer alignment is simply a timesaving

device, as any manual adjustment changes a computer alignment to a manual alignment. Clustal is easy to use, is available for multiple platforms, and permits multiple export formats. Export the alignment in a NEXUS format.

2. Open the Clustal NEXUS alignment with PAUP. From PAUP, go to the "file" pull down menus and "export data," using "file format, NEXUS," and clicking the box "interleave" with 130 characters per line. Close the original PAUP file, and reopen the new ".dat" file, which is now interleaved with 130 characters per line. This is just so that you have about the right number of characters in each block to be able to look at them all without scrolling back and forth on the screen.

3. Open the interleaved .dat file you just made with Microsoft Word (of course, there are other text editors; yet, not all of them allow for easy manipulation of the data). Format with courier bold 9 point font. You may have to format the document in "landscape" view, and reduce to 25% so that the lines do not wrap over. Setting a custom page size (22" by 22") may help. Now color the nucleotides by going to the "Edit" menu; go to "Replace" (select "more options" to find the "font" option, under "format"). Change all of the A's into green A's, the C's into blue C's, the U's into red U's, and leave the G's black. Now you have something that looks like Figure 7.11, except that yours would be in color (this example can be found on Kjer's website, http://rci.rutgers.edu/~insects/kjer.htm):

4. Add a *palette* to each of the rows. A palette contains a variety of symbols that you may wish to insert, as a column, into the data

```
AY037172  UUAUUAGAUCAAAGCCAAUCGAACUUUCGGGUU------------------------------CGUUUUAUUGGUGACUCUGAAUAAC
U61301    UUAUUAGAUCAAAGCCAAUCGAGUUUCGGCUC-------------------------------GUUUUGUUGGUGACUCUGAAUAAC
Z36893    UUAUUAGAUCAAAGCCAAUCGAACUCUCGGGU------------------------------UCGUUUAAUUGGUGACUCUGAAUAAC
X89485    UUAUUAGAUCAAAGCCAAUCGGACUCUCGGGU------------------------------UCGUAUUGUUGGUGACUCUGAAUAAC
Z26765    UUAUUAGAUCAAAGCAAAUCGGACCUUCGGG-------------------------------UUCGUUUUGUUGGUGACUCUGAAUAAC
AF173233  UUAUUAGACCGAAACCAACCUGGUCGUGUCUCAC----GGCACGGUCCGGUCUCUGGCUUUGCCCAGGGGUUUGGUGACUCUGAAUAAC
AY037170  UUAUUAGACCGAAAUCAACCUGGUCGUUCGCUU-----GCGAGCGGUCCGGUCUCUGGAUCUUCCAGGGGUUUGGUGACUCUGAAUAAC
AY037169  UUAUUAGACCGAAACCAACCUGGUCGUGUCUCUG-----GCACGGUCCGGUCUCUGGCUUUGUCCAGGGGUUUGGUGACUCUGAAUAAC
AF173234  UUAUUAGCUCAAAGCCGAUCGGGUCCUUGUGGCCC----------------------------GCAACUUGGUGACUCAAACGAAC
AY037168  UUAUUAGCUCAAAGCCGACCGGGCUUAGCCCGCGCUU-----CCGUUCGCGGUGCGCGGGCGGCCCGCCUCUCGGUGAAACGGACGAAC
AY037167  UUAUUAGUUCAAAGCCGAUCGGGUCCUUUGUG-------------------------------GCCCGCUACUUGGUGACUCAAACGAAC
AF005456  UUAUUAGCUCAAAGCCGACCGGGCUUCAACCCUUCGUCCCCUCGCGGGGCGUUGGGGCGGCCCGUUUCCACUCGGCGAAUCGAAAGAAC
AF005455  UUAUUAGCUUAAAGCCAAUCGGGUCCUUGUGGCC----------------------------CGCUUAUUGGUGACUCAAACGAAC
AF005454  UUAUUAGCUUAAAGCCAAUCGGGUCCUUGUGGCC----------------------------GCUUAUUGGUGACUCAAACGAA
```

Figure 7.11. Algorithmically aligned sequences imported into Word, awaiting manual refinement.

```
AY037172  UUAUUAGAUCAAAGCCAAUCGAACUUUCGGGUU-----------------------------CGUUUUAUUGGUGACUCUGAAUAAC [ ( ) | * - ]
U61301    UUAUUAGAUCAAAGCCAAUCGAGUUUCGGCUC------------------------------GUUUUGUUGGUGACUCUGAAUAAC [ ( ) | * - ]
Z36893    UUAUUAGAUCAAAGCCAAUCGAACUCUCGGGU-----------------------------UCGUUUAAUUGGUGACUCUGAAUAAC [ ( ) | * - ]
X89485    UUAUUAGAUCAAAGCCAAUCGACUCUCGGGU------------------------------UCGUAUUGUUGGUGACUCUGAAUAAC [ ( ) | * - ]
Z26765    UUAUUAGAUCAAAGCAAAUCGACCUUCGGG-------------------------------UUCGUUUUGUUGGUGACUCUGAAUAAC [ ( ) | * - ]
AF173233  UUAUUAGACCGAAACCAACCUGGUCGUGUCUCAC-----GGCACGGUCCGGUCUCUGGCUUUGCCCAGGGGUUUGGUGACUCUGAAUAAC [ ( ) | * - ]
AY037170  UUAUUAGACCGAAAUCAACCUGGUCGUUCGCUU------GCGAGCGGUCCGGUCUCUGGAUCUUCCAGGGGUUUGGUGACUCUGAAUAAC [ ( ) | * - ]
AY037169  UUAUUAGACCGAAACCAACCUGGUCGUCUCUG-------GCACGGUCCGGUCUCUGGCUUUGUCCAGGGGUUUGGUGACUCUGAAUAAC [ ( ) | * - ]
AF173234  UUAUUAGCUCAAAGCGAUCGGUCCUUGUGGCCC---------------------------GCAACUGGUGACUCAAACGAAC [ ( ) | * - ]
AY037168  UUAUUAGCUCAAAGCCGACGGGCUUAGCCCGCGCUU---------CCGUUCGGGUGCGCGGGGCCCCGUUCCGUGAAACGGACGAAC [ ( ) | * - ]
AY037167  UUAUUAGUUCAAAGCCGAUCGGUCCUUUGUG----------------------------GCCCGUACUGGUGACUCAAACGAAC [ ( ) | * - ]
AF005456  UUAUUAGCUCAAAGCGACCGGCUUCAACCCUUCGUCCCCUCGCGGUCCGCGGGGGGCGGCCCGUUCCGCGAAUCGAAAGAAC [ ( ) | * - ]
AF005455  UUAUUAGCUUAAAGCCAAUCGGGUCCUUGUGCC----------------------------CGCUAUUGGUGACUCAAACGAAC [ ( ) | * - ]
AF005454  UUAUUAGCUUAAAGCCAAUCGGGUCCUUGUGCCC----------------------------GCUAUUGGUGACUCAAACGAA [ ( ) | * - ]
```

Figure 7.12. The sequences with a symbol palette added onto the end of each row. The symbols will be used to describe structural elements. The palette should be contained within square brackets [] so as to keep the sequences in valid NEXUS format.

matrix. As shown in Figure 7.12, the palette starts and ends with brackets, so that NEXUS will ignore the contents after they are eventually reimported into PAUP.

5. Go to Gutell's (http://www.rna.ccbb.utexas.edu) or Gillespie's (http://hymenoptera.tamu.edu/rna/index.php) website and download the most recent secondary structure diagram.

6. Microsoft Word permits three essential things you need to be able to do. First, you want to see colors. Second, you must be able to move columns. To move columns in Word, you simply depress the "option" key as you drag down a column with the mouse. Finally, underlines are essential (more about them in step 9). So the next step is to find a stem from a structural model, and paste in the structural symbols from Kjer (1995) to indicate the putative boundaries of the stems. (Note: since 1995, I have replaced the bracket symbols with the "|" symbol to indicate long range stems, because the brackets have meaning in NEXUS that I had not considered in 1995.)

7. Apply structural symbols as in Kjer (1995) to the reference sequence, and fit them, one by one, onto each of the other sequences. Attempt to subdivide long single-stranded regions by looking for covariation, as in Kjer (1995). As a first pass, assume the structural model you have is correct, but if the data contradict it, then do what the data tell you. Structural models are inferred from comparative evidence, which is exactly what you have before you for a more specific set of taxa. These structures may evolve. If you see that your model does not fit your taxa, then alter it to a model that is supported by the evidence presented by the sequences. The signal in these regions comes from universal and covariable inferred hydrogen bonds (compensatory base changes). If all of the taxa can bond in a thermodynamically stable stem that is supported by compensatory base changes, and would also be unlikely to exist by chance, then this stem should be inferred and used in an alignment. You may propose modifications to structural models this way. It is not your task to construct a perfect secondary structural model, but, rather, to use the structure to infer homology. A portion of the stem for which the structure is ambiguous from the data cannot be used to define homology beyond what you can infer from the nucleotides (primary structure). So you should freely contract

Structural Considerations for RNA MSA

stems to the minimum common supported size, and let others whose primary goals are to develop structural models worry about the differences.

8. Consult Hickson et al. (1996) and "Phylogenetic conservation superimposed onto the *E. coli* SSU rRNA" on Gutell's website for conserved motifs.

9. Pasting the structural symbols provides only an initial rough hypothesis of base pairing. The next step is to confirm the hydrogen bonds. Since this is an iterative process, you MUST be able to trace what you have looked at, and differentiate those regions from the regions you have not yet finished. One way to do this is with underlines. Underlines indicate confirmed hydrogen bonds. They mean that you have looked at those individual nucleotides, and their partners, and a Watson–Crick, or G-U, base pair is possible. Laziness is the biggest problem at this point, because it is easy to drag entire columns of nucleotides, and simply underline them all without checking. A sloppy alignment is full of non-Watson–Crick pairs that are mistakenly underlined. Note in Figure 7.13 the bulge indicated by the lack of underlines in Z26765 GCAAA...UUGGU. If you cannot trust the underlines, you cannot trust the alignment. This is why we do not use some of the fancier phylogenetic data editors—because they do not offer the opportunity to visualize individual hydrogen bonds (or if they do, Kjer did not know). There may be a better way to do this. But any system must have confirmation of bonding at each site, as opposed to a mask applied to the top sequence.

10. Line up the stems. If the stems do not line up because there are alternative lengths, slippage, or a lack of structural conservation, pull back on the stems, and consider them *unaligned*. Put an empty space to mark the unaligned regions, as above. Use empty spaces to help you break up the alignment, so that you can get a better look at it. Think carefully about data exclusion. For example, can you justify aligning the above UUUUG with CAAC? If not, then eliminate this region, and code it as you see fit with some other method. The structure will define the aligned regions, and delimit the unaligned regions. If there is no length variation in the single-stranded region, keep it in, as in the region below "V2": AGAUCAAA. If there is length variation, without con-

```
[                    V2                                                       Region 4                                                                    ]
AY037172  (UUAUUAGAUCAAAGCCAA U CGAA -------------CUUCGGG UUCG UUUUA-- UUGGUGACUCUGAAUAA)C[ ()] *-]
U61301    (UAAUUAGAUCAAAGCCAA U CGAG U------------UUCGG-- CUCG UUUUG-- UUGGUGACUCUGAAUAA)C[ ()] *-]
Z36893    (UUAUUAGAUCAAAGCCAA U CGAA CU-----------CUCGGG  UUCG UUUAA-- UUGGUGACUCUGAAUAA)C[ ()] *-]
X89485    (UAAUUAGAUCAAAGCCAA U CGAA C------------UCUCGGG UUCG UAUUG-- UUGGUGACUCUGAAUAA)C[ ()] *-]
Z26765    (UUAUUAGAUCAAAGCAAA U CGGA CC-----------UCUGGG  UUCG UUUUG-- UUGGUGACUCUGAAUAA)C[ ()] *-]
AF173233  (UUAUUAGACCGAAACCAA C CUGG UCGUGUCUCACGGCA-CGGUCCGGUCUCUGGCUUUGC CCAG GGGU--- UUGGUGACUCUGAAUAA)C[ ()] *-]
AY037170  (UUAUUAGACCGAAAUCAA C CUGG UCGUUCGGUUGCGAG-CGGUCCGGUCUCUGGAUCUU- CCAG GGGU--- UUGGUGACUCUGAAUAA)C[ ()] *-]
AY037169  (UUAUUAGACCGAAACCAA C CUGG UCGUGUCUC-UG3CA-CGGUCCGGUCUCUGGCUUUGU CCAG GGGU--- UUGGUGACUCUGAAUAA)C[ ()] *-]
AF173234  (UUAUUAGCUCAAAGCCGA C CGGG UCCUU---------GUGG                  CCCG CAAC--- UUGGUGACUCAAACGAA)C[ ()] *-]
AY037168  (UUAUUAGCUCAAAGCCGA C CGGG CUUAGCCCCGCUUC--CGUUCGGGUGCGGGCGG    CCCG CCUC--- UCGGUGAAACGACGAA)C[ ()] *-]
AY037167  (UUAUUAGUUCAAAGCCGA U CGGG UCCU----------UGUGG                  CCCG CUAC--- UUGGUGACUCAAACGAA)C[ ()] *-]
AF005456  (UUAUUAGCUCAAAGCCGA C CGGG CUUCAACCCUUCGUCCCUCGGGGGCGUUGGGCGG   CCCG UUUCCAC UCGGCGAAUCGAAAGAA)C[ ()] *-]
AF005455  (UUAUUAGCUUAAAGCCAA U CGGG UCCUUGU-------------                CCCG CUUA--- UUGGUGACUCAAACGAA)C[ ()] *-]
AF005454  (UUAUUAGCUUAAAGCCAA U CGGG UCCUUGU---------------GG            CCCG CUUA--- UUGGUGACUCAACGAA)-[ ()] *-]
```

Figure 7.13. Underlined text is used to indicate confirmed hydrogen bonds (Watson–Crick and G–U base pairings) and allows one to keep track of which sites have been examined.

Structural Considerations for RNA MSA

served nucleotide motifs, throw it out, put it into INAASE, or try some iterative tree-based computer alignment on these regions.

11. Think about dots. When applied to sequences to indicate identity with a reference sequence (on top in Figure 7.14), they help you to visualize compensatory base changes, as well as synapomorphies. They will also make misalignments stand out so that you can see them. You can insert them by moving a section into MacClade, and selecting the "matchar" option, but in doing so, you would lose your structural symbols.

12. Define regions of ambiguous alignment. A candidate for a region of sequence that may be considered as "ambiguously aligned" is initially any region containing length variation among taxa. Objectively subdividing this assignment becomes the more important task because the initial definition applies to the whole sequence. There are three types of information that help to designate regions into aligned and ambiguously aligned classes. First, an ambiguously aligned region is any region containing length variation among taxa that is flanked by hydrogen-bonded stems, in which there is more than one equally plausible alignment. This assignment alone will subdivide the whole gene into multiple fragments. Once the secondary structure has defined the boundaries of ambiguity, additional information comes from the nucleotides. Attempt to manually align the region. Consult both Gutell and Hickson et al. for conserved motifs, and if all taxa have them, align the conserved motifs together to further subdivide the region. Ask yourself if a panel of judges were to look at every gap in your alignment, whether or not you could defend your decisions to the point where no other placement would be equally parsimonious. Consider transitions to be more likely than transversions, and also, make consistent decisions about how heavily to consider one or a few aberrant taxa in an otherwise length-homogeneous region. Decide the degree of nucleotide similarity among taxa that is required to expand into the regions defined by the flanking hydrogen bonds. Remember, each decision you make is a hypothesis of homology that can be reviewed and overturned. Therefore, you do not need to be perfect, because if you publish your hypotheses, they can be repeated and or contested.

```
                    [         V2                                                              Region 4                                                                       ]
                    [UUAUUAGAUCAAAGCCAA U CGAA ------------------------CUUUCGGG UUCG UUUUA-- UUGGUGACUCUGAAUAA)C[()| *-]
AY037172           (UUAUUAGAUCAAAGCCAA U CGAA ------------------------CUUUCGGG UUCG UUUUA-- UUGGUGACUCUGAAUAA)C[()| *-]
U61301             (..................  . .... U---------------------------UUCGG- C.... UUUUG- ...G..............).[0]| *-]
Z36893             (..................  . ...G CU-------------------------CUCGGG ..... UUUAA- ...................).[0]| *-]
X89485             (..................  . .... C--------------------------UCUCGGG ..... UAUUG- ...................).[0]| *-]
Z26765             (..............A...  . ...G. CC-------------------------UUCGGG ..... UUUUG- ...................).[0]| *-]
AF173233           (........C.G..A....  C ..UGG UCGUGUCUCACGGCA-CGGUCCGGUCUCUGGCUUUGC CCA. GGGU- ...................).[0]| *-]
AY037170           (........C.G..AU...  C ..UGG UCGUUCGCUUGCGAG-CGGUCCGGUCUCUGAUCUU- CCA. GGGU- ...................).[0]| *-]
AY037169           (........C.G..A....  C ..UGG UCGUGUCUC-UGGCA-CGGUCCGGUCUCUGGCUUUGU CCA. GGGU- ...................).[0]| *-]
AF173234           (........C.....G..  . ..GG UCCUU---------------------------GUGG CC... CAAC--- ..........AA.CG...).[0]| *-]
AY037168           (........C........  . ..GG CUUAGCCCGCGCUU----CGUUCGCGGGUGCGCGGGCGG CC... CCUC-- .C....AA.G..CG..).[0]| *-]
AY037167           (........U........  . ..GG UCCU-----------------------UUGUGG CC... CUAC--- ..........AA.CG...).[0]| *-]
AF005456           (........C........  C ..GG CUUCAACCCUUCGUCCCCUCGGGGGCGUUGGGGCGG CC... UUUCCAC .C..C..A.GA..G...).[0]| *-]
AF005455           (........C.U......  . ..GG UCCUUGU------------------GG CC... CUUA--- ..........AA.CG...).[0]| *-]
AF005454           (........C.U......  . ..GG UCCUUGU------------------GG CC... CUUA--- ..........AA.CG...)-[0]| *-]
```

Figure 7.14. The use of dots to indicate identity with a reference sequence can ease the identification of compensatory base changes, as well as synapomorphies.

13. Once you have finished the alignment in Word, import the whole thing back into PAUP, and use the "Replace" option in the Edit menu to change all the parentheses, and lines "(,)" and " | " into blank spaces. The NEXUS file should be an exact match to the Word file, except that it will lack color, and the structural symbols.

CHAPTER 8
Constructing Alignment Benchmarks

JULIE D. THOMPSON
Institut de Génétique et de Biologie Moléculaire et
Cellulaire, France

Criteria for Benchmark Development 153
Multiple Alignment Benchmarks. 155
    Benchmark Specification. 155
    Alignment Benchmarks. 158
    Comparison of Multiple Alignment Benchmarks 169
Multiple Alignment Revolution. 170
    Analysis of Program Strengths and Weaknesses 173
Conclusions. 175

Multiple sequence alignment is one of the most fundamental tools in molecular biology. It is used not only in evolutionary studies to define the phylogenetic relationships between organisms, but also in numerous other tasks ranging from comparative multiple genome analysis to detailed structural analyses of gene products and the characterization of the molecular and cellular functions of the protein. Many of these applications are discussed in detail in Chapter 11. The accuracy and reliability of all of these applications depend critically on the quality of the underlying alignments. Errors in the initial alignment will be propagated and further amplified in the subsequent analyses, leading to improper analyses and false predictions.

    An important application of multiple sequence alignments is in evolutionary studies, where protein and nucleotide sequence signatures are used to infer family relationships between species (Morrison and Ellis 1997; Ogden and Rosenberg 2006). For example, the accepted universal tree of life, in which the living world is divided into three domains (bacteria, archaea, and eucarya), was constructed from comparative analyses of ribosomal RNA (rRNA) sequences. However, the comparison of phylogenetic trees derived from selected proteins and from rRNA has frequently revealed substantial discrepancies between the respective trees. These discrepancies have often been interpreted as providing evidence for a high frequency of lateral gene transfer (Gogarten et al. 2002; Nesbø et al. 2001; Raymond et al. 2002), implying that the "tree of life" might actually resemble an interwoven mesh or network, rather than a hierarchical tree. Nevertheless, other investigators (Daubin et al. 2003; Eisen and Fraser 2003; Kurland et al. 2003; Yang et al. 2005) have argued that some of the discordant phylogenies are possibly due to methodological problems, such as sequence misalignments.

Multiple sequence alignments have also been widely exploited in the prediction of protein three-dimensional (3D) structure. Knowledge of a protein's structure is crucial to answering many biological questions, such as the understanding of functional mechanisms, the characterization of protein–protein and protein–ligand interactions (including drug design), or the correlation of genotypic and phenotypic mutation data. Despite considerable progress in *ab initio* structure prediction (Bradley et al. 2005), comparative modeling methods still provide the most reliable and accurate protein structure models, when a suitable prototype structure is available (Baker and Sali 2001). The traditional comparative modeling procedure consists of several consecutive steps usually repeated iteratively until a satisfactory model is obtained (Marti-Renom et al. 2000): identification of a suitable template protein related to the target, alignment of the target and template sequences, identification of structurally conserved regions, and construction and refinement of the structural model. Two important factors have been identified that influence the ability to predict accurate models: the extent of structural conservation between target and template, and the correctness of the sequence alignment (Kryshtafovych et al. 2005). For models based on distant evolutionary relationships, it has been shown that multiple sequence alignments generally improve the accuracy of the structural prediction (Moult 2005).

In the face of the growing number of alignment applications, a vast array of multiple alignment programs has been developed based on diverse algorithms, ranging from traditional optimal dynamic programming or progressive alignment strategies to the application of iterative algorithms, such as simulated annealing, hidden Markov models, or genetic algorithms (reviewed in Lecompte et al. 2001). Since the year 2000, thanks to the new high-throughput genomic and proteomic technologies, the size and complexity of the data sets that need to be routinely analyzed are increasing. In particular, with the arrival of numerous new genome sequences from eukaryotic organisms, large multi-domain sequences are becoming more and more prevalent in the sequence databases. In the face of these highly complex proteins, state-of-the-art multiple alignment methods now often use a combination of complementary techniques, such as local/global alignments or sequence/structure information (reviewed in Thompson and Poch 2006). When new methods are introduced, it is important to objectively measure the quality of the alignments produced by the program, and to assess the improvement obtained compared to existing methods. Ideally, a detailed evaluation should be performed to determine the performance of the different methods under various conditions. Such studies would allow the selection of the most appropriate program for a given alignment problem. In-depth analyses of the strengths and weaknesses of the different programs would also indicate guidelines for improvement of alignment algorithms.

In computer science, the quality of an algorithm is often estimated by comparing the results obtained with a predefined benchmark or "gold standard." The benchmark provides a set of standard tests that can be used to determine the relative performance of different programs. For multiple sequence alignment algorithms, several benchmarks are now available. These benchmarks will be described in detail in the section on multiple alignment benchmarks, but first we discuss the most important criteria involved in the development of a successful benchmark. The final section will describe the evolution of multiple sequence alignments since the introduction of benchmarking and, in particular, the improvements achieved by the most recent methods.

CRITERIA FOR BENCHMARK DEVELOPMENT

Benchmarking is widely used in computer science to compare the performance characteristics of computer systems, compilers, databases, and many other technologies. Within a scientific discipline, a benchmark captures the community consensus on which problems are worthy of study, and determines what are scientifically acceptable solutions. A benchmark generally consists of two components: a task sample used to compare the performance of alternative tools or techniques, and some kind of performance measure that evaluates the fitness for purpose. The task sample is used to reduce variability in the results, because all tools and techniques are evaluated using the same tasks and experimental materials. Another advantage of benchmarking is that replication is built into the method, and the results of the benchmarking experiments are therefore reproducible. Since the materials are designed to be used in different laboratories, people can perform the evaluation repeatedly on various tools and techniques, if desired. Also, some benchmarks can be automated, so the computer does the work of executing the tests, gathering the data, and producing the performance measures. The resulting evaluations allow developers to determine where they need to improve and to incorporate new features in their programs, with the aim of increasing specific aspects of performance. Consequently, the creation and widespread use of a benchmark within a research area is frequently accompanied by rapid technical progress (Sim et al. 2003). Nevertheless, continued evolution of the benchmark is necessary to prevent researchers from making changes to optimize the performance of their contributions on a particular set of tests. Too much effort spent on such optimizations indicates stagnation, and can hold back later progress.

The process of constructing a benchmark implies the rigorous definition of both what is to be measured (for example, the quality of a solution or the time required to produce it) and how it should be measured. This implies that the correct solutions to the tests in the benchmark are known in advance and that the exactitude of the results obtained by different tools can be quantitatively estimated. A number of requirements for successful benchmarks have been identified previously (Sim et al. 2003), which can be used as design goals when creating a benchmark or as dimensions for evaluating an existing one:

- *Relevance.* The tests in the benchmark need not be exhaustive, but should be defined by the community as being representative of the problems that the system is reasonably expected to handle in a natural (i.e., not artificial) setting. In addition, the performance measure used should be pertinent to the comparisons being made.
- *Solvability.* It should be possible to complete the task sample and to produce a good solution. A task that is too difficult for all or

most tools yields little data to support comparisons. A task that is achievable, but not trivial, provides an opportunity for systems to show their capabilities and their shortcomings.

- *Accessibility*. The benchmark needs to be easy to obtain and easy to use. The test materials and results need to be publicly available, so that anyone can apply the benchmark to a tool or technique and compare the results with those of others.
- *Evolution*. Continued evolution of the benchmark is necessary to prevent researchers from making changes to optimize the performance of their contributions on a particular set of tests. Too much effort spent on such optimization indicates stagnation, suggesting that the benchmark should be changed or replaced.

Benchmarks that are designed according to these conditions should lead to a number of benefits, including a stronger consensus on the community's research goals, more rigorous examination of research results, and faster technical progress.

MULTIPLE ALIGNMENT BENCHMARKS

Benchmark Specification

There are three main issues involved in the definition of a multiple alignment benchmark. First, what is the "correct" alignment of the sequences included in the tests? Second, which alignment problems should be represented in the benchmark, and how many test cases are needed? Finally, the benchmark needs to include a means of comparing the alignment produced by a program with the reference alignment.

Definition of the Correct Alignment The goal of a multiple sequence alignment is to identify equivalent residues in nucleic acid or protein molecules that have evolved from a common ancestor. However, the true evolutionary history cannot usually be reconstructed and, therefore, reference sequence alignments are generally constructed based on comparisons of the corresponding 3D structures. For proteins, the 3D structure is generally more conserved than the sequence, and a reliable structural superposition is normally possible between very divergent proteins sharing little sequence identity (Koehl 2001). Structural superposition is carried out between two known structures, and is typically based on the Euclidean distance between corresponding residues, instead of the distance between amino acid "types" used in sequence alignment. Thus, the structure alignment can provide an objective reference that is built independently of the sequences. However, there are inherent ambiguities in deriving a sequence alignment from structure superposition, and there is no unique way to align two 3D structures (Godzik 1996). To illustrate the problem, Figure 8.1 shows three pairwise sequence alignments derived from different structure superposition programs: SSAP (Taylor 2000), DALI (Holm and Sander 1998), and CE (Shindyalov and Bourne 1998).

On the other hand, most algorithms yield similar conclusions as to what is superposable and what is not. Even for the divergent sequences shown in Figure 8.1 (*1gky* and *1uky* share 16% residue identity), the three programs produced the same alignment for a large portion of the sequences. Ultimately, however, at very low sequence identity, structures diverge significantly enough that alignment of some parts of the target and template structures is not even meaningful. To overcome this problem, many multiple alignment benchmarks define "core block" regions

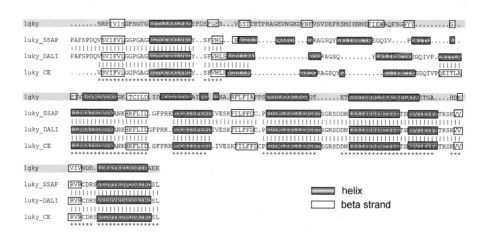

Figure 8.1. Comparison of alignments derived from three different structure superposition methods. The first row shows the *1gky* sequence from the PDB database; the subsequent rows represent different pairwise superpositions of *1gky* and *1uky* produced by the programs SSAP, DALI, and CE. Secondary structure elements are identified by shaded boxes. Vertical lines indicate positions that are superposed identically by the different programs. Asterisks below the alignments indicate positions that are aligned identically by all three programs.

in the reference alignments, where the sequences are assumed to be reliably aligned.

Definition of the Alignment Test Cases As mentioned in the previous section, a benchmark does not need to include all possible test cases and, in general, it is sufficient to provide enough representative tests in order to be able to differentiate between the different methods tested. In the case of a multiple sequence alignment benchmark based on 3D structure comparisons, the largest source of protein structures is the Protein Data Bank (PDB) database (Berman et al. 2000), although this set contains a certain amount of bias due to over-represented structures (Brenner et al. 1997). Protein structure databases, such as SCOP (Hubbard et al. 1999) or CATH (Pearl et al. 2005), provide useful resources, by classifying proteins at various levels of structural similarity and allowing the selection of as many different types of proteins as possible. However, the complexity of a multiple sequence alignment does not depend only on the structural class, but also on the nature of the set of sequences to be aligned. Therefore, the reference alignments should also include example test cases representing the diverse problems encountered when multiple alignments are performed, such as different numbers of sequences, different sequence diversity or sequence length, and so forth.

Definition of an Alignment Score A benchmark should also include a means of comparing a test alignment with the gold standard reference alignment. In this way, the same benchmarking process can be performed in different groups, and the results of the studies will be directly comparable. Most multiple alignment benchmarks use either the percentage of pairwise residues aligned the same in both alignments (known as the sum-of-pairs score), or the percentage of complete columns aligned the same. The most appropriate score will depend on the sequences in the reference alignment. For example, for a set of very divergent sequences, it might be more useful to use the sum-of-pairs score, since many programs will not be able to align all of the sequences correctly. However, there are situations where the column score is more meaningful than the sum-of-pairs score. This is the case, for example, for reference alignments containing a set of closely related sequences and a small number of more divergent, or "orphan" sequences. Here, most alignment programs will align the closely related sequences correctly and will obtain a high sum-of-pairs score even if the more divergent sequences are misaligned. In this case, the column score will better differentiate the programs that successfully align the orphan sequences.

Alignment Benchmarks

One of the first studies to compare the quality of different methods (McClure et al. 1994) used four sets of sequences from the hemoglobin, kinase, aspartic acid protease, and ribonuclease H protein families. Several progressive alignment methods, including both global and local algorithms, were compared by measuring their ability to identify highly conserved motifs involved in the function of the four protein families. Protein sequences with >50% amino acid residue identity were usually unambiguously aligned by most of the multiple alignment methods. For sequences with ~30% identity, the motifs were detected correctly only if the motifs were well conserved and when few indels existed. They concluded that, in general, global methods performed better than local methods. However, the number of suitable test sets available at that time was somewhat limited, and this was therefore not a comprehensive test. Since then, a number of larger, more comprehensive benchmark test sets have been developed (see Table 8.1).

These benchmarks are used to compare and evaluate the different methods, to select the most suitable method for a particular alignment problem (e.g., more efficient, more correct, more scalable), to evaluate the improvements obtained when new methods are introduced, and to identify the strong and weak points of the different algorithms.

TABLE 8.1. MULTIPLE SEQUENCE ALIGNMENT BENCHMARKS

	Creation Date	Sequence Type	Pairwise Alignments	Multiple Alignments	Number of Test Cases
Homstrad	1998	protein	×	√	1032
BAliBASE	1999	protein	×	√	217
OxBench	2003	protein	√	√	672
Prefab	2004	protein	×	√	1932
SABmark	2005	protein	√	×	29759
IRMBASE	2005	artificial	×	√	180
BRAliBASE	2005	RNA	×	√	388

BAliBASE One of the first large-scale benchmarks specifically designed for multiple sequence alignment was BAliBASE (Bahr et al. 2001; Thompson et al. 1999a). The alignment test cases in BAliBASE are based on 3D structural superpositions that are manually refined to ensure the correct alignment of functional residues. The alignments are organized into reference sets that are designed to represent real multiple alignment problems. The first version of BAliBASE consisted of five reference sets representing many of the problems encountered by multiple alignment methods at that time, from a small number of divergent sequences, to sequences with large N/C-terminal extensions or internal insertions. In version 2 of the benchmark, three new reference sets were included, devoted to the particular problems posed by sequences with transmembrane regions, repeats, and inverted domains. In each reference alignment, core blocks are defined that exclude the regions for which the 3D structure is unreliable, for example, the borders of secondary structure elements or in loop regions.

In order to assess the accuracy of a multiple alignment program, the alignment produced by the program for each BAliBASE test case is compared to the reference alignment. Two scores are used to evaluate the alignment. The SPS (sum-of-pairs score) is the percentage of correctly aligned pairs of residues in the alignment produced by the program. It is used to determine the extent to which the program succeeds in aligning some, if not all, of the sequences in an alignment. The CS (column score) is the percentage of correctly aligned columns in the alignment, which tests the ability of the program to align all of the sequences correctly. These scores are calculated in the core block regions only.

For the latest release of BAliBASE (Thompson et al. 2005b), many of the steps in the development were automated, including the search for homologous proteins, the 3D structure superposition, the definition of the core blocks, and the integration of structure/functional information for alignment annotation and display. This semiautomatic protocol allows the construction of a larger number of alignments, as well as the integration of more sequences in each alignment, compared to previous releases of the benchmark (see Table 8.2). In the future, the new protocol should also allow more frequent releases of the benchmark by the incorporation of new protein families and new sequences as they become available.

For BAliBASE version 3, representative protein families were selected from the SCOP protein domain database (Hubbard et al. 1999). Protein domains in SCOP are hierarchically classified into families, superfamilies,

TABLE 8.2. NUMBER OF TEST CASES IN VERSION 3 OF THE BALiBASE ALIGNMENT BENCHMARK

		Subtotal
Reference Set 1	Small number of equidistant sequences	
V1 (< 20% identity)	short: 14; medium: 12; long: 12	38
V2 (20–40% identity)	short: 14; medium: 16; long: 15	45
Reference Set 2	Family with one or more "orphan" sequences	41
Reference Set 3	Divergent subfamilies	30
Reference Set 4	Large N/C terminal extensions	48
Reference Set 5	Large internal insertions	16
Total		217

NOTE: Reference Set 1 contains alignments of equidistant sequences and is divided into six subsets, according to three different sequence lengths and two levels of sequence variability. Reference Set 2 contains families aligned with one or more highly divergent "orphan" sequences, Reference Set 3 contains divergent subfamilies, Reference Set 4 contains sequences with large N/C-terminal extensions, and Reference Set 5 contains sequences with large internal insertions. Reference Sets 6–8, containing transmembrane sequences, repeats, and circular permutations, were maintained in this version, although they were not updated.

folds, and classes. Example folds were selected for inclusion in BAli-BASE from each of the five main classes (excluding membrane and small proteins), provided that at least four different protein structures were available. The protein structures were superposed using the SSAP structural alignment program (Taylor 2000), and the sequence alignments were manually verified to ensure that annotated functional residues were aligned correctly. If too few sequences with known structures existed, the reference alignment was augmented with sequences from the Uniprot database (Wu et al. 2006). Figure 8.2 shows part of an alignment from the Reference Set 3 set of divergent subfamilies.

For each test case, two different reference alignments were constructed, containing either the homologous regions only or the full-length sequences. The alignment of homologous regions only is widely used in the construction of protein domain databases, whereas the alignment of full-length, complex sequences, such as those detected by the database searches, is routinely performed in automatic, high-throughput genome analysis projects.

OxBench The OXBench benchmark suite (Raghava et al. 2003) provides multiple alignments of protein domains that are built automatically using both structure and sequence alignment methods. The fully automatic

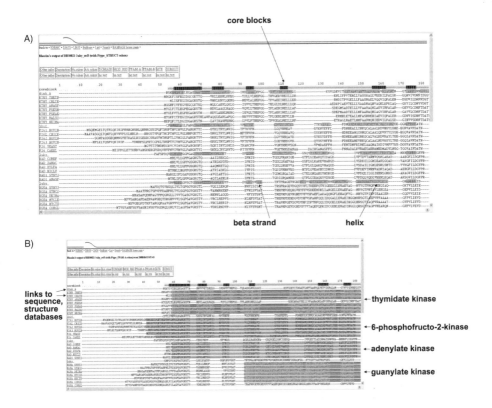

Figure 8.2. Example BAliBASE reference alignment. (A) Part of a multiple alignment of four divergent subfamilies of P-loop-containing kinases. Secondary structure elements are highlighted in black (helix) or grey (beta strand). Core blocks are indicated by black boxes above the alignment. (B) The same alignment shaded according to Pfam database entries. Links to the Uniprot and PDB databases are provided by clicking on the sequence names on the left.

construction means that a large number of tests can be included; however, the benchmark results will be biased toward sequence alignment programs using the same methodology as that used to construct the reference.

Reference proteins for alignment were drawn from the 3Dee database of structural domains (Siddiqui et al. 2001). The domains in 3Dee are organized into a hierarchy of structurally similar protein domain families classified by the "S_c score" from the automatic multiple structure superposition program STAMP (Russell and Barton 1992). The regions with STAMP scores of $P_{ij} > 6.0$ for three or more residues were considered as reliably superposed, structurally conserved regions (SCRs). From the complete 3Dee data set, 1,168 domains in 218 families were selected, such that no two sequences in a family shared ≥ 98% identity. The benchmark was then divided into three data sets. The *master set* contains only sequences of known 3D structure, and currently consists of 672 alignments from 218 distinct protein families, with from 2 to 122 sequences in each alignment. The *extended data set* was then constructed from the master set by including sequences of unknown structure. Finally, the *full-length data set* was built by including the full-length sequences corresponding to the domains in the master data set. Figure 8.3 shows an example alignment from the master data set.

A number of smaller data sets were also derived from the master data set:

- *Pairwise families.* The set of 273 families containing only two sequences was extracted from the master data set. This set is used to evaluate alignment methods that work for only two sequences (pairwise methods).
- *Multiple families.* The set of 399 families with more than two members was extracted from the master data set. This set allows the study of alignment algorithms and parameters on families having more than two sequences.
- *Test and training sets.* The master data set was split into two sets in such a way that there was no domain in one set that shared

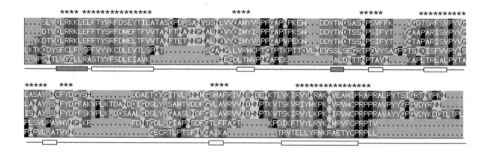

Figure 8.3. Part of an alignment from the Oxbench master data set. An alignment of five viral coat protein sequences from the PDB database. Asterisks above the alignment indicate structurally conserved regions (SCR). Secondary structure elements are indicated below the alignment by boxes shaded in grey (helix) or white (beta strand).

sequence similarity with domains in the other set. The separate, independent testing and training sets currently contain 334 and 338 alignments, respectively.

A number of different scores are included in the benchmark suite, so that the accuracy of multiple alignment programs may be evaluated. The average number of correctly aligned positions is similar to the column score (CS) used in BAliBASE. This can be calculated over the full alignment or over SCRs only. The position shift error (PSE) measures the average magnitude of error, so that misalignments that cause a small shift between two sequences are penalized less than large shifts. Two other measures are also provided that are independent of the reference alignment; the structure superposition implied by the test alignment is computed, and the quality of the test alignment is then estimated by computing the RMSD (Mclachlan 1972) and S_c values of the superposition.

PREFAB The PREFAB (Edgar 2004b) benchmark was constructed using a fully automatic protocol, and currently contains 1932 multiple alignments. Pairs of sequences with known 3D structures were selected and aligned using two different 3D structure superposition methods: FSSP (Holm and Sander 1998) and the CE aligner (Shindyalov and Bourne 1998). To minimize questionable and ambiguous structural alignments, only those pairs for which FSSP and CE agreed on at least 50 positions were retained. Each sequence in the aligned pair was then used to query a database using PSI-BLAST (Altschul et al. 1997), from which high-scoring hits were collected. Finally, the queries and their hits were combined to make test sets of 50 sequences.

Three test sets selected from the FSSP database were used, called SG, PP1, and PP2. These three sets vary mainly in their selection criteria. PP1 and PP2 contain pairs with sequence identity <30%. PP1 was designed to select pairs that have high structural similarity, requiring a z-score of >15 and a root mean square deviation (r.m.s.d.) of <2.5 Å. PP2 selected more diverged pairs with a z-score of >8 and <12, and an r.m.s.d. of <3.5 Å. SG contains pairs sampled from three ranges of sequence identity: 0 ± 15, 15 ± 30, and $30 \pm 97\%$, with no z-score or r.m.s.d. limits. PREFAB version 3.0 has 1932 alignments averaging 49 sequences of length 240, of which 178 positions in the structure pair are found in the consensus of FSSP and CE.

The accuracy of an alignment program is estimated by comparing the alignment of the structure pair in the test multiple alignment with the

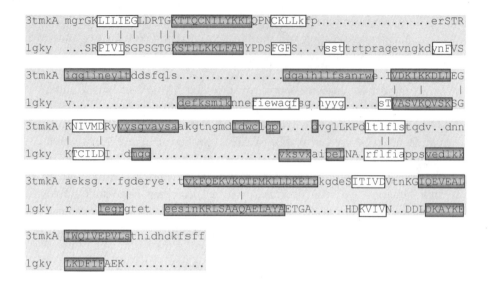

Figure 8.4. Example PREFAB alignment of *3tmkA* and *1gky*. Uppercase-aligned positions represent reliable regions in the alignment, whereas lowercase positions are not aligned. Secondary structure elements are identified by shaded boxes (grey = helix, white = beta strand). Identical residues are indicated by vertical lines.

reference superposition in each test case. An example pairwise alignment is shown in Figure 8.4. Only positions that are aligned the same by the two different superposition methods are considered. The quality of a multiple alignment program is thus evaluated by calculating the percentage of these positions that are correctly aligned in the multiple alignment.

SABmark SABmark (Van Walle et al. 2005) contains reference sets of sequences derived from the SCOP protein structure classification, divided into two sets, twilight zone (Blast E-value >= 1) and superfamilies (residue identity <= 50%). For each alignment problem, pairwise reference structure alignments are constructed, derived as a consensus from SOFI (Boutonnet et al. 1995) and CE (Shindyalov and Bourne 1998). For example, Figure 8.5A shows three pairwise reference alignments from the twilight test set.

However, the pairwise sequence alignments are not necessarily entirely consistent with each other and, as a consequence, no unique multiple alignment solution can be constructed from the pairwise reference alignments (see Figure 8.5B). Instead, to evaluate the quality of

Constructing Alignment Benchmarks

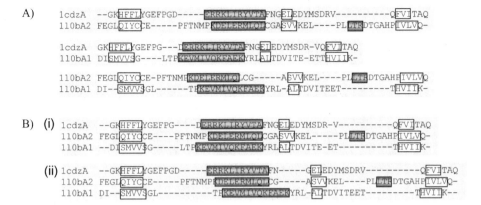

Figure 8.5. Example SABmark pairwise alignments. (A) Pairwise alignments of three sequences from a twilight test set. Secondary structure elements are identified by shaded boxes (grey = helix, white = beta strand). (B) Alternative multiple alignments of the same three sequences. *1cdzA* and *1lobA2* are aligned identically in both multiple alignments. The sequence *1lobA1* is aligned to *1cdzA* in (i) and to *1lobA2* in (ii).

a multiple alignment program, multiple alignments of each reference set are constructed, and pairwise alignments are then extracted from the multiple alignment and compared to the reference structure superpositions. Two different scores are provided to measure the quality of the alignment program. The first, f_D, is similar to the SPS score and is defined as the ratio of the number of correctly aligned residues divided by the length of the reference alignment, and may be thought of as a measure of sensitivity. The second, the f_M score, measures the specificity and is defined as the ratio of the number of correctly aligned residues divided by the length of the test alignment.

The benchmark includes test sets for most of the known protein fold space (see Table 8.3).

In addition, SABmark contains two test sets that include sequences that are not actually related to the other family members. In these sets, each group of N sequences is expanded with at most N other structurally unrelated, yet apparently similar, sequences. These false positives were selected from Blast searches of the original sequences (true positives) against a 70% identity subset of SCOP. Each false positive belongs to a different fold than the true positives and has at least one E-value to a true positive that is lower than at least one E-value of that true positive to another true positive.

TABLE 8.3. NUMBER OF TEST CASES IN SABMARK BENCHMARK

Set	Groups	Sequences	Sequence pairs
Twilight	209	1740	10667
Twilight with false positives	209	3458	44056
Superfamily	425	3280	19092
Superfamily with false positives	425	6526	79095

NOTE: The benchmark is divided into two test sets: twilight zone (Blast E-value ≥ 1) and superfamilies (residue identity ≤ 50%). The corresponding sets "with false positives" include sequences that are structurally unrelated to the original family members.

Homstrad HOMSTRAD (Stebbings and Mizuguchi 2004) is a database of protein families, clustered on the basis of sequence and structural similarity. Most of the families have, on the average, more than 30% sequence identity, although the lowest pairwise percentage identity may be less than 20%. Although HOMSTRAD was not specifically designed as a benchmark database, it has been employed as such by a number of authors (e.g., O'Sullivan et al. 2004; Roshan and Livesay 2006). The database provides combined protein sequence and structure information extracted from the PDB (Berman et al. 2002), and relies heavily on other databases, especially Pfam (sequence-based families) (Berman et al. 2002) and SCOP (structure-based families) (Lo Conte et al. 2002). The underlying principals for generating family-specific data are superposition of homologous protein structures, and an alignment produced using COMPARER (Sali and Blundell 1990) followed by a revised superposition using MNYFIT. The alignment is then annotated using JOY (Mizuguchi et al. 1998a), and a family profile is generated for use with the FUGUE (Shi et al. 2001) search engine. Finally, evolutionary trace data are provided by TRACE SUIT II (Innis et al. 2000), homologous sequences are appended using CLUSTALW (Thompson et al. 1994), and PROSITE (Sigrist et al. 2002) motifs are incorporated into the alignments. The result is a database of multiple alignments of protein domains containing 1032 families with 3454 structures and 5972 single-member families.

IRMBASE IRMBASE (Subramanian et al. 2005) uses simulated sequence data to test the ability of multiple sequence alignment programs to detect and align locally conserved motifs. Groups of artificial conserved-sequence motifs were produced using the ROSE software tool

Constructing Alignment Benchmarks

(Stoye et al. 1998), which simulates the process of molecular evolution. A set of "phylogenetically" related sequences is created from a user-defined "ancestor" sequence according to a phylogenetic tree. During this process, sequence characters are randomly inserted, deleted, and substituted under a predefined stochastic model. This way, a sequence family with known "evolution" is obtained, so the "correct" multiple alignment of these sequences is known. Note that these alignments contain mismatches as well as gaps. These families of conserved motifs created by ROSE were inserted at randomly chosen positions into nonrelated random sequences to produce three reference sets, *ref1*, *ref2*, and *ref3*, of artificial protein sequences. Sequences from *ref1* contain one motif each, and sequences from *ref2* and *ref3* contain two and three motifs each, respectively. The accuracy of a multiple alignment program is then scored by counting the number of columns in the ROSE motifs that are aligned correctly by the program.

Each reference set consists of 60 sequence families, 30 of which contain ROSE motifs of length 30, while the remaining 30 families contain motifs of length 60. Twenty sequence families in each of the reference sets consist of four sequences each, another 20 families consist of eight sequences, and the remaining 20 families consist of 16 sequences. In *ref1*, random sequences of length 400 are added to the conserved ROSE motif, whereas for *ref2* and *ref3*, random sequences of length 500 are added. Figure 8.6 shows part of an example IRMBASE reference alignment from the *ref2* data set.

However, the relationship of these simulations to evolutionary models of real biological sequences is not well understood, and the performance of alignment programs on artificial sequences should not be overestimated as the design of such data sets is necessarily somewhat arbitrary.

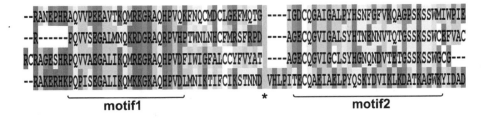

Figure 8.6. Part of an IRMBASE reference alignment. The alignment contains four random sequences with two conserved ROSE motifs. * 170 columns have been deleted at this position to save space.

BRAliBASE All of the above benchmarks contain protein sequence alignment test sets. However, the use of multiple sequence alignments is also an essential step for many RNA sequence analysis methods, for example, RNA structure analysis, RNA homology search, noncoding RNA (ncRNA) detection, and RNA-based phylogenetic inference. BRAliBASE (Gardner et al. 2005) is the only benchmark specifically designed to evaluate and compare multiple alignment programs on RNA data sets. BRAliBASE includes four diverse structural RNA datasets of Group II introns, 5S rRNA, tRNA, and U5 spliceosomal RNA. The sequences and the reference alignments were obtained from the Rfam v5.0 database (Griffiths–Jones et al. 2005). Approximately 100 subalignments were generated for each of the four families. The subalignments contained five sequences each and encompassed a range of sequence identities. This large dataset was divided into high (≥75% sequence identity, 73 alignments), medium (55–75% sequence identity, 73 alignments), and low (<55% sequence identity, 242 alignments) sequence homology groups. Figure 8.7 shows an example alignment from the low homology group. An additional tRNA dataset was also generated with just two sequences in each alignment.

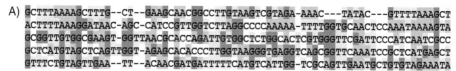

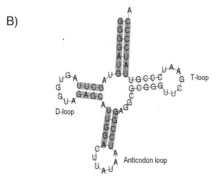

Figure 8.7. Example alignment from the BRAliBASE benchmark. (A) Alignment of 5 tRNA sequences from the low homology group (from Rfam:RF00005). The base-pairing structure is represented by colored sets of < and > tags. (B) Consensus secondary structure for the tRNA family. Base pair coloring is the same as for the multiple alignment tags.

Two quality measures are provided to evaluate alignment methods on structural RNAs. The first is the traditional sum-of-pairs score (SPS) employed in many of the previous alignment benchmarks. The second measure, known as the structure conservation index (SCI), provides an estimate of the conserved secondary structure information contained within the alignment (Washietl et al. 2005). It is a derivative of the score calculated by the RNAalifold consensus folding algorithm (Hofacker et al. 2002), based on the sum of a thermodynamic and a covariance term and is independent of the reference alignment.

The RNA datasets were used to evaluate a variety of both sequence and structure alignment methods. It was found that the multiple sequence alignment algorithms performed well on the high- to medium-homology data sets, whereas ClustalW (Thompson et al. 1994), ProAlign (Löytynoja and Milinkovitch 2003), and POA (Raphael et al. 2004) consistently ranked in the top ten across all homology ranges. Nevertheless, it was observed that sequence alignment alone was generally inappropriate for <50–60% sequence identity. Below this limit, algorithms incorporating structural information, for example, Dynalign (Mathews and Turner 2002), Foldalign (Havgaard et al. 2005), PMcomp (Hofacker et al. 2004), and Stemloc (Holmes 2005), outperformed the purely sequence-based methods. Unfortunately, these algorithms are computationally demanding, which severely limits their use in practice.

Comparison of Multiple Alignment Benchmarks

The goal of an alignment benchmark is not to provide a comprehensive database of all possible alignments, but it should include sufficient test cases in order to differentiate between the different alignment programs being compared. All of the benchmarks described above have large numbers of alignments and have been used in a number of studies to demonstrate the relative accuracy of various alignment methods. SABmark is by far the largest of the benchmarks and boasts full coverage of the known fold space. However, there are only pairwise references for each sequence group, and multiple alignment assessment becomes complicated depending on how the results are treated.

Another important criterion for the benchmarks is the quality of the reference alignments. Errors in the benchmark will lead to biased or erroneous conclusions about a program's performance. Two different approaches have generally been used to ensure the quality of the reference alignments in the benchmarks described above.

BAliBASE alignments are based on a structural superposition, with a manual verification and correction of the results, whereas most of the other benchmarks use several complementary structure superposition methods and consider only the alignments where the methods are in agreement.

However, a comprehensive multiple alignment benchmark should also include alignment tests that represent the most common problems encountered when real sequences are aligned. Most of the benchmarks described above contain real protein or RNA sequences, except IRMBASE, which contains artificial sequences created by the authors. Many of the benchmarks are organized into a number of test sets according to the similarity of the sequences, which is clearly a major factor affecting the accuracy of a multiple alignment. Nevertheless, other characteristics of the sequences also influence the final alignment quality, including their lengths, their phylogenetic distribution, the presence of nonhomologous regions, or a nonlinear domain organization. A comprehensive multiple alignment benchmark should specifically address these diverse complex issues.

A comparison of a number of benchmarks for protein sequence alignment algorithms, including those described above, has been performed recently (Blackshields et al. 2006). The authors demonstrated that categorizing HOMSTRAD, OxBench, and PREFAB test cases by percent sequence identity yielded similar results when they were used to rank alignment programs, as might be expected. Each of these datasets showed that the ability to accurately align sequences is largely dependent on the diversity of the sequences. Oxbench, PREFAB, and BAliBASE all contain difficult cases containing full-length sequences of low sequence identity. BAliBASE has the additional advantage that several distinct problem areas are explicitly addressed, which also makes the benchmark more difficult to over-train on. Therefore they concluded that BAliBASE is one of the best benchmarks available, although it may be advantageous to use several benchmarks for program assessment.

MULTIPLE ALIGNMENT REVOLUTION

The objective evaluation of alignment quality and the introduction of large-scale alignment benchmarks have clearly had a positive effect on the development of multiple alignment methods (as shown in Figure 8.8).

Constructing Alignment Benchmarks

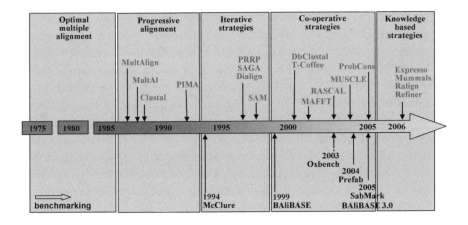

Figure 8.8. The simultaneous development of multiple alignment algorithms and alignment benchmarks. The development of multiple alignment algorithms (above the dateline) is shown in the context of the analogous developments of alignment benchmarks (below the dateline). The shaded boxes represent the introduction of a novel approach to the multiple alignment problem.

From their beginnings in 1975, until 1994 when McClure first compared different methods systematically, the main innovation was the introduction of the heuristic progressive method that allowed the multiple alignment of larger sets of sequences within a practical time limit, for example, MultAlign (Barton and Sternberg 1987a), MultAl (Taylor 1988), or Clustal (Higgins and Sharp 1988).

Soon after this initial comparison, various new methods were introduced that exploited novel algorithms, such as genetic algorithms (SAGA, Notredame and Higgins 1996), iterative refinement (PRRP, Gotoh 1996), or hidden Markov models (SAM, Karplus et al. 1998). These new approaches significantly improved alignment quality, as shown in the comparison of these methods (Thompson et al. 1999b). Nevertheless, this study highlighted the fact that no single algorithm was capable of constructing high-quality alignments for all test cases. For the first time, the study revealed a number of specificities in the different algorithms. For example, while most of the programs successfully aligned sequences sharing >40% residue identity, an important loss of accuracy was observed for more divergent sequences with <20% identity. Another important discovery was the fact that while global alignment methods in general performed better for sets of sequences that were of similar length, local algorithms were generally more successful at identifying the most conserved motifs in sequences containing large extensions and insertions. Of the local methods, DIALIGN (Morgenstern et al. 1996) was the most successful. The iterative methods, such as PRRP or SAGA, were generally more accurate than the traditional progressive methods, although at the expense of a large time penalty.

As a consequence, the first methods were introduced that combined both global and local information in a single alignment program, such as DbClustal (Thompson et al. 2000), T-Coffee (Notredame et al. 2000), MAFFT (Katoh et al. 2002), MUSCLE (Edgar 2004b), or Probcons (Do et al. 2005). Table 8.4 shows the scores obtained using these new methods for three different multiple alignment benchmarks described in the previous section. These benchmarks all contain alignments of protein sequences, where the reference alignment is constructed based on a 3D structural superposition.

The new combined strategies clearly improve overall alignment quality, compared to the more traditional methods, such as ClustalW or DIALIGN. Although there are some inconsistencies in the ranking of the different methods, ProbCons generally ranks as one of the highest scoring programs.

The IRMBASE benchmark contains a different type of alignment test. The goal here is to test the ability of multiple alignment programs to detect locally conserved motifs in otherwise unrelated sequences. In

TABLE 8.4. CURRENT STATE OF THE ART FOR MULTIPLE SEQUENCE ALIGNMENT METHODS

	SABmark		PREFAB		OxBench		
	0–25%	25–50%	0–20%	20–40%	0–20%	20–40%	40–60%
ClustalW 1.83	13.1	41.2	41.1	78.3	7.3	38.7	66.2
DIALIGN-T 0.1.3	13.7	41.9	42.0	78.1	4.4	34.0	68.1
T-Coffee 1.37	15.8	44.4	44.8	81.8	5.9	38.3	73.3
MAFFT 5.531	16.2	45.2	48.6	83.8	5.3	38.4	71.6
MUSCLE 3.6	20.8	50.6	46.1	83.0	5.0	39.9	72.1
Probcons 1.09	**21.1**	48.4	**49.0**	**85.2**	**9.6**	39.7	**74.4**

NOTE: All scores shown are percentage column scores (maximum score = 100). For PREFAB, the score is calculated in the superposable regions. For OxBench, the full alignments were used and the scores were calculated in structurally conserved regions only (from Blackshields et al. 2006). The top score in each data set is shown in bold.

TABLE 8.5. COMPARISON OF SIX MULTIPLE
ALIGNMENT PROGRAMS USING IRMBASE

Method	1 motif	2 motifs	3 motifs	Total
DIALIGN-T	82.3	78.4	79.7	80.1
T-Coffee 1.37	75.3	66.6	69.2	70.2
Probcons 1.09	33.1	37.9	51.3	40.8
POA 2	73.0	12.5	7.45	31.0
MUSCLE 3.5	9.4	10.9	22.4	14.2
ClustalW 1.83	0.0	0.8	5.1	1.9

NOTE: Values indicate the percentage of correct alignment columns of the ROSE motifs (from Subramanian et al. 2005).

these tests, the DIALIGN-T program (Subramanian et al. 2005) was shown to outperform the other programs evaluated (see Table 8.5).

Analysis of Program Strengths and Weaknesses

In order to provide helpful insight into the strengths and weaknesses of the many different methods that are now available, we need to use specific test sets that represent the typical problems encountered when aligning large sets of complex sequences. Version 3 of BAliBASE provides a number of such test sets that can be used to evaluate and compare the most recent multiple sequence alignment programs. The goal of this comparison is not to determine which program is the "best" for all alignments, but to measure the improvement in quality achieved by the programs and to identify their strong and weak points.

Table 8.6 shows the scores obtained by the different methods for five different BAliBASE reference sets, based on the alignments containing only the homologous regions. Reference Sets 6–8, containing transmembrane sequences, repeats, and circular permutations, were not used in this study, because the alignment programs have not been designed to handle these special cases.

In all of the reference tests, there is a significant difference between the traditional methods (ClustalW and DIALIGN) and the most recent developments. Nevertheless, in Reference Set 1, for all the programs tested, a decrease in accuracy of the alignments with decreasing residue identity is clearly demonstrated, with a significant difference between V_2 (20–40% identity) and V_1 (<20% identity), which corresponds to the "twilight zone" of evolutionary relatedness.

TABLE 8.6. SCORES FOR BAliBASE VERSION 3 REFERENCE SETS CONTAINING ALIGNMENTS OF HOMOLOGOUS REGIONS ONLY

	Reference 1: Equidistant sequences		Reference 2: Family with orphans	Reference 3: Divergent subfamilies	Reference 4: Large extensions	Reference 5: Large insertions	Time (sec)
	V1:<20%	V2:20–40%					
ClustalW 1.83	63/42	90/78	91/42	76/52	75/41	75/38	902
DIALIGN 2	50/31	86/71	89/37	70/39	79/45	78/43	6043
MAFFT 5.531	71/54	91/83	94/55	84/60	85/49	87/57	327
MUSCLE 3.6	71/52	91/82	94/50	85/58	84/46	86/54	523
T-Coffee 1.37	67/47	93/84	94/50	84/64	**87/54**	88/58	46335
Probcons 1.09	**79/63**	**94/87**	**95/60**	**87/65**	86/54	**90/63**	19035

NOTE: The scores shown in each column are percentage sums of pairs/column scores (maximum score = 100). The top score in each data set is shown in bold.

Of the two traditional methods, the global alignment program, ClustalW, performs better for the tests involving sequences of similar length in Reference Sets 1–3, while the local alignment program, DIALIGN, is more successful in Reference Sets 4–5, containing large N/C terminal extensions or internal insertions. This result confirms the observations made previously (e.g., Blackshields et al. 2006; Thompson et al. 1999b). By combining different complementary approaches, the more recent programs, TCoffee, MAFFT, MUSCLE, and Probcons are more reliable in all of the tests. The best alignments in all of the reference tests were achieved by Probcons, although a significant time penalty was incurred. The iterative programs, MAFFT and MUSCLE, represent a reasonable compromise between alignment accuracy and efficiency.

Table 8.7 shows the scores obtained by the different programs when the full-length sequences are aligned, instead of just the homologous regions. By comparing these scores with the scores in Table 8.6, it is clear that the inclusion of "noise," in the form of nonhomologous regions, represents a serious problem for all the programs tested, although the difference between the scores is smaller for the local alignment program, DIALIGN. Also, the difference is less for the more related sequences in Reference Set 1, V2, which suggests that, in general, all of the programs are capable of detecting and aligning domains with >20% residue identity, even for sequence sets that do not share the same overall domain structure. In these tests, Probcons again produces the most accurate alignments, but requires significantly more CPU time than most of the other methods.

CONCLUSIONS

Benchmarking can highlight the strong and weak points of multiple alignment methods and has had a positive effect on the development of new multiple sequence alignment methodologies. Considerable effort has been applied recently to the development of benchmarks aimed at different aspects of the multiple alignment problem. These benchmarks are generally designed to evaluate alignment quality, although the criteria for selection of an alignment program are often more complex, taking into account ease-of-use, efficiency, stability, robustness, and so forth, as well as alignment accuracy. Nevertheless, the objective evaluation of new algorithms should lead to more robust multiple alignment algorithms, which, in turn, will lead to more reliable results for the many applications that rely on accurate multiple alignments.

TABLE 8.7. SCORES FOR BAliBASE REFERENCE SETS CONTAINING ALIGNMENTS OF FULL LENGTH SEQUENCES

	Reference 1: Equidistant sequences		Reference 2: Family with orphans	Reference 3: Divergent subfamilies	Reference 5: Large insertions	Time (sec)
	V1:<20%	V2:20–40%				
ClustalW 1.83	46/24	85/72	86/20	62/27	61/34	2227
DIALIGN 2	47/26	85/70	85/29	64/31	77/42	12595
MAFFT 5.531	57/35	90/80	88/40	78/50	84/53	1409
MUSCLE 3.6	56/34	90/79	88/36	76/39	83/46	3608
T-Coffee 1.37	59/35	92/82	91/40	75/49	87/57	156373
Probcons 1.09	**65/43**	**93/86**	**90/41**	**79/54**	**88/57**	58488

NOTE: The scores shown in each column are percentage sums of pairs/column scores (maximum score = 100). The top score in each data set is shown in bold. Reference 4 is not included in this table because the full-length alignments in these tests are the same as in Table 8.6.

The comparison of the most recent multiple alignment programs using BAliBASE version 3 has shown that, despite significant progress, none of the available methods is capable of producing reliable alignments for the complex, divergent proteins that are detected by today's advanced database search algorithms. As a result, new multiple alignment methods are now being developed that exploit other information, for example, Expresso (Armougom et al. 2006b), Mummals (Pei and Grishin 2006), or RAlign (Sammeth and Heringa 2006), aimed at the construction of a high-quality multiple sequence alignment, even in the difficult case of complex, multi-domain proteins. Another area of growing interest is the development of methods for the refinement of an initial multiple alignment, for example, RASCAL (Thompson et al. 2003) and Refiner (Chakrabarti et al. 2006).

These new methods, together with the ever-increasing amount of sequence and structure information being generated by the new high-throughput technologies, mean that the size and complexity of the data sets that need to be routinely analyzed are increasing. As a consequence, the alignment benchmarks will also need to evolve in order to provide new larger test cases, which are representative of the new alignment requirements. It may also be useful in the future to address other factors that determine the suitability of an alignment program for a particular application. For example, most of the current benchmarks evaluate the ability of the programs to correctly align the most conserved segments of the sequences. Nevertheless, the accurate alignment of the regions between these "core blocks" is often essential for subsequent applications, such as 3D structure modelling or the identification of important functional sites. It is hoped that the creation of new benchmarks with larger, more complex test sets will encourage the continued development of new alignment algorithms, and vice versa.

CHAPTER 9

Simulation Approaches to Evaluating Alignment Error and Methods for Comparing Alternate Alignments

MICHAEL S. ROSENBERG
Arizona State University

T. HEATH OGDEN
Idaho State University

Simulation. 180
    Sequence Simulation Software 182
How Should One Compare Alignments?. .183
    Terminology 184
    Pairwise Alignment. 185
    Inclusion of Gapped Sites 188
    Multiple Alignments. 191
    A Comment on Global vs. Local Alignment 196
    A Comparison 196
Downstream Effects of Alignment Inaccuracy
Investigated Through Simulation. 199
    Evolutionary Distance Estimation. 200
    Phylogeny Reconstruction. 203
Conclusion 206

As this book demonstrates, sequence alignment is an important tool for biological research and may be used for a variety of purposes ranging from secondary structure identification (Coventry et al. 2004; Dowell and Eddy 2004; Holmes 2005; Knudsen and Hein 1999), noncoding functional RNA (ncRNA) detection (di Bernardo et al. 2004; Rivas and Eddy 2001), and phylogenetic inference. Generally speaking, the goal of multiple sequence alignment is to hypothesize site homology for a string of characters that represent evidence from data (DNA, amino acids, morphological data, etc.). While the goal is to hypothesize the correct or "true" site homologies, the reality is that identification of the true homologies is difficult for sequences that are at least moderately diverged. Failure to correctly predict site homologies is known as alignment error. Exploration of the effects of these alignment errors on downstream analysis in biological research has only recently begun in a serious fashion. This chapter will discuss the use of simulation approaches in the evaluation of alignment accuracy (in contrast to empirical benchmarks, as discussed in the preceding chapter), examine a variety of approaches (both established and novel) for directly comparing alternate alignments, and summarize some of the early findings on the effects of alignment error on downstream analysis, particularly with respect to phylogeny reconstruction.

SIMULATION

A simulation is an experiment designed to model reality. Simulations can take many different forms, but in this chapter we our focused on stochastic modeling of sequence evolution through computers. In its simplest form, the evolution of a DNA sequence could be modeled through a straightforward Markov process by taking an initial starting sequence and emulating the mutation process by randomly choosing sites within the sequence and allowing them to mutate to another nucleotide with a probability based on a theoretical model of nucleotide change, such as the Jukes–Cantor model (Jukes and Cantor 1969), the Kimura 2-parameter model (Kimura 1980), or the HKY model (Hasegawa et al. 1985), among many others. Protein sequence evolution could similarly be simulated using amino acid change matrices, such as the Dayhoff (Dayhoff et al. 1978) or JTT (Jones et al. 1992). More complex experiments generally include simulation of sets of sequences across a phylogenetic tree; this tree could be fixed *a priori* or could itself be generated from a random model or process, such as the Yule tree (Yule 1924), which assumes equal branching rates on all branches at any given time point. More complex simulations and models may include different rates of evolution at different sites, different models of evolution for different segments of a sequence (e.g., separate modeling of functional domains, exons, or introns), or different models of evolution across different parts of a tree (nonstationarity of the evolutionary process).

There are a number of advantages of the simulation approach. First, we can know the exact and precise truth behind the data. For the alignment, we know with absolute certainty which specific sites in one sequence are homologous to specific sites in other sequences. We know with absolute certainty the phylogenetic relationships and the history of the sequences. Thus, with a properly designed simulation system, we can track and distinguish every mutational event, whether it is a point substitution, insertion or deletion, translocation, transversion, gene duplication, or any other mutational mechanism causing variation in the sequence that we choose to model as part of the simulation. For even the best empirical data sets, we never know the absolute truth and can only infer it from the data. In some circumstances, this can be done with a reasonable amount of certainty, but under most conditions our "truths" are merely hypotheses. Because we have complete control over the simulation, a second advantage of simulation is that we can do very directed and specific tests of the effects of different variables or conditions. Every aspect of the simulation can be held constant but the one we want to test, allowing us to directly measure the effects of that variable on our outcome.

The primary disadvantage of simulation is realism. Our understanding of mutational processes is incomplete, and the simulation can only be as good as the model of the evolutionary process which underlies it, a model that can only be a simplification of the true molecular process. Also, the model consists of many parts, not simply the process by which a sequence changes, but also the specific parameters controlling the process (e.g., substitution rates, transition–transversion biases, nucleotide or amino acid compositions, and insertion and deletion rates and size distributions, just to name a few). Complex models have more parameters than simple models, but that leads to more parameter values that must be specified as part of the simulation. Usually, specific values are chosen to be representative of those estimated from empirical data, but occasionally there may be reasons to choose unrealistic values as part of the simulation experiment.

Therefore, in simulation studies, one must always question whether the simulation conditions were realistic enough to allow one's conclusions to have any applicability to questions involving empirical data. As an example, protein function is highly dependent on tertiary structure;

evolution of protein sequences must therefore be constrained by how mutations would affect this structure. Almost no work has been done that efficiently includes awareness of secondary or tertiary structure into protein sequence simulation, an extremely difficult problem that deserves more attention in the future.

Simulation has been widely used in phylogenetic studies for many years (e.g., Hillis 1995; Huelsenbeck and Rannala 2004; Nei 1996; Rosenberg and Kumar 2003; Takahashi and Nei 2000); it is only recently that simulation has begun to be regularly applied to the examination of sequence alignment. Part of the reason for this is that, to examine issues of sequence alignment, one must have a model of indel formation; until recently, little work has been done in developing such models (advances in such modeling are discussed in other chapters within this book).

Sequence Simulation Software

Numerous programs have been designed for simulating sequence evolution, but only a few include the simulation of insertions and deletions. The first widely used software for simulating alignments that included indels was the program ROSE (Stoye et al. 1998). This program allows the simulation of a family of related sequences guided by a known evolutionary tree. From this first study, the authors showed that the most optimal alignment is not necessarily the correct alignment. This program was created explicitly for benchmarking and comparative analyses. Numerous studies have used this software for testing alignment accuracy, particularly for protein sequences (Blackshields et al. 2006; Fuellen et al. 2002; Grasso and Lee 2004; Hollich et al. 2005; Hudek and Brown 2005; Lassmann and Sonnhammer 2005a; Pollard et al. 2004; Storm and Sonnhammer 2002; among others). Additional sequence simulation programs that incorporate indels have been developed recently, including Dawg (Cartwright 2005), MySSP (Rosenberg 2005c), Indel-Seq-Gen (Rambaut and Grassly 1997; Strope et al. 2007), and EvolveAGene3 (Hall, 2008). All of these programs are based on similar principles, but they may incorporate different models of sequence evolution, have different options with respect to recording information on ancestral sequences or specific mutational outcomes from the simulation process, or even run on different operating systems. Many studies that have used sequence simulation to examine alignment do not specify software and most likely made use of an in-house program (e.g., Blanchette, Green, et al. 2004; Blanchette, Kent, et al. 2004; Keightley and Johnson 2004; Wang et al. 2006). Some simulation programs are designed to model very specific situations or gene regions, such as CisEvolver (Pollard et al. 2006), which models the evolution of the transcription factor binding site, and EdiPy (Picardi and Quagliariello 2006), which simulates plant mitochondrial genes under RNA editing.

HOW SHOULD ONE COMPARE ALIGNMENTS?

For all of the work that has been done on alignments, one issue that has rarely been discussed in much detail is how one should actually compare alternate potential alignments. Researchers have taken a variety of approaches (Gardner et al. 2005; Karplus and Hu 2001; Lassmann and Sonnhammer 2002; Notredame et al. 2000; Ogden and Rosenberg 2006; Pollard et al. 2004; Rosenberg 2005a,b; Thompson et al. 1999b; Wallace, O'Sullivan, et al. 2005), which have most commonly been performed by comparing a hypothesized alignment from a program or algorithm to some form of benchmark alignment. In general, two primary measures have been used to assess alignment similarity: the sum-of-pairs score (SPS) and the column score, although variants do exist (Karplus and Hu 2001; Lassmann and Sonnhammer 2002). SPS and the column score were described by Thompson et al. (1999b) for comparing a hypothesized alignment to a reference alignment. SPS is the proportion of all paired residues from the reference alignment found in the hypothesized alignment. The column score is the proportion of all columns in the reference alignment found in the hypothesized alignment. These measures treat gaps differently: SPS explicitly ignores all paired comparisons containing a gap, whereas the column score implicitly includes gapped sites in columns by basing the measure only on residues within the column.

While there are clear fundamental differences between these and other potential measures, it is not clear how much difference it makes to use one or the other. Also, although some researchers are interested in including gaps in their measure, no one has described or explored the various approaches that would allow one to include them. In the following, we describe all of the logical metrics one could use for comparing alternate alignments, including different approaches to scaling, methods for the exclusion or inclusion of gapped sites, and approaches for both paired and multiple sequence alignments. Previously described metrics, such as SPS and the column score, are specific variants of the described system.

These metrics are equally applicable to DNA or protein sequences or any other type of character alignment one can imagine, because they are not at all dependent on the nature of the characters, just the order. Approaches for comparing structural alignments (Briffeuil et al. 1998; Gardner et al. 2005; Hickson et al. 2000) will not be discussed. Many of the alignments in our examples and figures were manually constructed to demonstrate specific characteristics and are not representative of optimal alignments that would be found from an alignment algorithm.

Terminology

In the following discussion, different alignments of the identical sequences will be referenced by roman numerals, for example, alignment *I* and alignment *II*. The two sequences within a pairwise alignment (Figure 9.1) are X and Y (additional letters will be used for a multiple alignment of more than two sequences). X_i will refer to the ith non-gapped character in sequence X; the total number of characters in a sequence is n_X (in Figure 9.1, $n_X = 7$ and $n_Y = 9$). The notation $I : X_i Y_j$ is used to indicate pairs of aligned characters, specifically that alignment I

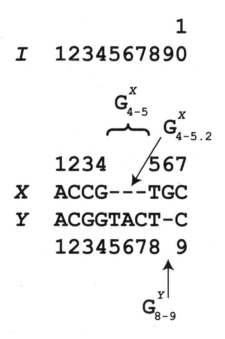

Figure 9.1. Illustration of alignment notation.

has the ith site in sequence X aligned with the jth site in sequence Y. For the alignment shown in Figure 9.1, for example, $I : X_5Y_8$ indicates that alignment I has the 5th non-gap character of sequence X (a T) aligned with the 8th non-gap character of sequence Y (also a T). Gaps in an alignment are referenced as $I : G^S_{i-j}$, where S refers to the sequence containing the gap and i and j are the non-gapped characters that bracket the sequence (j will always equal $i + 1$, but we find it convenient to list both bracketing characters). Thus, in Figure 9.1, there are two gaps: $I : G^X_{4-5}$ and $I : G^Y_{8-9}$. To refer to a specific single-base gap within a larger gap, we use $I : G^S_{i-j.k}$, where k represents the kth gapped site within the gap. Therefore, in Figure 9.1, the specific gapped site aligned with the 7th character in sequence Y (a C) would be $I : G^X_{4-5.3}$, and we can show that they are "aligned" using $I : G^X_{4-5.3}Y_7$. The total number of non-gapped aligned sites in an alignment, N_I, is the count of all $I : X_iY_j$; in Figure 9.1, $N_I = 6$ ($I : X_1Y_1$, $I : X_2Y_2$, $I : X_3Y_3$, $I : X_4Y_4$, $I : X_5Y_8$, and $I : X_7Y_9$). We will also refer to the total length of an alignment (all sites, gapped or ungapped) as n_I ($n_I = 10$ in Figure 9.1), and the kth site of the alignment as I_k; this notation can be combined with any of the above; thus, $I_{10} = X_7Y_9$ and $I_6 = G^X_{4-5.2}Y_6$. Subscripts of alignments and sequences may also refer to ranges of consecutive sites; thus, $I_{5...7} = G^X_{4-5} Y_{5...7}$.

Pairwise Alignment

We will begin by discussing similarity measures for alignments of pairs of sequences. The simplest way to compare two alignments is to count the proportion of non-gapped sites that are identical in the two alignments. There are actually three different ways to do this, and which one is chosen may depend on the purpose of the comparison. In the first two cases, one is comparing a hypothesized alignment to a benchmark alignment, perhaps from a database or an alignment known to be correct from a simulation (for consistency, the benchmark alignment will always be alignment I and the hypothesized alignment will always be alignment II). In this comparison, one may be interested in testing the accuracy of alignment II or how similar it is to alignment I. To measure this, one first would count the number of non-gapped sites in alignment II that are found in alignment I, C_U, that is, the count of all $II : X_iY_j$ that are also $I : X_iY_j$. Note that it is not necessary for the position of X_iY_j in alignment II be identical to the position of X_iY_j in alignment I (that is, if $II_p = X_iY_j$ and $I_r = X_iY_j$, p does not have to equal r), only that X_iY_j is found in both alignments. To get the

```
    I
    X    ATCT---TGTGGAACTG
    Y    A-GTAGCCGTAG---CG

    II
    X    ATCTTG---T-GGAACTG
    Y    A-GTAGCCGTAG---C-G
```

Figure 9.2. Example of a pair of alternate alignments for a pair of sequences. The similarity of these alignments varies from 56 to 33%, depending on the metric one chooses.

similarity, one divides this count by the total number of non-gapped aligned sites in sequence $II = N_{II}$; thus,

$$\omega_U = \frac{C_U}{N_{II}}.$$

Figure 9.2 shows an example of two alternate alignments. Using this measure, we find $N_{II} = 9$ (there are nine non-gapped sites in alignment II), $C_U = 5$ (X_1Y_1, X_3Y_2, X_4Y_3, X_7Y_9, and $X_{14}Y_{13}$ are found in both alignments), and $\omega_U = 0.56$. This measure makes sense when one wishes to compare one alignment to a benchmark or reference. It is less sensible if one is simply comparing two alternate alignments because the measure is directional; whereas the numerator is constant, the denominator may vary depending on whether one uses the count of non-gapped sites in alignment I or that of alignment II, N_I or N_{II}. For the example in Figure 9.2, if we use N_I (= 10) rather than N_{II}, we find $\omega_U = 0.5$. The commonly used SPS of Thompson et al. (1999b) is C_U/N_I; that is, it is the measure of common aligned pairs divided by the number of aligned pairs in the reference. It measures what proportion of the reference alignment is found in the proposed alignment, that is, what proportion of the reference alignment is described by the proposed alignment. In contrast, C_U/N_{II} is the proportion of the proposed alignment found in the reference alignment; it is a measure of what proportion of the proposed alignment is correct and is the measure used by Rosenberg (2005a).

However, if one simply wants to compare two alternate alignments (perhaps neither is a reference), one could instead use the average N:

$$\omega_U = \frac{C_u}{\frac{1}{2}(N_I + N_{II})} = \frac{2C_u}{N_I + N_{II}}.$$

This is equivalent to the overlap score described by Lassmann and Sonnhammer (2002). To distinguish among these variant measures, we will reference the alignment length that makes up the denominator as $\omega_{U'}$, $\omega_{U''}$, and $\omega_{\bar{U}}$. In the example in Figure 9.2, we find $\omega_{\bar{U}} = 0.526$. These measures are based on counting similarities only at non-gapped sites. The advantages of this are twofold: first, because gaps are inferred and not observed, one is measuring similarity based only on observed data and not on secondary inferences; second, many bioinformatic analyses only make use of non-gapped sites (e.g., in the estimation of evolutionary distance); therefore these are the sites whose similarity or accuracy is most important to those analyses. A disadvantage of these measures is they do not account for differences in sites that are aligned without gaps in one sequence but aligned with gaps in the other; thus they can make very different alignments seem very similar. Figure 9.3 A shows an extreme case. If alignment II were a hypothesis and alignment I a benchmark and we chose to use N_{II} for the denominator, we would find that $\omega_{U''} = 1$; that is, every pair of aligned characters in II is found in I. This is correct, but one

A)

```
I
X   ACTCGCTG
Y   AGTAGCCG

II
X   ACTCGCTG-------
Y   A-------GTAGCCG
```

B)

```
I
X   ACTCGCTGCGTCATCGCTAT
Y   AGTAGCCGCGTTACCGCGCC

II
X   ACTCGCTGCGTCATCGCTAT-
Y   -AGTAGCCGCGTTACCGCGCC
```

Figure 9.3. Alternate pairwise alignments illustrative of specific problems. (A) Extreme case of alignments that may be measured as very similar by non-gapped metrics, but that are actually very different. (B) Simple case where r_N is large but alignments have no similarity. Although $r_N = 0.95$, no sites are identically aligned.

can hardly claim the alignments are identical. Thus, when these metrics are reported, it will often be useful to also determine the ratio of the number of aligned sites in each alignment, that is, $r_N = N_{II}/N_I$. For the extreme case in Figure 9.3, we would see that $r_N = 0.125$; thus, while $\omega_{U^{II}} = 1$, it is only for a small proportion of sites. In the example in Figure 9.2, $r_N = 0.9$; for this example, these measures reflect similarity for a majority of the aligned sites. By itself, r_N is not particularly useful because pairs of alignments can have almost identical numbers of aligned sites but be completely different (Figure 9.3B).

Inclusion of Gapped Sites

Almost all previously described metrics for comparing alignments of pairs of sequences ignore gapped sites. However, one may wish to use similarity measures that include gaps. A possible advantage of a gap-based measure is that it would represent a greater proportion of the alignment than when gapped sites are ignored. It also is a sensible measure when considering bioinformatic analyses that explicitly make use of gaps, for example, phylogenetic analysis using parsimony with gaps treated as characters. A potential disadvantage is that gaps are never observed, only inferred, and one may be skewing the similarity measure based on inferences rather than observations.

There are three basic approaches to including gaps in a similarity measure that we refer to as any gap (AG), correct gap (CG), and correct gap site (CS). For all three methods, we calculate C_U for non-gapped sites identically as above; the difference is we will now add a count based on gapped sites. In the first of these approaches, AG, we generate a count, C_{AG}, of all characters in either sequence that are aligned to a gap in both alignments. In Figure 9.2, for example, there are four sites that are aligned to gaps in both alignments I and II (X_2, Y_6, X_{10}, and X_{11}); thus, $C_{AG} = 4$. Our new metric is, thus,

$$\omega_{AG^{II}} = \frac{C_U + C_{AG}}{n_{II}}.$$

The denominator has changed to the length of the alignment rather than the number of non-gapped aligned sites, because we are now considering gapped sites in our metric. As before, one needs to choose whether to use n_I, n_{II}, or their average in the denominator. For the example in Figure 9.2, $C_U = 5$, $C_{AG} = 4$, and $n_{II} = 18$; thus, $\omega_{AG^{II}} = 0.5$.

This measure treats characters matched with gaps in both alignments as identical, regardless of where the gap occurs. This has the advantage of recognizing that both alignments failed to align the specific character in one sequence with a character in the other sequence; on the other hand, it potentially allows for alignments with very different patterns of inferred gaps to be considered similar.

The first alternate to this is the CG measure. In this instance, we generate a count, C_{CG}, of all characters in either sequence that are aligned to the same gap in both alignments. For a gap to be identical in both alignments, it must fall between identical characters in the gapped sequence; that is, the observed character in one sequence must be aligned across from the identical G^S_{i-1} in both alignments. In Figure 9.2, there are three sites that are aligned to the same gap in both alignments I and II: $G^Y_{1-2}X_2$, $G^Y_{11-12}X_{10}$ and $G^Y_{11-12}X_{11}$; thus, $C_{CG} = 3$. Character Y_6, which was included in C_{AG}, is not included in C_{CG}, because we find $I:G^X_{4-5}Y_6$ in the first alignment, but $II:G^X_{6-7}Y_6$ in the second alignment. Y_6 is not aligned with the identical gap and is not counted for this measure. Thus we estimate our alignment similarity as

$$\omega_{CG^{II}} = \frac{C_U + C_{CG}}{n_{II}},$$

and, for our recurring example, find $\omega_{CG^{II}} = 0.44$.

The final alternate, CS, is even stricter. For this measure, not only must an observed character be aligned with the correct gap (as in CG), it must also be aligned with the correct position within the gap to be counted within C_{CS}. In our example, two of the three matching sites from the CG case now drop out: only $G^Y_{1-2.1}X_2$ is identical in both alignments. Site X_{10} is aligned with gap site $I:G^Y_{11-12.1}X_{10}$ and $II:G^Y_{11-12.2}X_{10}$, and site X_{11} is aligned with $I:G^Y_{11-12.2}X_{11}$ and $II:G^Y_{11-12.3}X_{11}$; thus neither is included in C_{CS}. As before,

$$\omega_{CS^{II}} = \frac{C_U + C_{CS}}{n_{II}},$$

and we find $\omega_{CS^{II}} = 0.33$ for the example in Figure 9.2. This measure is very strict in its inference of similarity among alignments, but it has an additional complication not found with the other gap-based metrics. As we have initially described, this metric is asymmetrically susceptible to variation in gap length; that is, proposed gap extensions that occur upstream of a site have more effect on accuracy

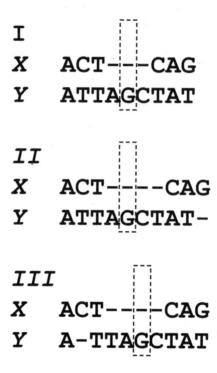

Figure 9.4. The correct gap site measure is asymmetrically susceptible to changes in gap length upstream and downstream from sites. In alignment *I*, the 5th site in sequence *Y* (a G) is aligned with the second gapped site in sequence *X*, that is, $I : G^X_{3-4.2}Y_5$. In alignment *II*, the gap has been extended because of changes at the 3′ end (downstream), having no effect on the alignment of Y_5, $II : G^X_{3-4.2}Y_5$. In alignment *III*, the gap has been extended because of changes at the 5′ end (upstream), changing the alignment of Y_5 to $III : G^X_{3-4.3}Y_5$. The correct gap and any gap measures would not be affected by either of the upstream and downstream changes. A solution to this problem is to count the position from both the 5′ and the 3′ ends of the gap.

than gap extensions that occur downstream of a site (Figure 9.4). To correct for this, one needs to examine the specific gap position relative to both the 5′ and the 3′ ends of the gap and to accept an aligned site as identical in two alignments if the position within the gap matches its distance from either end. (In our running example from Figure 9.2, this correction has no effect, but it would solve the variance seen in the alignment of the 5th site in the example shown in Figure 9.4).

The three gap-based metrics are hierarchical and thus related in a predictable manner, such that $\omega_{AG} \geq \omega_{CG} \geq \omega_{CS}$. As with the non-gapped metrics, for all of these measures, one may also want to report the ratio of aligned sites, based on the total alignment lengths rather than the non-gapped alignment lengths; thus, $r_n = n_{II}/n_I$. For the example in Figure 9.2, $r_n = 1.06$.

The $\omega_{AG'}$ term is similar to the variant of SPS described by Karplus and Hu (2001), except that they weighted the gapped sites half as much as the non-gapped sites. Weighting gapped and non-gapped sites differently is a possibility with all three of these gap-based measures; the obvious reason why one may choose to use weighting is so that matches based on observed data are given more weight than matches based on inferred data. Choice of weights is completely arbitrary, and we will use unweighted measures throughout the rest of our discussion.

One special case still needs to be discussed: what do we do about aligned sites that have gaps in both sequences? While this will never result from a typical pairwise alignment, it will frequently occur if a pair of sequences drawn from a multiple alignment is compared. We believe the best solution is to ignore them completely, to essentially remove them from any alignments prior to estimating similarity. They should not be counted as similarities or differences or be included in estimates of alignment length because they do not represent observed data, but are rather inferences drawn from an external source. For a pair of sequences, a double gap can neither be right nor wrong, because it represents the matching of nothing. One could add an infinite number of paired gaps to a pair of sequences without changing any aspect of the alignment. If one really wanted to include double gaps in the similarity metrics, one would be best using either the CG or the CS measure; weighting could also be included, perhaps counting these matches even less than the character–gap matches (e.g., character–character match = 1, character–gap match = 0.5, gap–gap match = 0.25).

Multiple Alignments

All of this discussion has dealt with alignments of pairs of sequences. Measures for comparing alignments of multiple sequences are derived directly from the pairwise metrics, with just a few additional complications. The first approach to comparing multiple alignments is to directly extend the pairwise comparisons to full columns of a multiple alignment. For example, for a five-sequence alignment with sequences V, W, X, Y, and Z, one would compare $V_i W_j X_k Y_l Z_m$ in the pair of alignments (just as one compares $X_i Y_j$ for a pair of sequences). All issues with ignoring or including gapped sites are treated identically as for pairwise alignments. One can calculate column-based similarities based only on completely ungapped columns, Ω_U; by allowing for any gap within a specific sequence in the column, Ω_{AG}; based only the correct gap (bracketed by the proper pair of observed sites), Ω_{CG}; or based on the specific gapped position within the gap, Ω_{CS}. One can also choose to base the measure on the number of aligned columns from alignment I, alignment II, or their average. The commonly used column score measure of Thompson et al. (1999b) is actually Ω_{AG}; it allows gaps in the correct sequences and bases the column correctness only on the observed characters. This full column approach is very conservative; it takes only one misaligned sequence to make the entire alignment look bad. Figure 9.5 shows an example of a pair of eight-sequence multiple alignments. Looking at just aligned sites 2–7, one can see that sequences A to G are aligned identically in both alignments, but sequence H is aligned differently; using this full column measure would cause all of these columns to be measured as misaligned. In general, the full column measure fails to capture blocks of similarity among sets of sequences in the alignments that are not shared across all sequences. For example, in Figure 9.5, sites 13 through 19 are aligned identically to each other for sequences A, B, C, and D in both alignments, and for sequences E, F, G, and H in both alignments, but the ABCD block and the EFGH block are aligned differently relative to each other across the two alignments.

An alternate approach is to choose one sequence as a standard to which all of the remaining sequences will be compared. This allows one to recognize overall similarity across alignments without potentially sacrificing entire columns due to one bad sequence. This metric can be calculated as the average of all pairwise similarity measures that include the standard sequence; that is, this metric is the average ω between the standard and all other sequences. Thus,

$$\Omega_{U \cdot i} = \frac{\sum_{j=1, j \neq i}^{n} \omega_U^{i \cdot j}}{n-1},$$

where n is the number of sequences, i indicates the standard sequence for comparison, and $\omega_U^{i \cdot j}$ represents the alignment similarity metric

```
              1111111111 2222
     I    123456789012345678 90123
     A    ACTCGCTGCGTCATTATGGCTAT
     B    AGTCGCTGCGTCATTATGGCTTT
     C    ACTCGCCGCGTCATTATGGCTAT
     D    ACTCGCCGCGTCATTATGGCTAC
     E    ACTCGCCGCGTCATC--G-CTAC
     F    GCTCGCCGCGTCATC--G-CTAT
     G    ACTCGCCGCGTCATC--G-CTAT
     H    ATTGGC-GCGTCATC--G-CTAT

              1111111111 2222
     II   123456789012345678 90123
     A    ACTCGCTGCGTCATTATGGCTAT
     B    AGTCGCTGCGTCATTATGGCTTT
     C    ACTCGCCGCGTCATTATGGCTAT
     D    ACTCGCCGCGTCATTATGGCTAC
     E    ACTCGCCGCGTC---ATCGCTAC
     F    GCTCGCCGCGTC---ATCGCTAT
     G    ACTCGCCGCGTC---ATCGCTAT
     H    A-TTGGCGCGTC---ATCGCTAT
```

Figure 9.5. Examples of some issues when comparing multiple alignments. Sites 2–7 are aligned identically in all sequences, except for sequence H. Sites 13–19 are aligned identically for sequences A through D in both alignments, and sequences E through H are aligned identically in both alignments, but the A–D and E–H sets are not aligned identically in both alignments.

measured between sequences i and j. (Again, one could base this on any of the ω measures, not just ω_U).

The disadvantage of this approach is that the metric may vary greatly depending on what sequence is chosen as the standard. It is well known that alignment accuracy between pairs of sequences decreases as their

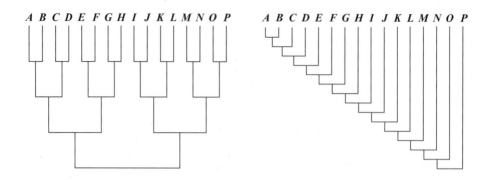

Figure 9.6. Alternate tree topologies used to illustrate the affect of tree shape on using a single sequence as a reference sample in a comparison of multiple alignments.

evolutionary distance increases (Pollard et al. 2004; Rosenberg 2005a). Figure 9.6 shows a pair of alternate 16-taxon trees of identical evolutionary depth. In the first tree, it may not matter which sequence is chosen as the standard because the balanced nature of the topology leads to an even distribution of branch lengths; the set of distances from all taxa to a standard is the same, no matter which sequence is chosen as the standard. However, in the second tree, one would expect a very big difference in measured similarity if sequence A were used as the standard versus that found if sequence P were used as the standard; the distribution of evolutionary distance between these sequences and the remaining sequences is very different. In a pair of simulated data sets based on these trees, a Clustal alignment was compared to the alignment derived directly from the simulation, using each sequence as a possible standard and the $\Omega_{U \cdot i}$ metric. (These simulations are explained in more detail below; in this specific example, the expected evolutionary distance between sequences A and P is 2.0 for both trees). Table 9.1 shows the estimated similarities for each potential choice of standard for each tree shape. As expected, there is little difference among choice of the standard for the balanced tree, but standard choice has a large affect for the pectinate tree. In general, this approach to estimating the similarity of multiple alignments may be reasonable if one has a good reason for picking a single sequence to serve as a standard (e.g., perhaps one wishes to compare how similar alternate alignments are specifically with respect to the human genome), but otherwise it should be avoided due to the arbitrary nature of choosing a standard.

Simulation Approaches to Evaluating Error

TABLE 9.1. SIMILARITIES BETWEEN
PAIRS OF ALIGNMENTS USING
ALTERNATE SEQUENCES AS A STANDARD

Standard	Balanced	Pectinate
A	0.533	0.723
B	0.543	0.724
C	0.560	0.722
D	0.555	0.717
E	0.540	0.702
F	0.546	0.709
G	0.559	0.695
H	0.556	0.686
I	0.563	0.604
J	0.553	0.622
K	0.566	0.643
L	0.553	0.412
M	0.564	0.433
N	0.550	0.275
O	0.550	0.196
P	0.550	0.140

NOTE: Similarities were estimated using $\Omega_{U''\cdot i}$, where i refers to the standard sequence.

The final alternative is to use all possible sequences as the standard; this is functionally equivalent to calculating similarity as the average of all possible pairwise comparisons:

$$\overline{\Omega}_U = \frac{\sum_{i=1}^{n-1}\sum_{j=i+1}^{n} \omega_{U_{i,j}}}{\frac{1}{2}n(n-1)}$$

Again, one could base this on any of the pairwise alignment measures (e.g., $\overline{\Omega}_{AG}$). The $\overline{\Omega}_{U'}$ term is equivalent to the multiple alignment version of SPS, because the number of aligned sites in the reference sequence is used in the denominator of individual $\omega_{U'}$'s. This approach maximally describes the similarity found between the alignments, but sacrifices some global measure: one can imagine having a very high value of $\overline{\Omega}_U$ without a single column being completely identical among the alignments, $\Omega_U = 0$. The potential for these metrics to differ will likely increase with the number of sequences in the multiple alignment; specific circumstances and questions may control which of these measures should be preferred.

A Comment on Global vs. Local Alignment

These measures have been described essentially for global alignments, where one assumes that all sites among the sequences are aligned. Local alignments, where many of the sites may remain unaligned, can be compared using the exact same metrics and measures. One needs to be careful to estimate alignment lengths (whether N or n) based only on aligned sites and not all sites. The aligned length ratios, r_N and r_n, potentially become very important for local alignments, because alignment lengths may be much more variable in local alignments than in global alignments.

A Comparison

In order to see how these metrics perform on a larger scale, we estimated these measures of alignment accuracy for 15,400 data sets simulated for a variety of 16-taxon trees (all simulations started with sequences of 2000 ungapped nucleotides). These data sets come from previous studies (Ogden and Rosenberg 2006, 2007a,b); we use them here for illustrative purposes because (1) for each data set we have a "true alignment" (TA), which came directly from the simulation, and a "hypothesized alignment" (HA), which came from aligning the sequences in ClustalW (Thompson et al. 1994); and (2) we know from our previous work that the similarity of TA and HA (estimated using $\overline{\Omega}_{U''}$) varies from extremely similar to extremely dissimilar across the full breadth of these data sets. Details on the simulations can be found in Ogden and Rosenberg (2006).

Because each of these data sets has a reference alignment (TA) and an alternate alignment (HA), we have chosen to always measure the similarity of the alternate to the reference (in this specific case, we are actually estimating the accuracy of the ClustalW alignments). For each data set, we estimated the similarity in the alignments for all possible pairs of sequences (120 pairs per data set × 15,400 data sets = 1.85 million pairs) using each of the pair metrics: $\omega_{U''}$, $\omega_{AG''}$, $\omega_{CG''}$, and $\omega_{CS''}$. We also estimated the similarity in the full multiple alignment of all 16 sequences for each of the 15,400 data sets using $\Omega_{U''}$, $\Omega_{AG''}$, $\Omega_{CG''}$, $\Omega_{CS''}$, $\overline{\Omega}_{U''}$, $\overline{\Omega}_{AG''}$, $\overline{\Omega}_{CG''}$, and $\overline{\Omega}_{CS''}$. (We do not include the full alignment metrics based on a standard sequence in this comparison because there is no simple way to choose a pair of contrasting standards that makes sense across the wide variety of tree shapes).

To be sure any conclusions are not an artifact of the simulation conditions, we used these same metrics to analyze alignments from the

TABLE 9.2. CORRELATIONS AMONG ALIGNMENT METRICS FOR THE SIMULATED DATA

						$\omega_{AG''}$	$\omega_{CG''}$	$\omega_{CS''}$	
$\Omega_{AG''}$	0.99813					0.99873	0.99751	0.99747	$\omega_{U''}$
$\Omega_{CG''}$	0.99575	0.99885					0.99948	0.99945	$\omega_{AG''}$
$\Omega_{CS''}$	0.99572	0.99883	1.00000					0.99999	$\omega_{CG''}$
$\bar{\Omega}_{U''}$	0.91082	0.91488	0.91514	0.91513					
$\bar{\Omega}_{AG''}$	0.90497	0.91032	0.91116	0.91118	0.99953				
$\bar{\Omega}_{CG''}$	0.89541	0.90151	0.90296	0.90301	0.99898	0.99979			
$\bar{\Omega}_{CS''}$	0.89530	0.90141	0.90287	0.90291	0.99895	0.99977	1.00000		
	$\Omega_{U''}$	$\Omega_{AG''}$	$\Omega_{CG''}$	$\Omega_{CS''}$	$\bar{\Omega}_{U''}$	$\bar{\Omega}_{AG''}$	$\bar{\Omega}_{CG''}$		

NOTE: The lower left triangle displays correlations among multiple alignment metrics for all 15,400 data sets; the upper right triangle displays correlations among pairwise alignment metrics for all 1.85 million sequence pairs.

BAliBASE database (Thompson et al. 1999a), consisting of high-quality structure-based core-block protein sequence alignments with a wide variety of properties. We used the 82 multiple alignment datasets that make up reference set 1, comparing the BAliBASE standard alignment to hypothesized alignments derived from ClustalW.

While similarity estimates for specific data sets vary quite a bit by metric (as much as 11% for the pairwise alignment metrics and 76% for the multiple alignment metrics), the correlations among the metrics (Tables 9.2 and 9.3) are extremely high. Not surprisingly, the weakest

TABLE 9.3. CORRELATIONS AMONG ALIGNMENT METRICS FOR BAliBASE DATA

						$\omega_{AG''}$	$\omega_{CG''}$	$\omega_{CS''}$	
$\Omega_{AG''}$	0.98940					0.99522	0.99013	0.99023	$\omega_{U''}$
$\Omega_{CG''}$	0.98492	0.99418					0.99435	0.99420	$\omega_{AG''}$
$\Omega_{CS''}$	0.98488	0.99415	1.00000					0.99993	$\omega_{CG''}$
$\bar{\Omega}_{U''}$	0.98303	0.98150	0.97763	0.97758					
$\bar{\Omega}_{AG''}$	0.97913	0.98600	0.98404	0.98382	0.99739				
$\bar{\Omega}_{CG''}$	0.97360	0.97933	0.98490	0.98484	0.99361	0.99645			
$\bar{\Omega}_{CS''}$	0.97355	0.97929	0.98488	0.98482	0.99363	0.99634	0.99998		
	$\Omega_{U''}$	$\Omega_{AG''}$	$\Omega_{CG''}$	$\Omega_{CS''}$	$\bar{\Omega}_{U''}$	$\bar{\Omega}_{AG''}$	$\bar{\Omega}_{CG''}$		

NOTE: The lower left triangle displays correlations among multiple alignment metrics for all 82 data sets; the upper right triangle displays correlations among pairwise alignment metrics for all 652 sequence pairs.

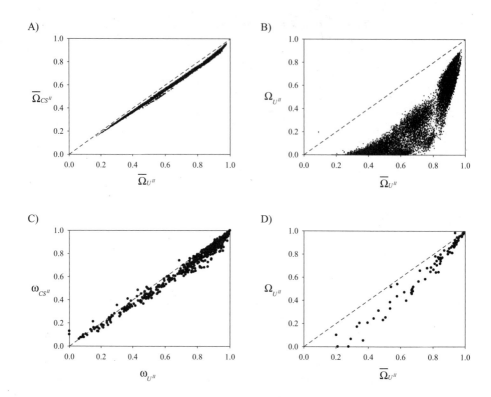

Figure 9.7. Comparison of multiple alignment metrics for the (A–B) 15,400 simulated data sets and the (C–D) BAliBASE data sets. (A) $\overline{\Omega}_{CS''}$ vs. $\overline{\Omega}_{U''}$ ($r = 0.99895$); (B) $\Omega_{U''}$ vs. $\overline{\Omega}_{U''}$ ($r = 0.91082$); (C) $\omega_{CS''}$ vs. $\omega_{U''}$ for 652 pairs of aligned sequences ($r = 0.99023$); (D) $\Omega_{U''}$ vs. $\overline{\Omega}_{U''}$ for 82 aligned data sets ($r = 0.98303$). The dashed lines indicate a 1:1 relationship.

correlations among the metrics are found between full column multiple alignment metrics and average pairwise multiple alignment metrics (Figure 9.7 and Tables 9.2 and 9.3). However, even with the apparent variation shown in Figure 9.7, these metrics are still correlated at $r > 0.9$. If one wishes to compare sets of alignments (e.g., those derived from different multiple alignment programs) to a standard or to use an estimate of alignment similarity as a covariate or variable in another analysis, it should not make any difference which metric is used, as long as one uses the same metric for all comparisons. Specific choice of the denominator of the metric (e.g., $\omega_{U'}$ vs $\omega_{U''}$ vs $\omega_{\overline{U}}$) may be used to answer slightly different questions (e.g., how much of a reference alignment is being recovered in a hypothesis or how much of a hypothesized alignment is found in the reference). Beyond consistency, ease and speed of calculation might be a factor for some researchers; in this case, pairwise metrics that ignore gapped sites (ω_U and $\overline{\Omega}_U$) are simpler and quicker to calculate than those that include gaps or are based on full column comparisons (e.g., ω_{CS} or Ω_U).

DOWNSTREAM EFFECTS OF ALIGNMENT INACCURACY INVESTIGATED THROUGH SIMULATION

The metrics described in the preceding section are designed to measure the similarity of alternate alignments based solely on properties of the alignments (primary measures). One can easily imagine other approaches if one considers results *derived* from the alignments. For example, one might consider a pair of alignments "similar" if evolutionary distances or phylogenetic trees or functional motifs estimated from the alignments were similar. These are secondary (or functional) measures of similarity. The relationships between primary and secondary measures of similarity have often been assumed, but only recently have researchers begun to examine them in detail (e.g., Ogden and Rosenberg 2006; Rosenberg 2005a,b). In the following, we describe some of these initial results.

Studies that use simulations with indels can take three basic approaches. First, the simulated data can be used directly in downstream analysis without alignment. In this case, the alignment (as output by the simulation program) is perfect, and the researchers are simply including sequences containing indels as part of their analysis. The presence of indels is the important feature of the simulation, and alignment is not being evaluated in any way. An example of this is part of a study by Ogden and Rosenberg (2007b) in which they examined whether certain approaches to coding gaps (e.g., treating them as missing data, as a fifth state character, or as separate presence/absence characters) were better than others in parsimony-based phylogenetic analysis. [A similar study has also been conducted by Simmons et al. (2007)]. In this case, they used the simulated alignments directly because they were interested in testing the efficacy of these approaches under the assumption that the alignments were perfect. They found that, for many cases (82%), there was no difference in topological accuracy among the different methods of gap coding. However, in cases where a difference was present, coding gaps as a fifth state character or as separate presence/absence characters outperformed treating gaps as unknown/missing data nearly 90% of the time.

Under the second approach to using simulations that include indels, the sequences resulting from the simulation are aligned using a standard approach after the simulation is complete, thus generating a hypothesized alignment logically similar to what one would use when working with empirical data (where the true alignment can never be known). This approach adds biological realism through the inclusion of uncertain alignment, but does not allow for direct testing of the effects of the alignment. Two examples of this approach are by Hall (2005, 2006), who used his EvolveAGene program to perform more realistic simulations that included sequence alignment for studying phylogenetic accuracy and ancestral protein prediction. The potential problem with this second approach is that while it includes the realism of the uncertain alignment, it does not allow for direct evaluation of the effect of the alignment on the downstream study, since only the hypothesized alignment is being evaluated.

The third approach allows for direct evaluation of the effect of alignment by combining the first two approaches. Both the true homologies derived directly from the simulation and a hypothesized alignment derived from algorithmic analysis of the sequences are evaluated for the downstream analysis being questioned; if all other factors are held constant, differences in these results must be due to differences in the alignments. Empirical studies have often taken a similar approach by comparing the results generated from two (or more) alternate alignments (Aagesen et al. 2005; Goldman 1998; Laamanen et al. 2005; Morrison 2006; Morrison and Ellis 1997; Mugridge et al. 2000; Ogden and Whiting 2003; Sharkey et al. 2006; Terry and Whiting 2005; Whiting et al. 2006); this approach allows one to show if the alignment has an effect. The difference is that in simulation studies we know what the true outcome should be and can directly compare the differences in results derived from the different alignments to this truth, rather than to each other (which can sometimes be very misleading). In the remainder of this chapter, we highlight the results of some studies that use both true and hypothesized alignments derived from simulation to explore the effects of alignment error on downstream analysis.

Evolutionary Distance Estimation

While it has been long understood that distantly related sequences are harder to align than closely related sequences, the exact nature of this relationship (e.g., what constitutes distant versus close) had never really been examined for DNA sequences until Pollard et al. (2004) and Rosenberg (2005a) undertook the study. Simulating pairs of noncoding DNA sequences separated by fixed divergences under a variety of simple to complex evolutionary models, Rosenberg (2005a) found that alignment accuracy is largely dependent on the proportion of homologous sites containing identical nucleotides.

A distinction needs to be made between the true identity of the sequences (the proportion of truly homologous sites in a pair of sequences containing identical nucleotides) and the aligned identity (the proportion of hypothesized homologous sites from an alignment that contain identical nucleotides). When true variation among sequences is large, algorithms can be quite efficient at incorrectly inferring identity; the theoretical minimum identity for a pair of sequences under the simulated conditions in Rosenberg's study (2005a) was 25–26% (depending on the specific substitution model), yet the hypothesized alignments yielded sequences with a minimum identity of 44%, even for random data. (Similar results have been reported by others, for example, Fleißner et al. 2000; Shabalina and Kondrashov 1999). The inflation in observed identity is predominantly found in sequences that truly differ by more than 50% of their sites; sequences with true identity of 50% or more have less than a 1% absolute increase in observed identity after alignment.

Up to a point, evolutionary distance estimation was somewhat robust to alignment error (Rosenberg 2005a). The relative difference between evolutionary distances estimated from the true and hypothesized alignments was less than 10%, even when up to 50% of the sites were aligned incorrectly. The robustness of these estimates appears to be related to the inflation of sequence identity. As long as the true identity was greater than 50%, there was little inflation in the estimated identity due to alignment (even when the alignment is largely wrong). This translated to relatively little error in the estimation of distance, because distance estimates are based solely on the observed proportions of sites that differ among the sequences. Since these counts are being estimated reasonably accurately (even though the specific sites are wrong), the distance estimates are also reasonably accurate. The 50% barrier for distance estimate accuracy was also reported by Fleißner et al. (2000) using a least-squares approach to estimating P-distance.

The effects of specific choice of evolutionary parameters on alignment and distance estimation accuracy have also been examined (Rosenberg 2005a), including sequence length, indel size and rate, nucleotide frequencies, and variation in substitution rate among sites. While the specific effect varied by parameter, one general observation was that when a parameter had an effect, the effect was amplified when the true distance between the sequence pair was larger.

Multiple Sequence Alignment It has generally been believed that multiple sequence alignments are more accurate than pairwise alignments (e.g., Duret and Abdeddaim 2000; Lesk 2002). Rosenberg (2005b) examined how well the pairwise analyses translated to multiple sequence alignment, by looking at (1) whether the accuracy of alignment of a pair of sequences is improved by the addition of a third sequence during the alignment process, (2) how the position of this additional sequence effects the accuracy, and (3) what effect, if any, this third sequence has on evolutionary distance estimation. He found that the third sequences could improve alignment accuracy and that the relative position of the sequence made a large difference. Specifically, the multiple alignment was maximally more accurate than the pairwise alignment when the third sequence was half as divergent from one of the target sequences than they were from each other. Analyses with more than three sequences yielded similar results, with the addition of more sequences improving alignment but with diminishing returns. Topological structure of the tree relating the sequences was important, with balanced topologies proving a greater increase in accuracy than pectinate structures.

Surprisingly, evolutionary distance estimation was decoupled from alignment accuracy; as the position of the third sequence moved toward the root, the estimate of the evolutionary distance between the other two sequences increased (Rosenberg 2005b). This bias was hypothesized to be a consequence of the greedy progressive alignment heuristic algorithm implemented in ClustalW (Thompson et al. 1994) and many any other alignment programs.

The above-mentioned studies were primarily conducted using ClustalW, since it has historically been one of the most widely used and consistently accurate alignment programs. Many of the analyses have been repeated with additional alignment programs, including DIALIGN (Morgenstern 1999), T-Coffee (Notredame et al. 2000), MAFFT (Katoh et al. 2002), DCA (Stoye 1998; Stoye et al. 1997; Tönges et al. 1996), PCMA (Pei et al. 2003), and MUSCLE (Edgar 2004a,b); while the vast majority of the results generally hold for all programs tested thus far, there is some interesting variation (unpublished). Specifically, the relationship between alignment accuracy and evolutionary distance estimation is not completely consistent. Under the simulated conditions, T-Coffee and DCA consistently produced the least accurate alignments of any of the programs tested; however, the T-Coffee alignments always led to the smallest distance estimates, whereas the DCA alignments always led to the largest distance estimates. Thus, there are fundamental differences in the characteristics of the resulting alignments from these programs that go beyond simple accuracy, highlighting the importance of secondary/functional measures of alignment accuracy and a potential problem with relying solely on a primary measure.

Phylogeny Reconstruction

It has been long recognized that changing an alignment can have a large effect on the final results of comparative sequence analysis, especially in phylogenetics (Cantarel et al. 2006; Fleißner et al. 2005; Hall 2005; Hall 2006; Liu et al. 2005; Ogden and Rosenberg 2006; Pollard et al. 2004; Rosenberg 2005a,b). The obvious prediction was that, in general, as alignment error increases, phylogenetic topological accuracy decreases. Ogden and Rosenberg (2006) directly addressed this question using simulation by estimating phylogenetic trees from both the true alignment and a hypothesized alignment. These predicted trees were compared to the true phylogenetic tree used in the simulation (rather than to each other as is often done with similar empirical studies); the differences in these accuracies were then compared to alignment accuracy in order to estimate the actual effect of alignment error on phylogenetic reconstruction (Figure 9.8). In general, while very large amounts of alignment error had a negative impact on phylogenetic accuracy, there were many cases where even moderate amounts of alignment error (30% or more) had no affect (on average) on phylogenetic accuracy. To be specific, it was not the case that the same phylogeny was generated from both alignments, but rather that the hypothesized alignment was just as likely to produce a tree that was as good, or better, than the true alignment. For even moderate levels of alignment error, the average difference in phylogenetic accuracy between the true and hypothesized alignments was essentially zero; stochastically, a wrong alignment often produced a better tree than the correct alignment. Only when alignment error became strong (> 40%) did it begin to have a consistently negative effect on phylogeny reconstruction.

Previous simulation studies have shown that, even given perfect alignments, it is often difficult to accurately reconstruct trees (e.g., Hillis 1995;

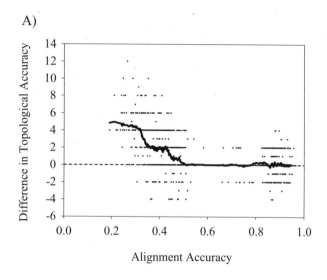

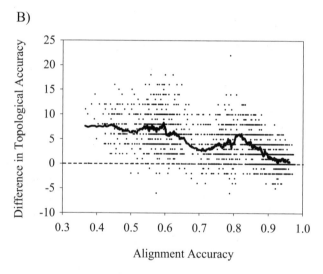

Figure 9.8. Relationship of alignment accuracy and the difference in topological accuracy (maximum likelihood) between trees reconstructed using the true and a hypothesized alignment for (A) fully balanced and (B) fully pectinate tree topologies (see Figure 9.6), including a variety of branch length and depth combinations. Points to the far right are the most accurate alignments, whereas points to the left are the least accurate alignments. The dashed line indicates where trees reconstructed from the true and hypothesized alignments were equally accurate. Points above the line are cases where the tree obtained from the true alignment was more accurate than that from the hypothesized alignment; the opposite is true for points below the line. The solid lines are moving averages based on overlapping sliding windows of 50 consecutive points.

Huelsenbeck and Rannala 2004; Nei 1996; Rosenberg and Kumar 2003; Takahashi and Nei 2000), so the smaller than expected effect of alignment accuracy on phylogenetic accuracy under many conditions is not entirely surprising. Further investigation is needed to more fully understand, and even predict, how any specific dataset will perform. Next we discuss a number of additional factors that have been identified as playing a significant role in the interaction between alignment and phylogenetic accuracy.

Tree Shape In general, pectinate trees are more difficult to reconstruct than balanced trees, regardless of alignment accuracy (Ogden and Rosenberg 2006). For example, phylogenetic reconstruction of simulated data sets on balanced topologies were less affected by error in sequence alignment than those on pectinate topologies. For balanced trees, alignment accuracy did not have an affect on phylogenetic accuracy until alignment accuracy dropped below 60% (Figure 9.8A). In contrast, for pectinate trees, increasing alignment error had an immediate effect on phylogenetic accuracy (Figure 9.8B). Although pectinate trees, on average, contain less alignment error than more balanced topologies (for a constant tree depth, pectinate trees have shorter average distances among sequences leading to less total alignment error), these errors have a larger effect on tree reconstruction then the same amount of alignment error in a balanced tree. Thus, balanced topologies are much less affected by alignment error than are pectinate topologies. Why phylogenetic estimation of balanced trees appears to be more robust in the face of alignment error is not yet understood.

Branch Length Branch length clearly plays an important role in both alignment and phylogenetic accuracy. The relationship between these values is not as intuitive as one might guess. In Ogden and Rosenberg (2006), phylogenetic accuracy of specific branches was estimated from the number of replicated simulations in which the specific correct branch was recovered. Similarly, the overall alignment error associated with a branch (the amount of alignment inaccuracy in the dataset that can be attributed to evolution along the branch) was estimated using a least-squares fitting of the overall pairwise alignment error matrix to the tree. As would be expected from the results on distance estimation (summarized earlier), longer branches were associated with more alignment error than shorter branches. In contrast, however, longer branches were associated with less phylogenetic error, leading to a negative correlation between alignment accuracy and phylogenetic accuracy. This counter-intuitive result is explained by the realization that extremely short branches (which would produce almost no alignment error) are very difficult to reconstruct phylogenetically, while longer branches (which would produce a lot of alignment error) are much easier to reconstruct phylogenetically because of the increased number of character changes (extremely long branches would lead to saturation of the phylogenetic signal, but these were outside the bounds of the simulation conditions). Thus, an understanding of the relationship between alignment and phylogenetic accuracy at the branch-by-branch level requires more complex analysis of neighboring as well as focal branches (Ogden and Rosenberg 2006).

Ancient Radiations Cantarel et al. (2006) point out that, although less divergent sequences produce more accurate topologies than highly divergent sequences, topological accuracy can be highly dependent on whether the radiation (divergence) was ancient or recent. In fact, they found only a few cases where having the true alignment improved the reconstruction of ancient radiations. This underscores the notion that some topologies are just inherently hard to reconstruct. Therefore, one must consider the overall branch structure across the tree (not just total tree depth or topological shape) when considering the role of alignment on phylogenetic accuracy.

CONCLUSION

The use of simulation to test the accuracy and effects of alignment is a fairly recent line of investigation in phylogenetics and bioinformatics. The advantages of the simulation approach is precise control of conditions and *a priori* knowledge of the truth; the disadvantage lies with the potential oversimplification and debatable applicability of results generated using these approaches. Alignments can be evaluated through primary or secondary (functional) measures, the specific use and choice of which is dependent on the details and questions of the experimental study. While errors in alignment can generally be expected to have an effect on downstream analysis in bioinformatics and genomics, these effects are neither as simple nor as straightforward as one would suspect, and much more work is needed to more fully understand the exact role and importance of producing accurate alignments in this research.

Acknowledgments

The authors thank Sudhir Kumar and Sonja Prohaska for comments on an early draft of this manuscript, as well as Ahmet Kurdoglu and Meraj Aziz for performing some of the computational work. This project was partially supported by the NIH R03-LM008637 and Arizona State University.

CHAPTER 10

Robust Inferences from Ambiguous Alignments

BENJAMIN D. REDELINGS
North Carolina State University
MARC A. SUCHARD
University of California, Los Angeles

Alignments, Inferences, and Uncertainty 210
    Multiple Sequence Alignments Represent
    Evolutionary History 211
    What Is Alignment Ambiguity? 213
    Inferences from Multiple Sequence Alignments 215
The Structure of Bioinformatic Inferences 216
    Robust Inference 216
    Sequential Estimation 217
    Joint Estimation 220
    Cost-Based Methods versus Statistical Estimation 223
    Sources of Alignment Uncertainty 225
Alignment Uncertainty: Pairwise Alignments 228
    Probabilistic Models and Pairwise Alignment Uncertainty 228
Alignment Uncertainty: Multiple Sequence Alignments 230
    Identification of Ambiguous Regions Based
    on a Single Alignment 231
    Multiple Sequence Alignment Ambiguity Resulting
    from Parameter Uncertainty 233
Statistical Inference under Alignment Uncertainty 236
    Joint Statistical Model 237
    Inference under a Statistical Model 241
    Improved Statistical Models of Insertion and Deletion 246

Limitations of Statistical Methods 249
Representing Alignment Ambiguity. 249
  Alignment Ambiguity for Use in Further Analysis. 250
  Graphically Representing Alignment Ambiguity 250
Example: 5S Ribosomal RNA 257
  Model and Priors 258
  Results 259
  Bias and Alignment Uncertainty 259
  Rate Heterogeneity. 261
Discussion. 266
  Speed and Sequential Estimation. 266
  Testing Alignment Reliability Measures 269
  Verifying Bioinformatic Inferences by Simulation 269

ALIGNMENTS, INFERENCES, AND UNCERTAINTY

Molecular sequence data have become an invaluable source of information for understanding evolutionary processes and for inferring evolutionary relationships between organisms. Molecular sequences provide a large number of separate characters (individual nucleotides, amino acids, or codons) that are easy to identify and distinguish. In addition, probabilistic models of how these molecular characters change over time allow us to estimate evolutionary process parameters and to quantify the evidence for evolutionary hypotheses. These parameters include phylogenetic trees, divergence times, insertion and deletion rates, and substitution rates. Probabilistic models of evolution also enable researchers to locate sequence motifs that are especially conserved or exhibit positive selection. These bioinformatic inferences depend not only on the observed molecular characters but also on the homology structure of the molecular characters, termed the "alignment" of the molecular sequences. This homology is not directly observed and can be difficult to assess, especially when the sequences have diverged for a long time so that sequence similarity is low. As a result, in many analyses, there is substantial uncertainty about the alignment. In this chapter, we examine methods for making robust inferences when the alignment is uncertain.

Traditional bioinformatic inference methods have usually assumed that the alignment was known with certainty. Ignoring alignment uncertainty when it is present can undermine bioinformatic inferences in several ways. For example, the use of only one plausible alignment estimate among many can lead to severely biased estimates of the phylogenetic tree (Lake 1991) and other parameters (Thorne and Kishino 1992), as well as unreliable or nonrepeatable results (Morrison and Ellis 1997). Furthermore, ignoring alignment uncertainty can lead to exaggerated measures of confidence in those results. Therefore, it is important for bioinformatic inferences to take alignment uncertainty into account. The scarcity of methods that are robust to uncertainty in multiple sequence alignments has been a significant obstacle to inferences based on highly divergent molecular sequences and, thus, to answering questions about ancient divergences in the Tree of Life.

A number of bioinformatic techniques have been developed to avoid the bias and exaggerated confidence that typically result from conditioning on a single alignment estimate. This is a difficult and multifaceted problem, and an ideal technique faces many challenges. First, methods should characterize alignment uncertainty in a rigorous and objective fashion. This involves taking all sources of uncertainty into account, as well as avoiding subjective and nonreproducible judgments of homology. Second, it is beneficial to report fine gradations of uncertainty that indicate the weight of evidence for homology, instead of dividing the alignment into "ambiguous" and "unambiguous" regions. Third, an ideal method would make full use of the information in the data set, including information in ambiguous regions.

Although such techniques for handling alignment uncertainty have been developed for pairwise alignments, methods for handling uncertainty in multiple alignments have been slower to develop. Recent improvements in statistical methodology have made it feasible to rigorously assess the degree of uncertainty in multiple sequence alignments, to represent finer gradations of uncertainty, and to make use of information in ambiguously aligned regions without bias. In this chapter, we will discuss several traditional methods of handling ambiguity in multiple sequence alignments, as well as new statistical approaches to this problem. We then illustrate how alignment uncertainty can affect phylogeny estimation by focusing on a small data example. This example demonstrates the negative effects of ignoring alignment uncertainty and illustrates the practical benefits of the new statistical methods.

Multiple Sequence Alignments Represent Evolutionary History

Homology-Based Alignments versus Function-Based Alignments In this chapter, we follow common practice in the field of molecular evolution in using alignments of molecular sequences to specify evolutionary homology. That is, multiple sequence alignments specify the homology of individual residues in a set of homologous sequences by arranging the residues in a matrix so that residues in a column (also called a "site") all descend from a single residue in the common ancestral sequence. Thus, the complete evolutionary history of individual residues in a set of homologous sequences is given by the combination of a multiple sequence alignment and a phylogenetic tree. However, we note that specifying homology by use of a matrix in which each row represents one sequence does not allow one to easily represent homologies within a single sequence. Thus, it is difficult to represent the within-sequence homology that occurs as a product of gene-internal duplications.

Researchers in other fields have sometimes conceived of alignment in different ways. For example, they may align residues that do not share a common ancestor but that perform a common function or occupy similar positions in the three-dimensional structure of a protein (Mizuguchi, Deane, Blundell, and Overington 1998b; Thompson et al. 1999a). While both kinds of information are useful, we note that functional and structural interpretations of alignments may not be well-defined, since residues may have multiple functions, and multiple residues may share one function. If a column corresponds to a function, then it could be necessary to place two adjacent residues from the same sequence in a single column or to place a single residue in several columns. Therefore, in this chapter, we use alignments only to represent evolutionary homology.

Commonly Observed Violations of Homology in Alignment Estimates These different interpretations of multiple sequence alignments can lead to different alignment matrices in practice. For example, it sometimes occurs that two sequences in a multiple sequence alignment experience independent insertions in the same location. These two insertions are not homologous, because they do not descend from any sequence in their common ancestor. Thus, under the evolutionary homology interpretation, these insertions must not be aligned in the same columns, even though they may share similar functions. Similarly, it may happen that a residue in an ancestral sequence is deleted and that one of its descendant sequences later experiences an insertion in the same location. Because the deleted residue is not ancestral to the inserted residue, these residues must not be aligned even though they may occupy similar positions in the protein's three-dimensional structure. These constraints, when observed, rule out many of the alignments that are commonly created by multiple sequence alignment programs today. However, without knowledge of the evolutionary tree relating the sequences and of the presence or absence of residues at ancestral sequences, it is difficult to determine conclusively if an alignment violates these constraints.

What Is Alignment Ambiguity?

Before we discuss sources of alignment ambiguity and methods of handling this ambiguity, we first seek to clarify the meaning of alignment ambiguity. We begin by defining alignment ambiguity to mean that there are two or more plausible alternative alignments. We claim that this definition of ambiguity is the correct one, and we note that it also applies to the estimation of other parameters, such as phylogenetic trees. However, it is common to think of alignment ambiguity simply in terms of the degree of sequence conservation in multiple sequence alignments. This conception of alignment ambiguity can be implicitly used in identifying ambiguous regions by visual inspection, and it is also explicitly used in some computerized procedures that exclude alignment regions based on the presence of gaps or of sequence variation between taxa. Although nonconserved regions are often a good indicator of alignment ambiguity, such indicators are successful only to the extent that they actually indicate the presence of plausible alternative alignments. Another class of methods seeks to characterize ambiguity in multiple sequence alignments by assessing the sensitivity of inferred alignments to perturbation in alignment parameters. Again, these methods are successful only to the extent that they indicate the presence of two or more plausible alternatives. Finally, we note that in considering alignment uncertainty, we are not addressing uncertainty about whether or not biological sequences are homologous. Instead, we are concerned with uncertainty about the homology of individual residues, but take as given that all sequences in a data set are homologous.

Second, homologies in an alignment are not simply "ambiguous" or "unambiguous," but they can be of varying degrees of certainty, depending on the strength of evidence in the observed data for the homology[1]. Many methods have been developed to improve phylogenetic inferences by discarding portions of the aligned data matrix and keeping the rest, and it might be natural to speak of the discarded portions as "ambiguous regions." However, it is important to note that this does not mean that the resulting portions are "unambiguous," because complete certainty about homology is not possible.

Third, it is useful to consider alignment ambiguity for parts of the alignment as well as the whole. When considering partial alignments, it is important to define them in a coherent fashion. For example, it is natural to talk about ambiguous "regions" in an alignment, but this is problematic, since the regions are defined by the alignment itself, which is not known with certainty. Indeed, if a "region" is ambiguous, then it may not actually exist. To avoid this problem, we define a partial alignment as a hypothesis about homology and note that the smallest possible hypothesis of homology is that two residues from different sequences are homologous or not homologous. Most homology hypotheses, including the hypothesis of a full alignment, can be decomposed into a collection of minimal homology hypotheses of this type. Now that we have more carefully defined what we mean by a partial alignment, we can extend the definition of alignment ambiguity to partial alignments. We note that many full alignments can conform to any partial alignment; for example, many full alignments satisfy the criterion that residue 1 of sequence 1 is homologous to residue 1 of sequence 2. With this in mind, we state that alignment ambiguity for partial alignments means that some plausible full alignments conform to the partial alignment and that some do not.

Fourth, plausibility is a relative term that always depends on the degree of the researcher's knowledge. Thus, the degree of ambiguity in the alignment depends on what is known, so that it is not meaningful to simply speak of an alignment region as "ambiguous" without first specifying what facts about evolutionary parameters should be taken as given. For example, it may be the case that two alternative phylogenies support alternative alignment estimates. In this case, if the phylogeny is unknown, the alignment must be considered ambiguous, but if one of the phylogenies is known to be correct, then the alignment can be well supported. Similarly, different values of other alignment parameters, such as the indel and substitution rates, may lead to different estimates of the alignment. In some cases, precise knowledge of parameter values could result in substantially less uncertainty in the alignment. The general idea that knowledge of parameter values affects the degree of alignment ambiguity can be conveniently expressed in terms of conditional probabilities. For example, if A and τ refer to the unknown true alignment and tree while A_0 and τ_0 are specific possibilities for the alignment and tree, then the probability

$P(\mathbf{A} = \mathbf{A}_0)$ that the alignment is $\mathbf{A}_0$ when the tree is unknown can differ substantially from the probability $P(\mathbf{A} = \mathbf{A}_0 | \tau = \tau_0)$ that the alignment is $\mathbf{A}_0$ when the tree is known to be τ_0.

The interaction of alignment ambiguity with bioinformatic inferences is most important when alignment estimates are largely determined by initial assumptions about parameter values. In this case, alignment estimates will reflect and reinforce assumptions embodied in tunable parameter values, such as the "guide tree" that is used to guide alignment construction in progressive alignment. In such cases, the alignment must be considered dangerously ambiguous, because the ambiguity may undermine the inference procedure. This observation can be clarified by noting that, when inferring an evolutionary parameter such as the tree or the insertion–deletion rate, it must not be treated as known during the inference[2]. For example, if we seek to infer the evolutionary tree topology, then we must not assume knowledge of a particular evolutionary tree during the estimation of the alignment. Nevertheless, this is exactly what is done when a guide tree is used. Likewise, when attempting to infer indel rates from an alignment, we must assume that there is a range of uncertainty about (at least) the gap-opening cost, because this corresponds to the indel rate.

Inferences from Multiple Sequence Alignments

Multiple sequence alignments are a prerequisite for inferring a number of different biological properties. Therefore, alignment ambiguity must be considered in each of these methods. Most of the properties inferred from multiple sequence alignments can be considered evolutionary parameters in that they specify the process of molecular sequence change or the historical relationships of the sequences.

First, modern methods for estimating phylogenetic trees relating molecular sequences require a multiple sequence alignment as input. By specifying which residues are homologous, multiple sequence alignments divide the set of sequences into a set of separate single-residue characters. This decomposition is a requirement for the most powerful and accurate modern phylogeny estimation methods, which infer clades based on the evidence of shared, derived characters. In this chapter, our central focus is on the estimation of phylogenetic trees, but we briefly consider implications of alignment uncertainty for other methods as well.

Second, alignments are necessary for the estimation of mutation rates, divergence times, and evolutionary distances. This includes the estimation of the rates of insertion and deletion, as well as substitution rates and other parameters of the evolutionary process.

Third, alignments are required in many methods for labeling sites as functionally divergent, conserved, positively selected, or hypervariable. This is because the alignment defines the sites that are given functional labels. Thus, alignment uncertainty means uncertainty about what the sites are. In addition, most methods for annotating sites use the pattern of different character values at each site to assign meaningful labels. Thus, alignments also are required for motif-discovery methods if those methods rely on conservation patterns to find motifs.

Lastly, we note that alignments also are used to infer whether or not two sequences are homologous. However, we do not focus on this kind of analysis in this chapter, because questions about the existence of sequence homology are not about which alignment is correct, but about whether any alignment is correct. Furthermore, determinations of homology are often based on the distribution of various statistics under the null model of nonhomology, whereas we are interested in alignment ambiguity under the assumption of homology. However, we note that improved methods that explicitly consider the alternative model of homology as well may improve the ability to detect homologs with very high sequence divergence (Csuros and Miklós 2005).

THE STRUCTURE OF BIOINFORMATIC INFERENCES

To characterize the quality of bioinformatic inference methods, we first describe the properties that we desire in a robust inference method. We then discuss the multistage pipeline structure that is common to many traditional estimation methods and how estimation errors can propagate through this pipeline. We then overview two different approaches to bioinformatic inferences: the cost-based paradigm and the statistical paradigm. Finally, we detail the two primary sources of alignment uncertainty and compare the two paradigms in terms of their ability to make robust inferences in the presence of alignment uncertainty.

Robust Inference

Inference methods must have a number of important properties to provide a sufficient basis for robust decisions. First, inference methods must avoid bias. Specifically, while inference methods do not always yield correct results because of insufficient information, we desire that results tend toward the correct answer as the amount of information increases and not toward any other value. Second, we desire an inference method to make the efficient use of all information in the data set, thereby decreasing the range of error and increasing power to detect things. Third, to be a useful basis for decisions, an inference method should provide accurate measures of confidence or precision that indicate the weight of evidence in favor of the inference or estimate. For example, if a researcher wishes to use molecular data to determine whether a clade is monophyletic, a tree estimate is of little value unless it is accompanied by the weight of evidence for clade monophyly. Bias and uncertainty in the alignment estimate naturally propagate to bias and uncertainty in estimates of trees, evolutionary rates, and other parameters, because these estimates are based on the alignment estimate. However, the way in which this happens depends on the way the inference is structured.

Sequential Estimation

The Structure of Sequential Estimation Bioinformatic inference methods often consist of a series of chained estimation steps that are performed in a particular order. Instead of estimating the parameter of interest directly from unaligned sequence data, such "sequential estimation" methods are characterized by a pipeline structure in which the output of each estimation step can be used as input to the following step. For example, phylogenies are traditionally estimated in a two-step process. In the first step, a multiple sequence alignment is constructed from unaligned sequence data. This alignment is often manually edited or trimmed to remove estimation artifacts. Then, in the second step, this alignment estimate is used to construct a phylogeny estimate. Estimation of other evolutionary parameters might follow the same pattern. For example, a researcher might first construct a single alignment estimate and then use that alignment to estimate branch lengths and the relative rates of indels and substitutions (Thorne et al. 1991). In both cases, the error in inaccurate alignment estimates might propagate through the estimation pipeline and affect downstream estimates, including finally the parameter of interest.

In addition to unaligned sequence data that are used as input, the output of the estimation pipeline is influenced by several tunable parameters that must be specified by the researcher. For example, an alignment estimate might be influenced by the value of tunable alignment parameters, such as the gap opening penalty (GOP), the gap extension penalty (GEP), and various mismatch penalties. When the alignment estimate is constructed by progressive alignment, these parameters additionally include a guide tree that determines the order in which pairs of partial multiple sequence alignments are combined to produce the full multiple sequence alignment estimate. Thus, alignment estimates might be influenced by the guide tree as well as by mutation costs. Although the guide tree can be specified by the researcher, it is frequently estimated from the unaligned sequence data using distance-based techniques, such as neighbor joining (NJ). In this case, phylogeny estimation becomes a three-step procedure in which the first step involves estimating the guide tree.

Sources of Error in Sequential Estimation Sequential estimation works well when the sequences are closely related, and there is little ambiguity about the true alignment. However, when the sequences are more divergent, two sources of uncertainty may lead to uncertainty in the alignment estimate. First, there may be a myriad of near-optimal alignments. In this case, the best-scoring alignment cannot be confidently preferred over the second best alignment, because both alignments have similar quality. In this case, the choice of a single alignment estimate to submit to the next stage is somewhat arbitrary and could result in an error that propagates through the estimation pipeline. We note that the degree of uncertainty here does not depend solely on prior beliefs about the alignment, but instead it depends on the strength of evidence in the observed unaligned sequence data for each alignment as assessed by a probabilistic model or objective function. Therefore, this uncertainty can be categorized as *a posteriori* uncertainty.

Second, when the input sequences are divergent, uncertainty about tunable parameter values may also lead to error. This is because small changes in the values of tunable parameters may result in a different estimate. For example, different gap penalties in the alignment stage may lead to different alignment estimates, which, in turn, lead to substantially different phylogeny estimates (Morrison and Ellis 1997). Thus, when the true value of these parameters is unknown or not precisely known, uncertainty in the parameters leads to uncertainty in downstream estimates that depend on the alignment estimate. In the sequential estimation paradigm, tunable parameter values are chosen initially based on prior knowledge or belief, instead of being estimated from the observed data that are used as input. Therefore, this uncertainty can be categorized as *a priori* uncertainty.

Circular Dependencies in Sequential Estimation Sequential estimation can lead to biased inferences, because accurate knowledge of tunable parameters for the alignment estimation process is not available during the alignment construction step. To demonstrate this, we note that tunable parameters of the alignment construction process in fact correspond to the parameters of the evolutionary process. For example, progressive alignment algorithms require phylogenetic information in the form of a guide tree to yield high-quality alignment estimates (Thompson et al. 1994). Likewise, the GOP is a proxy for the indel rate, the GEP corresponds to the mean indel length, and mismatch penalties depend on knowledge of the percent identity of the sequences. Therefore, the phylogeny, indel rate, and branch lengths should be known in advance before the alignment is estimated. This observation makes intuitive sense, in that knowledge of the evolutionary process should lead to better estimates of the alignment. However, it leads to problematic circular dependencies. For example, traditional approaches to phylogeny estimation require the alignment to be known in advance before the phylogeny can be estimated. However, high-quality estimates of the alignment require the phylogeny to be known in advance before the alignment can be estimated. Clearly, both of these conditions cannot be satisfied: in a sequential estimation framework, either the phylogeny or the alignment must be estimated first.

Circular dependencies lead to biased estimates when the alignment is ambiguous, because alignment estimates may simply reflect and artificially reinforce the initial guesses for tunable parameters in this case. For example, alignments constructed by progressive alignment tend to support phylogenies that are similar to the guide tree, which is a tunable parameter (Lake 1991; Thorne and Kishino 1992). Likewise, the number of indels in an alignment estimate can be largely determined by the GOP, so that estimates of indel rates strongly depend on that parameter. One possible way to handle circular dependencies is to alternate between estimating the alignment and the parameter of interest. However, the tendency of the alignment estimates to reinforce bad initial guesses when the alignment is ambiguous can lead to mutually reinforcing, but incorrect, estimates for the alignment and other parameters. In such a case, iteration will not step away from local optima to achieve global convergence.

Joint Estimation

Bioinformatic inference methods can avoid the bias and overconfidence introduced in sequential estimation by estimating mutually dependent parameters simultaneously. This "joint estimation" approach has two primary attractions. First, it eliminates the circular dependencies that plague the sequential estimation paradigm. This is done by means of a joint score function for all mutually dependent parameters, including the alignment, phylogeny, indel rates, substitution rates, and other parameters. For example, if the alignment **A** and phylogeny τ are jointly estimated, then a joint score function of the form $f(\mathbf{A},\tau)$ is necessary. Because an alignment is always available when scoring the tree, a previous alignment estimation step is unnecessary. Likewise, because a phylogeny is always available when scoring the alignment, a separate guide tree is unnecessary. We note that this score function may be either a cost function as in Wheeler (1996) or a probability function based on a joint probabilistic model as in Redelings and Suchard (2005).

Second, joint estimation allows all possible alignments to be considered when scoring each evolutionary parameter. For example, Thorne et al. (1991) estimate indel rates by maximum likelihood in a way that is not dependent on the choice of a single alignment. This involves computing the likelihood of parameters by summing the probability of all alignments conditional on those parameter values. Similarly, a joint score function can be used to estimate phylogenies in a way that considers all alignments. We also note that when the alignment is the parameter of interest, all possible values of other parameters can be considered.

Types of Joint Estimation There are a number of different ways that a joint score function can be used (Wheeler 2006), and we return to our previous example in which the alignment **A** and topology τ are to be estimated to illustrate this point.

First, the researcher can estimate the optimal parameter combination by finding the parameter values that optimize the joint score function. Wheeler (1996) defined a cost-based score function and followed this optimization approach to estimate phylogenies and alignments. Thus, the estimates $\hat{\tau}$ and $\hat{\mathbf{A}}$ are constructed such that

$$(\hat{\tau}, \hat{\mathbf{A}}) = \arg\min_{\tau, \mathbf{A}} f(\tau, \mathbf{A}) \qquad (1)$$

where the mathematical notation $\arg\min_{\tau, \mathbf{A}}$ denotes the value of the arguments $(\tau, \mathbf{A})$ at which the function $f(\tau, \mathbf{A})$ takes its minimum (equation 1). Nonoptimal parameter combinations make no contribution to parameter estimates in this approach. We also note that each topology τ can be compared to other topologies based on a reduced score function $\tilde{f}(\tau)$ in which the alignment argument **A** is optimized out (equation 2):

$$\tilde{f}(\tau) = \min_{\mathbf{A}} f(\tau, \mathbf{A}) = f(\tau, \arg\min_{\mathbf{A}} f(\tau, \mathbf{A})) \qquad (2)$$

Thus, each topology is scored using a different alignment that is optimally adapted to it. In contrast, the sequential estimation approach compares topologies based on a score function $f_{\mathbf{A}_0}(\tau)$, where the alignment $\mathbf{A}_0$ was computed in a previous step and remains the same for all topologies instead of being adapted to each topology.

Second, the researcher might construct the maximum likelihood estimate of τ. Here, the joint score function $f(\tau, \mathbf{A})$ is the probability Pr(**Y**, **A**|τ) of the data **Y** and alignment **A** given topology τ. The reduced score function $f(\tau) = \Pr(\mathbf{Y}|\tau)$ that is used to compare topologies is constructed by summing over the alignments instead of maximizing or minimizing over them (equation 3):

$$\tilde{f}(\tau) = \Pr(\mathbf{Y} \mid \tau) = \sum_{\mathbf{A}} \Pr(\mathbf{Y}, \mathbf{A} \mid \tau) = \sum_{\mathbf{A}} f(\tau, \mathbf{A}). \qquad (3)$$

This process of summing over all possible values of a variable is known as marginalization. Marginalization leads to less biased estimates than simple maximization, because it does not score a topology τ only against the alignment that is optimally adapted to it, but it also considers near-optimal alignments. This makes allowances for the fact that, conditional on a topology τ being correct, the optimal alignment for τ may still not be the correct alignment. For example, Thorne et al. (1991) showed that alignments must be summed out to give unbiased estimates of indel rates and evolutionary distances when the alignment is ambiguous. We note that in the maximum likelihood paradigm, the phylogeny, indel rates, and other evolutionary parameters are not summed out, because they are parameters and not missing data like the alignment.

Lastly, the researcher might use a Bayesian approach to estimate τ. In this case, the joint score function $f(\tau, \mathbf{A})$ is the posterior probability Pr(τ, **A**|**Y**). As in the maximum likelihood case, estimation of the topology is made by marginalizing over the alignment (equation 4):

$$\tilde{f}(\tau) = \Pr(\tau \mid \mathbf{Y}) = \sum_{\mathbf{A}} \Pr(\tau, \mathbf{A} \mid \mathbf{Y}) = \sum_{\mathbf{A}} f(\tau, \mathbf{A}). \qquad (4)$$

However, unlike the maximum likelihood case, inference on the alignment can be made in the same way by summing out the topology. This is made possible in the Bayesian paradigm by placing a prior distribution over the topology and other parameters to treat them as random variables. Thus, inferences about the alignment can take into account uncertainty in the topology and other parameters.

Joint Estimation and Bootstrap Fractions One of the most common ways of characterizing the strength of evidence for a phylogenetic hypothesis is the bootstrap fraction. We note that the bootstrap fraction should not be interpreted as a probability that a clade is correct. For example, a bootstrap fraction of greater than 0.7 is often considered to represent strong support. Nevertheless, the bootstrap approach is attractive, because it assesses the sensitivity of conclusions to data selection for estimators that do not provide this information themselves.

However, the bootstrap fraction cannot be used to characterize uncertainty of phylogenetic hypotheses when the alignment is coestimated with the phylogeny. This is derived from the fact that the bootstrap approach assumes that the data can be decomposed into a series of separate columns or sites that can be resampled in the bootstrap procedure. However, when the alignment is unknown, the sites are no longer defined; thus, the bootstrap fraction cannot be computed as usual.

Joint Bayesian Estimation Joint Bayesian estimation is our preferred method of analysis for bioinformatic inferences. One benefit of the Bayesian approach and the maximum likelihood approach is that all mutually dependent parameters can be jointly estimated. In contrast, score-based approaches, such as in Wheeler (1996), are able to coestimate the alignment and phylogeny but allow indel and substitution costs to remain as tunable parameters. Thus, circular dependencies still exist between the alignment and these costs. Another benefit of the Bayesian and maximum likelihood approaches is that alignments can be summed out, which is necessary to avoid bias. As noted by Wheeler (2006), optimization-based approaches can optimize either a cost-based score function or a probability expression. In the second case, a statistical model of the insertion–deletion process is necessary, but unobserved internal sequences can be averaged over instead of maximized over (Wheeler 2006). In other approaches, the letters of the internal sequences can be averaged over, but their homology can be maximized over (Fleißner et al. 2005). Both approaches lead to less accurate estimates that may be biased but can improve computational efficiency. Finally, we recall that Bayesian approaches are able to incorporate parameter uncertainty into posterior distributions of latent variables, such as the alignment, while it is not clear how to do this in the maximum likelihood framework.

Joint Bayesian estimation of pairwise alignments and evolutionary process parameters is not a new development (Allison and Wallace 1994). However, it is only recently that statistical techniques for estimating multiple sequence alignments have been developed (Holmes 2003; Holmes and Bruno 2001), leading to joint Bayesian estimation of alignments and phylogenies (Fleißner et al. 2005; Lunter et al. 2005; Redelings and Suchard 2005). Because many bioinformatic analyses require multiple sequence alignments, these new techniques have created new opportunities for robust statistical inference in the presence of alignment uncertainty. However, for the moment, the extreme amounts of computation time required for these analyses limits the size of data sets that they can be applied to.

Cost-Based Methods versus Statistical Estimation

Estimation in the Cost-Based Paradigm Both pairwise and multiple sequence alignments have traditionally been estimated by finding the alignment that optimizes some score function. In the cost-based paradigm, the score function is computed as a sum of penalties, or "costs," for each observed sequence change in the alignment. In many pairwise alignment algorithms, each aligned residue pair may incur a mismatch penalty if the residues are different, and each indel incurs a GOP as well as (in some algorithms) a separate GEP for each additional deleted or inserted character. The maximum parsimony method of inferring phylogenetic trees is an example of a cost-based method, because its score function is the sum of a number of penalty terms for observed sequence differences. In addition, some early (Sankoff 1973; Sankoff et al. 1976; Sankoff et al. 1973) and some later (Wheeler 1996, 2006) methods for multiple sequence alignment seek to minimize a score function that explicitly accounts for insertions and deletions on internal branches of the tree by including in the alignment unobserved ancestral sequences at internal nodes on the phylogeny.

However, most newer methods are willing to sacrifice some of this biological realism to substantially increase speed. First, most commonly used methods for multiple sequence alignment rely on an objective function that does not explicitly consider substitutions or indels occurring on each branch of the tree. Instead, they might use a less biologically motivated score function, such as a sum-of-pairs score (Edgar 2004b) or a tree-based, weighted-sum-of-pairs score (Thompson et al. 1994). Because the sum-of-pairs score does not depend on the evolutionary tree, it may double-count the cost for shared, derived changes that are observed in more than one leaf sequence. Second, most of these alignment algorithms rarely succeed in optimizing their score function, preferring instead to use progressive alignment to quickly discover a relatively high-scoring alignment.

Despite the variation in these methods, they all share a common drawback of the cost-based paradigm: cost parameters cannot be estimated by minimizing a cost function, since this would simply result in setting all the costs to zero. This is primarily a problem when attempting to determine gap penalties, since substitution costs can be determined from data using probabilistic methods, as in the PAM and BLOSUM matrices. However, even in this case, cost-based methods do not allow the cost parameters to be tuned to the data set at hand through optimization of the cost function. Lastly, we note that cost parameters do not represent a biological property, so it is unclear what it means for a cost parameter value to be called correct or incorrect.

Estimation in the Statistical Paradigm Alignment estimation in the statistical paradigm requires a probabilistic model of insertion and deletion. The alignment can then be inferred either by maximizing the likelihood (a frequentist approach) or by calculating the probability distribution of the alignment, given the observed sequence data (a Bayesian approach). If the frequentist approach is used, then the likelihood becomes a score function, although it is maximized instead of minimized like cost-based score functions. We note that log probabilities of accepted mutations are similar in meaning to the penalties for these mutations in the cost-based paradigm. This is because the log probabilities of independent events are added to compute the total probability, similar to the way penalties are added in the cost-based paradigm to compute the total penalty.

One benefit of the statistical paradigm is that statistical models of evolution allow the cost parameter for each type of mutation to be replaced with a biologically interpretable parameter that measures the rate of accepted mutations. For example, the GOP can be replaced by an insertion–deletion rate, and mismatch penalties can be replaced by substitution rates. One exception to this rule is that the gap-extension cost is not replaced with a rate, but with an extension probability that controls the mean length of gaps. This is because each gap extension is not a separate mutation, but it is simply a penalty for longer gaps that is separate from the penalty for gap creation. We note that rate parameters play the same role as penalties or costs—a high rate corresponds to a low cost, and a low rate corresponds to a high cost. However, modeling evolution in terms of mutation rates more naturally accounts for multiple changes on a branch of the tree and for the occurrence of more changes on branches of longer duration.

A second benefit of the statistical paradigm is that it is possible to estimate rate parameters from the data. This differs from cost-based estimation in that increasing the mutation rates, which corresponds to decreasing mutation costs, decreases the penalty for a mutation but also increases the penalty for not mutating. Therefore, the likelihood does not always increase with increasing mutation rates, allowing rates to be estimated by maximizing the likelihood. Because all parameters in a statistical model can be estimated from data, it is not necessary to specify the rate parameters based only on prior belief. For example, the relative rates of transitions and transversions can be estimated from the observed sequence data, as can the relative rates of indels and substitutions. This yields empirically driven parameter values, instead of parameter values chosen based on subjective or heuristic choices. Thus, the relative weight of different types of mutational events can be driven by the data.

Sources of Alignment Uncertainty

Alignment uncertainty comes primarily from two sources: uncertainty in parameter values and uncertainty due to near-optimal alignments. These sources of uncertainty are dealt with quite differently in the cost-based paradigm and the statistical paradigm. Additionally, there are smaller but important differences between maximum likelihood estimation and Bayesian estimation within the statistical paradigm.

Alignment Ambiguity from Parameter Uncertainty Parameter uncertainty leads to alignment ambiguity, because different values of tunable alignment parameters lead to different alignment estimates. This is to be expected, since these parameters characterize the evolutionary process and, therefore, determine how plausible each alignment should be. In the cost-based paradigm, such parameters include a cost for each type of mutation in addition to the guide tree, whereas in the statistical paradigm, cost parameters for each type of mutation are replaced with mutation rates. Parameter uncertainty is almost always *a priori* uncertainty in the cost-based paradigm and *a posteriori* uncertainty in the statistical paradigm. That is, uncertainty about cost parameters in the cost-based paradigm is based on prior belief and not on the data. This is because it is not clear how to estimate cost parameters from the observed data. Therefore, when determining alignment ambiguity via sensitivity analysis, it may be difficult to justify any particular range of parameters as being large enough. In the statistical paradigm, on the other hand, the amount of parameter uncertainty depends primarily on how much data are collected and how informative they are. Parameters can be estimated via maximum likelihood, and parameter uncertainty can be characterized in terms of confidence intervals. If a Bayesian approach is taken, prior information or belief can be incorporated into parameter estimates, but this *a priori* information has decreasing influence on estimates as the amount of data increases. In contrast, maximum likelihood estimates and confidence intervals do not incorporate prior information.

Alignment Ambiguity from Near-Optimal Alignments   Near-optimal alignments indicate alignment ambiguity, because they indicate the presence of plausible alternatives to the optimal alignment. Taking into account near-optimal alignments is difficult to do within a cost-based framework, because differences in cost scores have no intrinsic meaning. The scores can be multiplied by any positive scaling factor without changing the optimum; therefore, it is unclear how close an alignment must be to optimal before it is considered to be "near" the optimum. In addition, if one seeks to down-weight near-optimal alignments that are further away from the optimum, it is unclear how much a suboptimal alignment should be down-weighted. In contrast, when inference is carried out under a probabilistic model, it is possible to incorporate alignment uncertainty into a further analysis by weighting each alignment according to its probability. In addition, it is possible to determine (for example) a 95% probable set of alignments if a cutoff is needed.

Effects of Underestimating Ambiguity   Simultaneously accounting for both parameter uncertainty and near-optimal alignments is an important feature of any bioinformatic analysis. If one of these sources of uncertainty is ignored, then exaggerated confidence can be ascribed to the resulting estimates. Mathematical readers will appreciate that this is similar to the common ANOVA formula (equation 5) about the proportion of the variance in X that is explained by Y

$$\mathrm{Var}(X) = \mathrm{Var}[E(X|Y)] + E[\mathrm{Var}(X|Y)], \qquad (5)$$

where the first term is the variance in X that is explained by the variation in Y, and the second term is the variation in X that is not explained by the variation in Y. Ignoring either of these contributions to the uncertainty could lead to underestimation of the effects of alignment ambiguity and, hence, overconfidence in a bioinformatic inference. However, it is not clear how one can consider both sources of uncertainty when using cost-based models.

Alignment Ambiguity in the Frequentist Paradigm While both maximum likelihood and Bayesian methods attempt to simultaneously estimate model parameters Θ and alignment A from the data Y, the methods differ substantially in their handling of uncertainty in the alignment. In the maximum likelihood framework, a common method is to first construct an estimate $\hat{\Theta}$ of Θ by summing over all possible alignments in proportion to their probability (equation 6):

$$\hat{\Theta} = \arg\max P(Y|\Theta) = \arg\max \sum_{A} P(Y, A|\Theta) \qquad (6)$$

Approximate confidence intervals for Θ can then be obtained by assuming the asymptotic normality of $\hat{\Theta}$ and estimating the inverse of the Fisher information matrix. These confidence intervals then account for uncertainty in both the alignment and parameter values, because the alignment is summed out. However, uncertainty in the alignment is usually determined under the assumption that $\Theta = \hat{\Theta}$, by considering the alignment distribution (equation 7)

$$P(A|Y, \hat{\Theta}) = \frac{P(A, Y|\hat{\Theta})}{\sum_{A} P(A, Y|\hat{\Theta})}. \qquad (7)$$

Unfortunately, confidence intervals obtained from this distribution ignore uncertainty in the parameter estimates $\hat{\Theta}$ and, therefore, do not take into account parameter uncertainty. Thus, in estimating alignments, the common maximum likelihood method accounts for alignment uncertainty in parameter estimates, but it does not account for parameter uncertainty in alignment estimates.

Alignment Ambiguity in the Bayesian Paradigm   The Bayesian approach to statistical estimation naturally incorporates both sources of uncertainty simultaneously. Bayesian inference is based on the joint posterior distribution of the alignment **A** and parameters **Θ** given the data **Y**. This distribution represents the posterior uncertainty in both **A** and **Θ**, as well as representing the correlation between them. The posterior distribution for **A** is obtained by integrating over possible values of **Θ** (equation 8)

$$P(\mathbf{A} \mid \mathbf{Y}) = \int d\Theta P(\mathbf{A}, \Theta \mid \mathbf{Y}). \tag{8}$$

Thus, credible intervals for **A** that are based on this posterior alignment distribution take parameter uncertainty into account. As a result, joint Bayesian estimation is an attractive method for estimating alignments and other parameters, because it simultaneously accounts for uncertainty resulting from near-optimal alignments and parameter uncertainty.

ALIGNMENT UNCERTAINTY: PAIRWISE ALIGNMENTS

Before we consider methods of handling alignment ambiguity in multiple sequence alignments, we first must summarize the progress made for pairwise alignments. Methods for handling alignment uncertainty in pairwise alignments preceded methods for handling uncertainty in multiple sequence alignments by a significant time period. Nevertheless, the development and improvement of methods are parallel, so that many improvements can be illustrated with pairwise alignments and then extrapolated to multiple sequence alignments.

Probabilistic Models and Pairwise Alignment Uncertainty

Models of the Insertion–Deletion Process   Probabilistic approaches to pairwise alignment rely on a probabilistic model of the insertion–deletion process and the substitution process. For example, Bishop and Thompson (1986) specified a probability distribution on pairwise alignments in terms of the probabilities of gap opening and gap extension. Under such a model, parameters can be estimated via maximum likelihood without relying on the choice of a single alignment estimate. Instead, the likelihood for any set of parameter values is computed by using dynamic programming to sum the probabilities of all pairwise alignments, conditional on the given parameter values. Given an estimate of the evolutionary process parameters, a probabilistic model makes it possible to measure confidence for homology of two residues in terms of probabilities, by summing over all alignments that display the homology and weighting each alignment by its probability given the parameters. However, this maximum likelihood approach does not take parameter uncertainty into account when considering alignment uncertainty. A Bayesian approach to estimation allows the incorporation of prior beliefs about parameters and incorporates posterior parameter uncertainty into measures of posterior alignment uncertainty (Allison and Wallace 1994).

A further advance was the construction of stochastic process models for the insertion–deletion process (Thorne et al. 1991, TKF1). The TKF1 model specifies not only a distribution on pairwise alignments, but it also describes how insertion and deletion events accumulate over time to create a pairwise alignment between ancestor and descendant sequences. For example, previous probabilistic models did not distinguish between two adjacent deletion events and one long deletion, because they model only "gaps" and not the indel events that create them. Additionally, the TKF1 model replaces the gap probability of the Bishop and Thompson model with an insertion rate and a deletion rate that are biological meaningful parameters. However, the TKF1 model has the drawback of assuming that all indels are of unit length. The Thorne et al. (1992, TKF2) model extends the TKF1 model by allowing indel lengths to follow a geometric distribution. Therefore, the TKF2 model adds an additional parameter to specify the extension probability of this distribution.

As noted above, the use of probabilistic models can decrease bias in parameter estimates by performing a weighted average over all possible alignments instead of considering only the optimal alignment. Especially for divergent sequences, estimates of the number of substitutions are biased upwards and estimates of the number of indels are biased downwards when using the optimal alignment (Thorne et al. 1991; Yee and Allison 1993). Thus, when there is significant alignment uncertainty, the optimal alignment may not be typical of the set of plausible alignments. By instead averaging over all pairwise alignments, the bias is much decreased.

Extrapolation to Multiple Sequence Alignments Probabilistic models of insertion and deletion allow bioinformatic inferences to take alignment ambiguity into account by summing over all alignments. However, the number of possible pairwise alignments is astronomically large, growing faster than exponentially as sequence length increases. Sums over all pairwise alignments are computationally tractable because of the use of dynamic programming, an approach later formalized in terms of Hidden Markov Models (HMMs) (Durbin et al. 1998). The amount of computation time required by dynamic programming algorithms for pairwise alignments grows only as the square of sequence length for TKF models and the Bishop and Thompson (1986) model. This makes it possible to analyze sequence lengths of several thousand letters without approximations, and even greater lengths when approximations are made.

Unfortunately, there are several reasons that dynamic programming algorithms cannot be practically applied to directly sum over the possible alignments between three or more sequences. Firstly, the computation time and memory requirements increase as $O(L^n)$, where L is the sequence length and n is the number of sequences. Thus, dynamic programming can be performed for three sequences, provided the sequences are kept quite short, but it is impractical to extend to more than three sequences. Secondly, computation time and memory requirements grow exponentially in the number of observed sequences, because the number of states in the HMM increases exponentially. Thus, even if it were a simple matter to describe a dynamic programming algorithm for aligning many sequences simultaneously, this algorithm would be too computationally burdensome to carry out. Therefore, sums over all possible multiple sequence alignments must be approximate. Using MCMC techniques, these sums can be approximated without visiting every possible multiple sequence alignment. However, the development of practical techniques for performing MCMC on multiple sequence alignments was delayed until Holmes and Bruno (2001) introduced new strategies for sampling alignments on a fixed topology. These new developments were of course assisted by the fact that the speed of desktop computers has continued to increase.

ALIGNMENT UNCERTAINTY: MULTIPLE SEQUENCE ALIGNMENTS

In contrast with the advanced statistical methods for dealing with uncertainty in pairwise alignments, common procedures for dealing with uncertainty in multiple sequence alignments have until recently been much less developed. Methods to handle uncertainty in multiple sequence alignments must be able to handle many problems. First, they must be able to detect uncertainty in the alignment. In doing so, they must incorporate uncertainty from both near-optimal alignments and from parameter uncertainty, including the phylogeny.

Identification of Ambiguous Regions Based on a Single Alignment

When phylogenies are inferred by sequential estimation, alignment uncertainty is often handled by labeling some regions of the single alignment estimate as "ambiguous," leaving the remainder as presumably "unambiguous." The ambiguous regions are then thrown out, and the remaining alignment columns are then submitted as input for further analysis. However, identification of ambiguous alignment regions is in itself a challenging task. Therefore, the success of this approach depends critically on how well the researcher is able to identify unambiguous columns in the alignment. If the researcher fails to identify incorrectly aligned columns, then this failure could bias the rest of the analysis. On the other hand, the removal of correctly aligned columns decreases the power to distinguish between alternative hypotheses.

Unfortunately, alignment ambiguity is commonly identified by a subjective and *ad hoc* "visual inspection." Subjective determination of alignment ambiguity can lead to conflicting phylogeny estimates when researchers make different choices about including or excluding alignment regions (Lutzoni et al. 2000). In addition, subjective determination of alignment ambiguity makes it very difficult to reproduce reported results. Therefore, a significant challenge has been to identify ambiguous sites in an objective and repeatable way (Gatesy et al. 1993).

One method for avoiding this subjectivity is to use a computer program to remove all alignment columns that are near gaps and are not part of conserved blocks according to specified rules (Castresana 2000, GBLOCKS). This approach codifies some of the intuition that is commonly used in removing alignment columns to prepare for phylogenetic analyses. However, this approach does not fully address the problem of identifying ambiguity; thus, it has a number of drawbacks that make it difficult to apply to highly divergent sequences. First, the GBLOCKS method relies on a single alignment estimate to locate ambiguously aligned regions, and the method is sensitive to the specific placement of gaps in this alignment, which could be incorrect. Thus, the automatic censoring of alignments created using different methods can retain different columns (Talavera and Castresana 2007).

Second, the method attempts to assess ambiguity without any knowledge of evolutionary parameters, such as the frequency of indels or the phylogenetic tree. Because the tree is not known, columns are identified as "conserved," based on the frequency of the majority character value in the column, instead of the predicted number of substitutions in the column, which can be small or large. Third, the GBLOCKS method uses a stringent and conservative test for retaining alignment columns, so it may lead to the removal of a large number of phylogenetically informative characters. This results partly from the fact that the method is completely agnostic about alignment parameters and bases confidence in alignments on conservation. Thus, it may throw out blocks that would be unambiguously aligned if the guide tree or indel costs were known. The removal of phylogenetically informative characters also is partly by design, because the removal of phylogenetic informative but rapidly changing sites is sometimes considered to be a positive feature when the phylogenetic inference method does not account for rate heterogeneity between sites. However, if the phylogenetic inference method can handle rate heterogeneity, then removal of these sites could be a significant drawback.

Whether the approach is repeatable or not, the approach of categorizing columns as ambiguous or unambiguous has a few more drawbacks. First, the categorization is binary, but it would be preferable to have a degree of certainty in the homology of a column, because all alignment regions have some degree of uncertainty, no matter how small. Second, in identifying columns as ambiguous, the side knowledge of the researchers is not taken into account. For example, it is possible that a region could be ambiguous if the phylogeny is unknown, but it could be relatively unambiguous if the phylogeny is known. Third, methods that remove gaps or alignment columns are common in phylogenetic inference, where it is common to view each column as giving independent evidence about the phylogeny. However, methods that remove columns with gaps are inherently unable to estimate some interesting biological parameters, such as indel rates. Methods that remove ambiguously aligned columns can also be detrimental to motif detection, because they may remove the columns that contain a motif when one of the sequences is ambiguously aligned to the other sequences. However, despite these issues, censoring simulated alignments via the GBLOCKS algorithm has been shown to improve the accuracy of phylogenies inferred from ClustalW alignments using maximum likelihood, maximum parsimony, and neighbor joining (Talavera and Castresana 2007).

Multiple Sequence Alignment Ambiguity Resulting from Parameter Uncertainty

Many researchers have observed that the outcome of multiple alignment estimation depends on which values are chosen for parameters that characterize the evolutionary process (Gatesy et. al. 1993). These parameters include the phylogeny used as a guide tree in many alignment methods, as well as the relative rates or costs assigned to insertions, deletions, and different types of substitutions, such as transitions and transversions. When different values of evolutionary parameters are used, the estimated alignment could change. Therefore, uncertainty about the correct values of the parameters leads to uncertainty about the correct alignment.

Identifying Ambiguous and Unambiguous Alignment Regions   Gatesy et al. (1993) introduce a procedure called "culling" to objectively divide alignments into ambiguous and unambiguous regions. They propose to generate a collection of alignment estimates from a range of different costs for gaps and substitutions. Alignment columns that are not present in all of the resulting alignments are considered to be ambiguous and are removed or "culled" from the alignment. The remaining columns are considered to be unambiguously aligned and are retained for further analysis. This kind of method is called sensitivity analysis, because it attempts to identify alignment regions that are not "sensitive" to parameter values[3]. This method is one of the only methods available that can be used to account for alignment uncertainty in a *nonstatistical* (e.g., cost-based) framework.

Although this procedure is objective in the sense that it is repeatable, the range of parameter values is chosen based on the researcher's prior knowledge or subjective beliefs. This range includes all plausible values for these parameters, and it is the range of parameter values that determines which alignment columns are considered certain or uncertain. Thus, the degree of ambiguity in an alignment does not depend only on the sequences to be aligned, but it depends also on prior knowledge about the evolutionary process. For example, Gatesy et al. (1993) varied transition, transversion, and gap costs. For the gap costs, they used values of 2/3, 1, 2, 3, 4, 5, 6, 7, 8, 9, 10, 20, 50, 100, and 300 times the cost of a transition. However, if the gap cost was known to be between 2 or 3, then a much narrower range could be used, leading to a greater number of columns labeled as unambiguous. In addition, if the alignment of a region is sensitive to a guide tree parameter, which is the case in progressive alignment, the region would be considered "ambiguous" if the phylogeny is not known, but it would be considered unambiguous if the phylogeny is considered known. Clearly, if the resulting alignment estimates will be later used to estimate the phylogeny, the phylogeny must not be considered known. Instead, the range of parameters should include all plausible phylogenies, as well as all plausible values for substitution and indel process parameters. However, many sensitivity studies do not account for this dependence (Morrison and Ellis 1997).

Incorporating Ambiguous Information The culling method has the unfortunate downside that it throws out a large fraction of all informative features. For example, in the two data sets considered by Gatesy et al. (1993), only 91/250 and 12/250 sites were considered unambiguous. In addition, the culling method allows only two levels of confidence, ambiguous or unambiguous, instead of allowing various degrees of certainty about a column. Wheeler et al. (1995) address these concerns by a technique called "elision." Similar to culling, this technique also consists of generating a collection of alignments using a range of parameter values. However, instead of removing columns, the collection of alignments is concatenated end-to-end. When used as input for phylogeny estimation in a parsimony framework, this effectively weights each alignment column by the fraction of the total alignments that it occurs in.

We note that the use of several alternative alignments should not be seen as a contradiction (Lutzoni et al. 2000) but as an *ad hoc* way of considering several equally weighted alternatives. If the parsimony score for the concatenated alignment is divided by the number of alternative alignments, then the elision method simply maximizes the average of the parsimony scores over all alternative alignments. However, it is not usually necessary to carry out this division, because it simply scales the score function and, therefore, does not affect the optimal phylogeny estimate. We note that this approach could in principle be applied in a statistical framework by maximizing either the average likelihood or the average log likelihood, where the average is taken over all alignment alternatives. We recommend the first approach, since it corresponds to the hypothesis that all generated alignments are equally likely *a priori*, whereas it is not clear to what hypothesis the second approach corresponds. However, the second approach is the direct analog of the parsimony-based technique of Wheeler et al. (1995), because each alignment has equal influence in selecting the optimum tree, even though some alignments may lead to a lower score than others.

Because the elision method maximizes a score function that is obtained by averaging over a collection of alternative alignments, this collection can be seen as representing a probability distribution on alignments that results from a probability distribution over alignment parameters. From this perspective, the range of parameter values used in the sensitivity analysis then represents a prior distribution on parameter values. Unlike the culling procedure, multiple inclusion of a single parameter value leads to higher *a priori* confidence in that value. Although a discrete range of separate values can be used in practice, we note that denser sampling of values from a region leads to increased prior confidence that the true value lies in that region. For example, the choice of gap-opening costs used for culling means that 10/15 values lie between 1 and 10, corresponding to a weight of 2/3 on this possibility. In contrast, only three values lie between 10 and 300, indicating a sparser sampling and a lower prior weight on this region.

Shortcomings of Sensitivity Analysis Compared to statistical methods, there are several shortcomings of sensitivity analysis, especially as applied to cost-based methods. First, the sensitivity analysis methods described above do not consider uncertainty resulting from near-optimal alignments, but they only consider alignment uncertainty that results from parameter uncertainty. Because near-optimal alignments are ignored, the culling method may fail to identify some ambiguous regions, and the elision method may fail to down-weight them sufficiently. In addition, ignoring near-optimal alignments may exaggerate the effect of changing parameter values, because it is possible that the set of near-optimal alignments remains almost the same under new parameter values, but a different alignment is chosen from this set. However, we note that this shortcoming seems to be primarily an accident of implementation instead of an essential property of sensitivity analysis. We also note that if multiple equally optimal alignments are found, then the culling and elision approaches do consider these alternative alignments. While this does not seem to us to go far enough, it does seem to be a helpful feature and a step in the right direction.

Second, the prior information or subjective beliefs used in sensitivity analysis are not informed by the data. As a result, conducting a sensitivity analysis with a very broad range of values is likely to result in the detection of few unambiguous columns. Unfortunately, it is unclear how cost parameters in cost-based alignment methods can be informed by the sequences that they are aligning. In addition, it is unclear what it would mean for a value for a cost parameter to be "correct," since it does not correspond to a biological quantity. In contrast, given a statistical model of the evolutionary process, a Bayesian approach allows incorporation of prior knowledge or belief about parameter ranges in the form of a prior distribution and allows this information to be updated by the data. Thus, the posterior distribution may have a significantly narrower range, representing decreased uncertainty. Unlike sensitivity analysis, broad or "diffuse" prior distributions on evolutionary parameters do not necessarily lead to extreme uncertainty about the alignment. Therefore, jointly estimating alignments and mutation rates via statistical methods should commonly result in less alignment ambiguity than a sensitivity analysis, and it is less dependent on prior knowledge or subjective beliefs about parameter ranges.

STATISTICAL INFERENCE UNDER ALIGNMENT UNCERTAINTY

Sound statistical methods for inferring evolutionary parameters from molecular sequence data in the presence of alignment uncertainty all take as given only the observed, unaligned sequence data. These methods require a joint probabilistic model that describes how unobserved evolutionary parameters may combine to generate the observed sequences. These parameters include the alignment, the tree, and the rate of occurrence of different kinds of mutations. We note that when inferring the tree or mutation rates, the alignment is considered a nuisance variable that can be summed out. However, the same statistical model can be used when the alignment is the parameter of interest, to sum out uncertainty in the tree. In this section we give a brief introduction to the model, notation, and methods described in Redelings and Suchard (2007), because this paper describes how we analyze the data example presented later in this chapter. Then, using this approach as a reference, we describe other recent developments in statistical estimation of alignments, trees, and other parameters. We then discuss shortcomings and possible improvements.

Joint Statistical Model

Variables and Definitions We begin by describing the data and unobserved evolutionary parameters in the model. We consider a collection **Y** of n homologous molecular sequences. The individual sequences are labeled $\mathbf{Y}_i$, indexed by $i = 1...n$ and have lengths $|\mathbf{Y}_i|$. Each sequence $\mathbf{Y}_i$ has elements $\mathbf{Y}_i[j]$ indexed by $j = 1...|\mathbf{Y}_i|$ that take on values in a set called the alphabet. Each letter in the alphabet represents a monomer in the molecular sequences **Y**. For example, if the sequences are DNA sequences, then the alphabet consists of the nucleotides {A, T, G, C}, whereas if the sequences are protein sequences, then the alphabet is the set of amino acids. We note that the alphabet does not contain a gap letter "-" because gaps are not part of the observed data, but they must be inferred.

The data **Y** are generated from an evolutionary process that is characterized by a number of unobserved parameters that we seek to estimate. The alignment **A** and the phylogenetic tree combine to specify the complete evolutionary relationship of the sequences in **Y**. The alignment **A** is separable from the observed letters in **Y**, because it specifies the homology of these letters without mentioning their values. Therefore, the alignment specifies how the letters in **Y** can be arranged to form the aligned data matrix **f**. This matrix indicates which letters are homologous to each other by arranging groups of homologous letters into a single column. Each row of **f** contains one of the sequences of **Y**, so that each column contains one letter from each sequence, or possibly a missing value. We denote the unknown number of columns C. The phylogenetic tree can be separated into its topology τ and its branch lengths **T**. We assume that the tree is unrooted and define the topology as an undirected acyclic graph in which all internal nodes have three neighbors. The topology contains exactly n leaf nodes and $n - 3$ internal nodes, and each leaf node corresponds to one of the n observed sequences. The total number of nodes, which we denote by N, is therefore $2n - 3$. Each branch b is associated with a branch length $T^{(b)}$.

Evolutionary parameters also include parameters Θ that characterize the substitution process and parameters Λ that characterize the insertion–deletion process. The substitution parameters Θ include the rates of nucleotide or amino acid replacement and describe how these rates vary between different sites. The insertion–deletion parameters Λ include not only the rates of accepted insertions and deletions, but they also specify the length distribution of accepted indels. We assume that insertions and deletions have the same rate λ and that the length of indels follows a geometric distribution with extension probability ε, so that $\Lambda = (\lambda, \varepsilon)$. Taken together, the entire state space is composed of points $\omega = (Y, A, \tau, T, \Theta, \Lambda)$.

Probability Expression Traditional methods for estimating the tree or other bioinformatic parameters have assumed that the alignment was known with certainty. Therefore, these approaches implicitly condition on the alignment, leading to the probability expression (equation 8)

$$P(Y, \tau, T, \Theta, | A) = P(Y | A, \tau, T, \Theta) \times P(\tau, T) \times P(\Theta) \qquad (8)$$

where, following a common abuse of notation, we write $P(X)$ to represent $P(X = x)$ for any random variable X taking on a realized constant value x. The first term $P(Y | A, \tau, T, \Theta)$ in equation (8) is the likelihood for the model, and it is determined by the model of the substitution process. The other terms represent prior distributions on trees and on substitution process parameters, respectively.

In contrast with this traditional approach, a joint probability model allows the alignment to vary, yielding a modified probability expression

$$P(Y, A, \tau, T, \Theta, \Lambda) = P(Y | A, \tau, T, \Theta) \times P(A | \tau, T, \Lambda) \times$$
$$P(\tau, T) \times P(\Theta) \times P(\Lambda) \qquad (9)$$

We note that equation (9) is identical to equation (8) except for the addition of two new terms. Therefore, we can base the likelihood $P(Y | A, \tau, T, \Theta)$ on traditional substitution models, such as reversible, continuous-time Markov chains (CTMCs). The first new term, $P(A | \tau, T, \Lambda)$, is the prior distribution on alignments and is based on the insertion–deletion process. We describe a prior distribution on alignments below that is biologically realistic and penalizes alignments with more indels. The second new term, $P(\Lambda)$, is the prior distribution on insertion–deletion process parameters. We note that the likelihood and the alignment prior are separable in equation (9), because the substitution process and the insertion–deletion process are separate and operate independently. This is possible, because the alphabet does not include a "gap" letter "-", so that the substitution process is not responsible for insertions and deletions.

Substitution Model   The substitution model determines the likelihood $P(Y \mid A, \tau, T, \Theta)$ that the letters Y are observed. This probability can be expressed in terms of the aligned data matrix f that depends on both the data Y and the alignment A, as (equation 10)

$$P(Y \mid A, \tau, T, \Theta) = P(f \mid \tau, T, \Theta) \qquad (10)$$

This approach is useful, because we assume that observations in each column of the aligned data matrix are independent realizations from the substitution process, so that the full likelihood is the product of the likelihood of each column in the aligned data matrix f. We follow common practice in molecular phylogenetics by using the reversible CTMC models to describe the process of substitution in each column (Goldman 1993) and will not describe them here. Therefore, to express this probability, we must define the aligned data matrix f and describe how to construct it from the data Y and the alignment A.

The matrix f consists of rows indexed by $i = 1...N$ and columns indexed by $c = 1...C$. The letters in row i of f all come from sequence i and must occur in order. The matrix f represents the hypothesis that all the letters in each column c descend from a single residue in the sequence of the common ancestor. If no letter in sequence i is homologous to other residues in column c, then we place a gap "-" at f_{ic} to indicate a missing value. In addition to the observed leaf sequences, the matrix f also includes unobserved sequences at internal nodes as missing data. These sequences are composed of Felsenstein wildcards that are represented by "*" to indicate that a letter is present, but its value is unknown.

We introduce the matrix $M(A)$ to specify how the alignment A arranges the data Y into the matrix f while remaining separable from f. The matrix $M(A)$ has the same dimensions as f and specifies which letter of sequence Y_i belongs in column c through equation (11)

$$f_{ic} = Y_i[M_{ic}]. \qquad (11)$$

If no element of Y_i belongs in column c, then $M_{ic} = \text{'-'}$ and we define $Y_i[\text{'-'}] = \text{'-'}$. Figure 10.1 illustrates the relationship of Y, $M(A)$ and f for a 4-taxon example.

Alignment Prior   As described above, the multiple sequence alignment A includes alignment information for internal node sequences as well as leaf sequences. However, the observed data Y specifies the letters

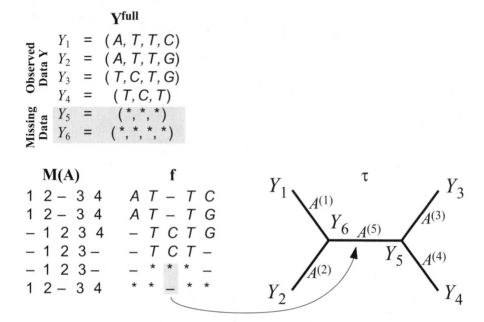

Figure 10.1. Construction of aligned data matrix **f** from data **Y** and alignment **A**. The first column shows some unaligned sequences Y^{full} of various lengths. These sequences include the observed leaf sequence data **Y** as well as unobserved sequences Y_5 and Y_6 at internal nodes of the tree. The lengths of the sequences Y_5 and Y_6 are unknown, and all possibilities must be considered; one possibility is shown. The second column shows **M(A)**, a matrix that parameterizes the multiple sequence alignment **A** by specifying where gaps appear in the aligned data matrix and which letters of **Y** appear in each column. Sequences at internal nodes are included in the multiple sequence alignment. The third column shows the aligned data matrix **f**, constructed by combining **Y** and **M(A)**. Letters that are present at internal sequences are unobserved and are drawn as Felsenstein wildcards. While the alignment is separable from the data, as shown in column 2, the aligned data matrix **f** is not. The fourth column contains the evolutionary tree topology τ. Each branch b is associated with a pairwise alignment $A^{(b)}$ between the two sequences at the endpoints of the branch. This is made possible by the inclusion of the missing data Y_5 and Y_6 in the alignment **A** and allows each indel to be localized to a particular branch. For example, the shaded indel in column 3 of **f** must occur on the internal branch of the tree.

only for leaf sequences (Figure 10.1). The alignment **A** also specifies the length of all sequences in addition to homology information, so that **A** must agree with **Y** on the length of observed sequences, but the length of unobserved sequences at internal nodes of the tree is unknown.

Augmenting the alignment to include homology about sequences at internal nodes is beneficial, because it allows us to decompose the multiple sequence alignment **A** into a collection of pairwise alignments (Holmes and Bruno 2001). Given a topology τ with B branches, **A** can be decomposed into a tuple of pairwise alignments $(A^{(1)}, \ldots, A^{(B)})$. Each branch b of the tree is associated with a pairwise alignment $A^{(b)}$ that specifies the homology of the sequences at each end of the branch. Representing the alignment **A** in this way allows us to construct a distribution on multiple sequence alignments from a distribution $v_{\lambda t_b}$ on pairwise alignments that is parameterized by indel rate λ and branch length t_b for branch b. Pairwise alignments on each branch will be independent conditional on the lengths of the sequences at the internal nodes, because evolution occurs independently on each branch of the tree. This leads to an alignment prior of the following form in equation (12)

$$P(\mathbf{A} \mid \tau, \mathbf{T}, \Lambda) = \frac{\prod_{b=1}^{B} P_{v_{\lambda t_b}}(\mathbf{A}^{(b)})}{\prod_{i \in I} \phi(|\mathbf{A}_i|)^2}, \qquad (12)$$

where $\phi(\cdot)$ indicates the time-independent sequence length distribution (Redelings and Suchard 2007). We note that augmenting the alignment to include internal sequences also is beneficial, because it specifies exactly on which branch of the tree each indel occurs, making it possible to model indel events instead of modeling gaps (Figure 10.1).

Inference under a Statistical Model

Three methods for joint estimation of alignment and phylogeny have been proposed to date, and we compare and contrast these methods here. Redelings and Suchard (2005, RS05) and Lunter et al. (2005, LMDJH05) draw inference in a Bayesian framework using Markov chain Monte Carlo (MCMC) integration, whereas Fleißner et al. (2005, FMH05) employ a maximization approach using simulated annealing. Both MCMC and simulated annealing are guaranteed to compute exact solutions given sufficient processing time. However, in practice, both approaches sacrifice some degree of precision to avoid considering all possible combinations of alignments, trees, and other parameters, which would be computationally prohibitive.

MCMC differs from the simulated annealing approach of FMH05 in two major ways. First, MCMC naturally leads to measures of uncertainty, such as posterior confidence intervals for continuous parameters and posterior probabilities for tree topologies, alignments, and other discrete parameters. As a maximization approach, the FMH05 method instead generates a single parameter estimate, but it does not produce a measure of confidence in the estimate that would reflect the degree of evidence for it in the data. However, the maximization approach may be significantly faster, since it only needs to find the optimum tree and alignment, instead of visiting a large number of other points to yield measures of confidence.

Second, the Bayesian approach involves summing over all possible alignments to evaluate the relative probability of evolutionary tree topologies and other parameters. In contrast, the FMH05 approach does not sum over all possible alignments, but instead it maximizes over them. This approach differs from the Bayesian approach and from the standard maximum likelihood procedure, which involves maximizing the likelihood $P(Y \mid \tau, T, \Theta, \Lambda)$ of the observed data. Computation of this likelihood naturally involves a sum over all alignments (equation 13), since

$$P(Y \mid \tau, T, \Theta, \Lambda) = \sum_A P(Y, A \mid \tau, T, \Theta, \Lambda). \qquad (13)$$

The FMH05 approach avoids computing or approximating this sum and instead maximizes the likelihood $P(Y, A \mid \tau, T, \Theta, \Lambda)$ of the observed data and an unobserved alignment. Therefore, the approach maximizes over the alignment A even though it is a latent variable representing missing data and should instead be summed out, perhaps using the expectation-maximization (EM) algorithm. Therefore, the choice to maximize over the alignment may lead to biased estimates of branch lengths and other parameters, as mentioned above. However, because it can lead to faster computation, it may allow analysis of larger data sets.

Computational Efficiency: MCMC and Simulated Annealing To be computationally efficient, both MCMC and simulated annealing require well-designed "transition kernels" that propose new alignments and trees. Proposed new alignments and trees should frequently be of similar or higher probability than the previous value or they will be rejected, leading to wasted computation time. In addition, proposed values must sometimes be substantially different than current values. If this is not the case, then simulated annealing methods will move only slowly away from the starting point, or fail to explore the full parameter space, and thus fail to find the optimum point in a short time. Similarly, MCMC proposal functions that fail to propose substantially different points could lead to a failure to converge to the equilibrium distribution or a high autocorrelation between adjacent samples and a low effective sample size. MCMC samplers can be used as simulated annealing search algorithms by raising the posterior probability density to successively higher powers in each iteration. This approach is roughly followed by FMH05, with the exception that certain details required for correct sampling from the posterior distribution were ignored, because they were not necessary to find the optimum.

Bayesian sampling of multiple sequence alignments via MCMC was pioneered by Holmes and Bruno (2001, HB01). The HB01 approach samples the alignment under the TKF1 model given a fixed tree topology. HB01 introduced the idea of augmenting the alignment to include homology information about internal node sequences and introduced two novel transition kernels that require this augmentation to resample parts of the multiple sequence alignment. While the augmentation enables the use of the new HB01 transition kernels, it also makes it difficult to change the tree topology. This is because the augmented alignment makes sense only on a given tree topology, since it specifies a pairwise alignment for each branch. Therefore, when changing the tree topology, a new indel history must be proposed that is consistent with the new topology.

To relax the constraint of a fixed tree, each of the three joint-estimation methods takes a different approach that has unique advantages and disadvantages. The RS05 and FMH05 methods retain alignment augmentation, while the LMDJH05 approach dispenses with it. The LMDJH05 approach is able to dispense with alignment augmentation, because it uses an "indel peeling algorithm" to evaluate the probability of unaugmented alignments on a tree by summing over all possible augmentations (Lunter, Miklós, and Jensen 2003). The indel peeling algorithm is currently restricted to the TKF1 model; thus, the LMDJH05 approach assumes that all gaps have unit length. Summing out the augmentation is beneficial to the LMDJH05 approach, because it means that new tree topologies can be proposed without being constrained to be consistent with the current indel history. In addition, summing over missing data generally leads to decreased autocorrelation between adjacent MCMC samples and, thus, improves mixing efficiency compared to augmenting with missing data if the time per sample is not increased excessively by the summation.

However, the lack of alignment augmentation has the drawback that the HB01 transition kernels for alignment sampling are not available. The second of these transition kernels (HB01-TK2) simply updates the alignment augmentation; therefore, it is not needed in the LMDJH05 approach. However, the loss of the first of the transition kernels (HB01-TK1) is an important drawback. The HB01-TK1 transition kernel divides the alignment into two subalignments along a branch of the tree and re-aligns the two subalignments with respect to each other. This proposal is an efficient method for resampling alignments and is never rejected, because the proposal is proportional to the target distribution. However, this is achieved by resampling the pairwise alignment of two (possibly unobserved) sequences at either end of the tree branch. LMDJH05 is not able to re-align subalignments in this way, because it does not store alignment information for internal sequences, and it cannot sample this information from the correct distribution to use temporarily. Thus, the LMDJH05 approach is instead forced to propose alternative alignments, which may be rejected. Because the probability of rejection grows with sequence length, only blocks of the alignment of randomly chosen lengths have their alignment resampled in this fashion.

In contrast, the RS05 approach extends the HB01 method to sample topologies without removing the alignment augmentation for internal node sequences. By summing out only the alignment of sequences at internal nodes that lose definition during a nearest-neighbor interchange (NNI), the RS05 approach is able to compare nearby topologies. The RS07 method extends this approach to subtree-prune-and-regraft (SPR) changes to tree topology as well. Because these summations are carried out using dynamic programming, the alignment information for internal node sequences that is summed out to change topologies can be resampled to be compatible with the topology that is chosen, thus reconstituting the alignment augmentation. In summary, traditional transition kernels for topologies, such as NNI and SPR, are modified in the RS05 approach to resample the indel history as well as the topology to maintain compatibility between the two. Unfortunately, this kind of procedure may be too computationally inefficient to use for larger topology changes, such as those induced by tree-bisection-and-reconnection (TBR). Thus, the RS05 approach is constrained in the topologies that it can propose compared to the LMDJH05 approach.

RS05 introduces two novel alignment transition kernels that can substantially improve mixing efficiency. The first one (RS05-TK1) resamples alignment information about unobserved sequences at two adjacent internal nodes, which keeping fixed implies the alignment between all other sequences. This transition kernel is helpful for efficient mixing of MCMC and is required for summing out local alignment augmentation to allow NNI proposals. The second of these novel alignment transition kernels (RS05-TK2) divides the alignment into two subalignments along a branch of the tree and simultaneously resamples the alignment of the two subalignments and the alignment information about an internal node sequence at one end of the branch. This transition kernel helps to avoid improbable intermediate states and may improve mixing efficiency dramatically. For example, this transition kernel improved convergence speed by more than 70-fold in a simple 12-taxon example (Redelings and Suchard 2005). The RS05-TK2 transition kernel is used to sum and resample alignment augmentation for SPR topology changes. We also note that the RS05-TK2 transition kernel subsumes both HB01-TK1 and HB01-TK2 in the sense that it resamples the alignment with fewer constraints than either of HB01-TK1 or HB01-TK2.

Insertion–Deletion Models The three approaches to joint estimation of alignment and phylogeny make use of three different indel models. The FMH05 approach makes use of the TKF2 indel model. The TKF2 model has the benefit of allowing multiple residue indels, as well as the benefit of being a continuous-time stochastic process that specifies how indel events occur along a branch. We note that the TKF2 model makes the biologically unrealistic assumption that each sequence is composed of unbreakable fragments, which are inserted or deleted as a unit, thus disallowing overlapping insertions and deletions. The drawbacks of this approach can be removed largely by allowing the fragment boundaries to be different and independent on each branch of the tree, so that indels on different branches can indeed overlap. However, in the FMH05 approach, these fragments' boundaries are not allowed to differ across branches of the tree. This choice results in simpler and faster dynamic programming algorithms, but it leads to unrealistically low predictions for the probability of overlapping indels.

In contrast to the FMH05 approach, the other two approaches each give up one of the advantages of the TKF2 model. For example, the LMDJH05 approach uses the TKF1 model. Like the TKF2 model, the TKF1 model is a continuous-time stochastic process, and it describes the generation of individual indel events along each branch of the tree.

However, it allows only single-residue indels and, therefore, may fail to cluster gaps in its alignment estimates. In addition, by treating a deletion of several residues as several independent deletions of one residue, the TKF1 model may over-weight the phylogenetic evidence in shared indels of multiple residues.

The RS05 approach uses a model that allows indels of multiple residues, but it is based on an HMM instead of a continuous-time stochastic process. As a result, the RS05 model places a distribution directly on pairwise alignments and does not describe the dynamics by which indel events accumulate along a branch of the tree to yield a pairwise alignment between ancestor and descendant sequences. Additionally, in the RS05 model, the probability of an indel occurring on a branch is independent of the branch length. This drawback was remedied in Redelings and Suchard (2007, RS07), which introduced an indel rate parameter, instead of just specifying the probability that an indel occurs on a branch. As a result, the RS07 model has biologically interpretable parameters similar to a TKF2 model in which the insertion and deletion rates are equal and lead to a distribution on pairwise alignments that is approximately the same as the distribution generated by the TKF2 model for short times.

Improved Statistical Models of Insertion and Deletion

Statistical methods for inferring alignments or for inferring other parameters in the presence of alignment uncertainty all rely on stochastic models of the insertion–deletion process to correctly down-weight alignments with more indel events. To yield high-quality inferences, such models must be biologically realistic in specifying the frequency of occurrence for different kinds of indel events. Unfortunately, there is often a trade-off between biological realism and computational efficiency, so researchers must seek a balance between the quality of inference and speed of inference, instead of employing the most realistic models that are known.

Since the initial development of the TKF1 model, several extensions have been proposed. The TKF1 model has just two additional parameters: the insertion rate λ and the deletion rate μ. The TKF1 model assumes an exponential distribution of sequence lengths at equilibrium with mean $\lambda/(\mu - \lambda)$. Only single residues can be inserted or deleted, leading to "linear gap penalties." To remedy this problem, the TKF2 model applies the TKF1 model to unbreakable sequence "fragments" containing multiple residues. The number of residues in a fragment is random and is distributed according to a geometric distribution with parameter ε and mean $1/(1 - \varepsilon)$. This leads to "affine gap penalties." Thus, insertion and deletion of fragments lead to multiple-residue insertions and deletions. The downsides of this approach are that (a) imaginary fragment boundaries remain after fragments are inserted, (b) it is impossible to delete parts of a fragment, and (c) indel rates are now per fragment instead of per nucleotide. Lastly, another difficulty that is less important, but still real, is that (d) probably neither indel mutations nor accepted indels have a geometric length distribution (Cartwright 2006). A geometric length distribution is significantly easier to deal with computationally, but it tends to underestimate the probability of seeing long indels. Therefore, there have been several attempts to improve on these models. The models are used at equilibrium so that the total forward and backward rates are equal. Length distribution is geometric (TKF1) or roughly geometric (TKF2). In both the TKF1 and TKF2 models, pairwise alignments can be estimated in $O(NM)$ time for two sequences of length N and M. One approach that avoids some problems with the TKF2 models is simply to assume that fragment boundaries can be different on each branch of the evolutionary tree. This allows a fragment that is inserted on one branch to be partially deleted on a different branch of the tree. This technique is attractive, because it does not substantially increase the computational burden. However, other, more accurate models have been proposed as well.

Miklós et al. (2004) introduce a new model that improves on the TKF1 model by allowing insertions and deletions of multiple residues without relying on unobserved fragments as the TKF2 does. Furthermore, this model allows an arbitrary length distribution for indels instead of requiring a geometric length distribution and is, therefore, called the "long indel model." However, conducting inference under this model is significantly more computationally burdensome than the TKF1 or TKF2 models. Firstly, the dynamic programming algorithm for this model is $O(M^2N^2)$ when no approximations are made, instead of $O(MN)$. Secondly, the dynamic programming recursion involves terms whose value cannot currently be calculated analytically. Instead, these terms are estimated using MCMC methods to approximately calculate "trajectory likelihoods," where all intermediate events on a branch must be explicitly considered. Additionally, certain reasonable assumptions about the number of overlapping indel events were made to ensure computational tractability. While this long indel model does yield improved estimates of pairwise alignments, it may be too computationally expensive to incorporate in multiple sequence alignment estimation. Despite this fact, the long indel model has been used to improve pairwise alignment accuracy.

One statistical issue that is not addressed by current models is the sequence length distribution. Most statistical models assume that over evolutionary time the equilibrium distribution on sequence length will converge to a geometric distribution whose mean is the same for all genes. The use of a geometric distribution implies that a shorter sequence length is always more likely than a longer sequence length. Although this assumption is mathematically convenient, it seems more biologically realistic that each protein family would have a separate mean length over evolutionary time and that the most likely length would not be zero. When analyzing a collection of homologous genes in a Bayesian framework, it would be desirable to use an informative prior on the mean sequence length that is based on an empirical distribution of protein family lengths in curated databases. However, the degree of variation around this mean could be estimated from the homologous genes themselves. Unfortunately, the failure of current models to separate the distribution of lengths within and between protein families prevents informative priors from being used.

We note that equilibrium sequence length distributions are determined entirely by mutation pressure and, thus, imply completely neutral evolution on sequence lengths. This unrealistic assumption could be replaced in two separate ways. First, it is possible to add to an indel model selection for sequence length *per se*. Second, it is possible to consider that some "conserved" residues are much less tolerant of deletion over evolutionary time and that the number of such residues within a protein remains roughly constant within the protein family. This second option would tend to counteract the assumption within current indel models that over a long evolutionary time period, all ancestral residues should be deleted and replaced with newly inserted residues. It would also be able to model heterogeneity in the indel process and handle "indel hot-spots" similar to site-heterogeneity models for substitution rates. As an extreme example of this type, one could consider a model in which some unknown fraction of residues may never be deleted, similar to the invariant sites assumption for substitution models (Thorne et al. 1992).

Limitations of Statistical Methods

Although statistical techniques are now available for making robust inferences in the presence of alignment ambiguity, these methods have important limitations. First, these techniques may require vast amounts of computer processing time, thereby severely limiting the size of data sets that can be analyzed in practice. For example, an MCMC analysis under the model described in equation (9) of 12 protein sequences with lengths of about 450 amino acids required seven days in 2005 (Redelings and Suchard 2005). While such analyses are feasible, they indicate that current approaches to joint estimation of alignment and phylogeny are not able to cope with either long sequences or a large number of taxa. In the current manuscript, we analyze a 25-taxon data-set, but we note that the length of the sequences is extremely short at about 130 nucleotides, indicating a trade-off between sequence length and number of taxa.

Second, the statistical models of indel formation are limited in the types of biological processes that they consider. For example, current models assume that insertion and deletion rates are independent of the DNA or amino-acid sequence where they occur and are distributed uniformly across the DNA or protein molecule. In addition, current models do not allow duplication or within-sequence homology, but they assume that inserted sequences are random and independent of the parent sequence. These assumptions can lead to inaccurate estimates of phylogenetic trees and other parameters when the sequence data have evolved according to indel processes that are not present in the model. For example, in phylogeny estimation, indel events are all assigned the same weight, regardless of whether or not they occur in an indel hotspot. Expansion and contraction of simple sequence repeats may be more rapid than other types of indels because of slipped-strand mispairing, but these changes will not be appropriately down-weighted.

REPRESENTING ALIGNMENT AMBIGUITY

Alignment ambiguity is more difficult to represent than ambiguity in continuous parameters, such as branch lengths or indel rates. This follows from the fact that alignments are discrete and unordered, so that one cannot summarize their distribution by a mean and variance or by a median and a confidence interval of the form (a, b). Alignment ambiguity must be represented either for visual display or for use as input to an inference procedure. When representing alignment ambiguity visually, it is often more important that the alignment distribution be summarized in a single figure than that all the information in the distribution is preserved.

Alignment Ambiguity for Use in Further Analysis

The simplest approach is to divide an alignment into certain and uncertain regions. This has the downside of assuming that homology is known with complete confidence, which is unlikely, or else not known at all. That is, there are no degrees of confidence in homology statements. We note that the issue of confidence is theoretically separate from how the confidence measure is calculated. Even the best method for assessing confidence will be of marginal usefulness if it is limited to declaring alignment regions as "certain" or "uncertain." However, the results of this method are easy to visualize. This representation of alignment ambiguity is used by the culling method and the GBLOCKS approach.

The next method of representing alignment confidence is by a list of alignments of equal weight, as in the elision method. For analyses that treat columns independently, columns that occur in many alignments effectively have a higher weight than columns that do not. If alignments themselves can be repeated multiple times, then we can replace the list by a set of unique alignments associated with integer weights that indicate the number of times these alignments occur in the previous list. This list may then indicate a collection of plausible alignments, along with their weights. However, we note that the elision method explicitly concatenates the alignment end-to-end and that the weights lead to a weighted sum of parsimony scores for each tree. We note that a list of alignments is difficult to visualize.

Finally, a general framework for dealing with certainty and uncertainty about alignments is to consider a probability distribution on alignments. This distribution may be represented by a sample, and this sample will most likely be unweighted, since the number of possible alignments makes it unlikely that a single alignment will recur. However, the sample conceptually represents a weighted collection of alignments, as opposed to an unweighted collection. This conception naturally arises in the Bayesian statistical framework; in which case, the posterior alignment distribution must be represented.

Graphically Representing Alignment Ambiguity

Pairwise Alignments A probability distribution on alignments between two sequences can be summarized on a 2-dimensional sheet of paper or computer screen using a path graph representation (Naor and Brutlag 1994). Given two sequences of length M and N, each pairwise alignment corresponds to a path through an integer lattice from (0, 0) to (M, N), such that diagonal edges represent match columns and vertical or horizontal segments represent gap columns. For example, the edge ($i - 1$, $j - 1$) → (i, j) indicates that character i of the first sequence and character j of the second sequence are homologous. Such paths are known as path graphs and correspond to a route through the dynamic programming matrix.

One method for using path graphs to indicate alignment ambiguity is to color each possible edge according to the probability that it occurs in the alignment (Figure 10.2). For example, edges can be colored black if they are certain to occur, white if they are certain not to occur, and shades of gray for intermediate degrees of certainty. We note that representation of the alignment distribution for two sequences may not capture the full information in the distribution, because it does not represent correlations in support of adjacent columns. Additionally, when an insertion is adjacent to a deletion, multiple paths may correspond to the same homology structure, leading to trouble with path graph interpretation. However, drawing weighted path graphs in this manner does successfully convey which regions of the alignment contain plausible alternatives, and it also conveys how plausible these alternatives are.

Although this method could in theory be used to represent alignment distributions of n sequences where $n > 2$, such representations are impractical, because they would require an n-dimensional plotting surface. Even a 3-dimensional version of this technique is not practical, because the most probable edges may be hidden behind less probable edges from every viewing angle. Thus, 2-dimensional projections of a 3-dimensional image are not sufficient. Another alternative for using path graphs to represent a distribution on multiple alignments is to draw all projected pairwise alignments separately. This technique can be useful when n is small, although information about the correlation between sequences is lost. However, the number of such graphs grows quadratically with n; therefore, it quickly becomes unmanageable for large n.

Selecting a Representative Multiple Sequence Alignment When summarizing multiple sequence alignment distributions, one common approach is to construct a representative alignment and then annotate it to indicate the degree of confidence in various regions. Selection of a representative

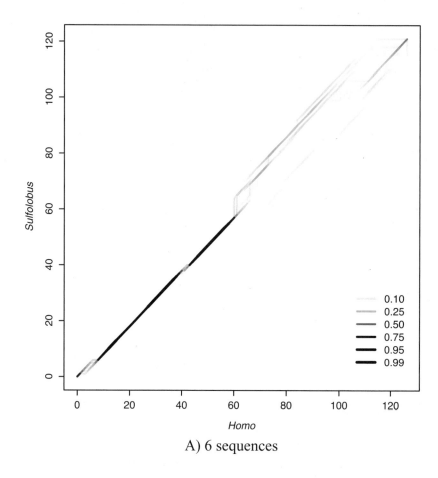

A) 6 sequences

Figure 10.2. Weighted path graph representation allows easy comparison of two pairwise alignment distributions. While only the alignments between the *Homo* and *Sulfolobus* sequences are shown, the full analyses are based on (A) 6 sequences and (B) 25 sequences, respectively. The pairwise alignment distribution based on 25 sequences contains substantially less alignment ambiguity, indicating that alignment ambiguity between a pair of sequences may decrease when additional sequences are included in the analysis. Line segments with high posterior probability are darker and thicker, while edges with lower posterior probability are colored with lighter shades of grey. The posterior probabilities in this figure are based on the 25-taxon 5S rRNA data set (B) described in the Results section and a 6-sequence subset of those sequences (A). The RS07 indel model was used, along with the GTR+gwF+log-Normal$_8$ model.

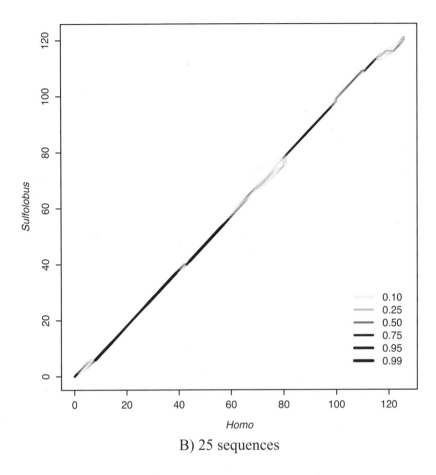

B) 25 sequences

Figure 10.2. *(continued)*

multiple sequence alignment is itself a challenging task. One criterion would be to select the multiple sequence alignment with the highest posterior probability. However, this criterion has two drawbacks: one theoretical and one practical. First, the most probable alignment may not be very probable, and individual columns may have low posterior probability. Second, it is common for each alignment sampled via MCMC to be unique. This indicates that the most probable alignment may not have been sampled and that it is not possible to determine the posterior probability of the alignments that have been sampled. One simple way around this problem is to select the multiple alignments from the sampled point with the higher posterior probability (Redelings and Suchard 2005).

Although such estimates are always at hand in an MCMC approach and give usable results, they are *ad hoc*, fail to marginalize properly, and are rarely repeatable. One way around these difficulties is to select a representative alignment by maximizing the sum of column probabilities or "posterior decoding" (Durbin et al. 1998; Lunter et al. 2005). This approach is practical, because each column often occurs enough that its posterior probability can be estimated from MCMC samples; however, a full alignment does not occur sufficiently often. However, it is not clear how such an alignment should be found if the researcher desires to look beyond the MCMC samples that have been generated to find the maximum. Dynamic programming can be used to find a maximum posterior decoding alignment for two or perhaps three sequences, but the speed and memory requirements grow exponentially in the number of sequences in a multiple sequences alignment, making such approaches impractical.

Annotating a Representative MSA with Column Probabilities Once a representative alignment is found, this alignment must be annotated to indicate the degree of confidence in various regions. For maximum posterior decoding alignments, a natural measure of confidence is the posterior probability of each column, which can be plotted above the chosen alignment. This method clearly indicates when a column is strongly aligned. However, when a column has a low posterior probability, there are a number of possible reasons. A low column probability could result, because two subgroups of characters are weakly aligned to each other, but strongly aligned within groups. Alternatively, a low column probability could result, because all characters are weakly aligned to each other. Thus, the use of only one value per column is not capable of capturing the confidence in partial columns. We also note that, if one sequence such as an outgroup is weakly aligned to all other sequences, then both alignment selection by posterior decoding and alignment annotation using column probabilities may be undermined.

Alignment Uncertainty (Au) Plots Alignment uncertainty (Au) plots offer an alternative annotation method to the use of column probabilities (Redelings and Suchard 2005). The underlying idea behind Au plots is to annotate each letter separately by shading or coloring it to indicate the posterior probability that it is placed in its correct column. Thus, when annotating a multiple sequence alignment with n sequences, an Au plot may use n values instead of just the one value of column probability.

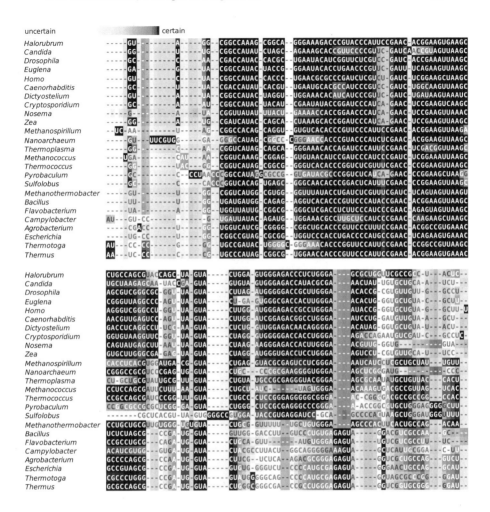

Figure 10.3. Au plot showing alignment uncertainty for 5S rRNA sequences. Each residue is shaded to indicate the probability that it is correctly aligned with other residues in its column. The probability distribution that is summarized in this figure is the posterior distribution on multiple sequences alignments that is calculated in section 10.7.

We note that Au plots do not depict exact probabilities. Instead, they approximate the posterior probabilities for the $n \times (n-1)/2$ pairs of characters in each column using a tree structure with only $2n-3$ branch lengths. Each character is then annotated with the probability that it aligns with a hypothetical character at the root of the tree and is shaded accordingly (Figure 10.3). This procedure can easily depict confidence in a partial column even when a few characters in the column are weakly aligned. Thus, when the exact position of a gap within a sequence is ambiguous, adjacent characters in the same sequence can be lightly shaded to indicate ambiguity.

Multidimensional Scaling Multidimensional scaling offers an alternative to Au plots and provides a starting point from which to assess convergence in Bayesian samplers that explore alignment uncertainty. Multidimensional scaling is a statistical technique commonly used for high-dimensional data visualization (Borg and Groenen 1997; Cox and Cox 2001; Young and Hamer 1987). Multidimensional scaling algorithms start with a sample-to-sample distance (or similarity) matrix $\mathbf{D} = \{D_{ij}\}$ and assign a location $\vec{x}_i$ in a low-dimensional, visualizable space to each sample. Optimal assignments proceed via minimizing a stress function, such as the Kruskal-1 function. Hillis et al. (2005) recommend multidimensional scaling to explore phylogeny distributions. Setting $D_{ij} = m(\mathbf{A}^{(i)}, \mathbf{A}^{(j)})$, where $m(\cdot,\cdot)$ is an arbitrary distance metric between two multiple sequence alignment samples $\mathbf{A}^{(i)}$ and $\mathbf{A}^{(j)}$, allows for multidimensional scaling projections of multiple sequence alignment distributions. In low-dimensional spaces, visual comparisons can assess differences between distributions. This is useful for assessing convergence and interactive displays.

Distance metrics for multiple sequence alignments remain underdeveloped (see Chapter 9). However, metrics on pairwise sequence alignments are well studied. Schwartz et al. (2005) provide a biologically interpretable metric for pairwise alignments. The metric counts the number of homology statements on which two pairwise alignments disagree. Schwartz et al. (2005) also suggest how to construct a metric $m(\cdot,\cdot)$ on multiple sequence alignments as the sum of all possible pairwise alignment metrics; while this metric over counts homology disagreements because the metric ignores the evolutionary correlation between pairs, the metric warrants consideration as the underlying phylogeny is unknown.

Figure 10.4 demonstrates the use of multidimensional scaling to visualize convergence of alignment samples. The 100 samples depicted in the figure are drawn from a posterior simulation involving 12 taxa that span the Tree of Life. After approximately 80 steps, the alignment samples appear to reach convergence.

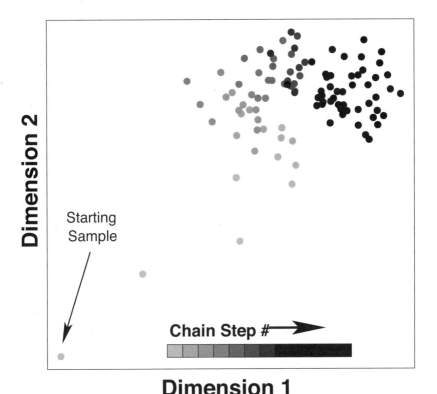

Figure 10.4. Alignment burn-in of a Bayesian sampler set visualized via multidimensional scaling. The data set consists of 12 taxa that span the Tree of Life. The data set contains sequences of EF-1α and EF-Tu (its bacterial homolog). Sequences are about 450 amino acids in length.

EXAMPLE: 5S RIBOSOMAL RNA

As an illustration of the above techniques, we examine a data set consisting of 25 5S ribosomal RNA gene sequences that displays substantial alignment uncertainty. The 5S ribosomal RNA is present throughout the Tree of Life except in a few basal eukaryotes, such as *Giardia*. Therefore, we included 5S sequences from organisms spanning the Tree of Life, including 7 Eubacteria, 9 Archaea, and 9 Eukaryotes. The 5S rRNA is relatively short, ranging from 111 nucleotides to 131 nucleotides in this data set. Although we have previously shown that alignment uncertainty is so high that there is little phylogenetic information in a data set of only 5 sequences (Redelings and Suchard 2005), we seek to show that this data set containing 25 sequences does indeed contain substantial phylogenetic information. However, this data set must be analyzed using joint estimation instead of sequential estimation to avoid conclusions with strongly supported phylogenetic errors.

Model and Priors

For the nucleotide substitution process, we employ the GTR+gwF+log-Normal$_8$ model. We also consider the simpler GTR+gwF model that assumes no rate heterogeneity between sites. The GTR+gwF model is a reversible CTMC model of nucleotide substitution. Nucleotide frequencies are specified by parameters $\pi = (\pi_A, \pi_T, \pi_G, \pi_C)$. This yields 3 degrees of freedom, because of the constraint that the sum of the frequencies must be 1. The GTR+gwF model also contains parameters $\psi = (\psi_{AT}, \psi_{AG}, \psi_{AC}, \psi_{TG}, \psi_{TC}, \psi_{GC})$ specifying the symmetric exchangeability of each pair of nucleotides independently from their frequencies. Because only the relative rates can be estimated from data, we follow common practice in scaling the exchangeabilities so that the mean substitution rate is 1. Because of this constraint, these 6 parameters yield only 5 degrees of freedom. Finally, the GTR+gwF model contains an additional parameter $f \in [0,1]$ that specifies whether common letters occur more frequently because they are more highly conserved (low f) or because they are more often proposed as replacements for other letters (high f). The +gwF formulation (Goldman and Whelan 2002) can be reduced to the more common +F formulation (Cao et al. 1994) by fixing $f = 1$. We model rate heterogeneity between sites using a log-Normal distribution that is approximated by a discrete distribution with 8 bins. We set the mean of the distribution to 1 and parameterize it by its standard deviation V, so that there is no heterogeneity when $V = 0$. Therefore, the substitution model is fully characterized by $\Theta = (\pi, \psi, f, V)$ with 10 degrees of freedom.

For the insertion–deletion process, we use the RS07 indel model from Redelings and Suchard (2007). This model contains two parameters: the insertion–deletion rate λ and the gap extension probability ε. Therefore, the insertion–deletion model is characterized by $\Lambda = (\lambda, \varepsilon)$.

We place a uniform Dirichlet prior on π. We place a nonuniform Dirichlet prior on ψ with weight 4 on transversions and weight 8 on transitions (Zwickl and Holder 2004). We place a Uniform(0,1) prior on f, and we place a Laplace prior on log V with scale 1 that is centered at -3. This leads to a prior median for V of about 0.05. We place a

Laplace prior on log λ that is centered at -5 with scale 1.5 and an Exponential(5) prior on the mean indel length minus one. For the prior on phylogenies, we place a uniform distribution on tree topologies and an Exponential(μ) prior on branch lengths, where μ is a hyperparameter. We then place an Exponential(1) prior on μ.

Results

Inference under the model was conducted using the MCMC sampling program BAli-Phy. We ran two chains from different starting positions and obtained 400,000 samples from each chain. This analysis took about 2 weeks to complete on a Pentium 4 processor. We discarded the first 40,000 samples from each chain as burn-in. We additionally analyzed the same data with the alignment held constant to either the ClustalW estimate or the MUSCLE estimate. For these latter analyses, we used the same burn-in period and the same number of samples. However, we did not make use of an indel model, thus ignoring the phylogenetic evidence of shared indels and considering only shared substitutions.

We report estimates of model parameters for both models in terms of the posterior median and a 95% Bayesian credible interval. As shown in Table 10.1, the log indel rate ln λ was estimated as -4.24 with rate heterogeneity and as -3.41 without. However, we note that because the branch lengths are defined in terms of substitutions, the indel rate here is defined relative to the substitution rate; thus, in this case, the difference in indel rate seems to indicate only that the substitution rate scale has changed. This is indicated by the fact that the total tree length $|T|$ is estimated as 16.7 with heterogeneity, but 7.57 without. Additionally, the mean branch length μ was estimated as 0.365 with rate heterogeneity and 0.165 without. Thus, $\lambda \cdot \mu$ remains roughly constant.

Bias and Alignment Uncertainty

The 5S rRNA data set exhibits alignment uncertainty in two major ways. First, the posterior alignment distribution is diffuse, placing similar support on a large number of distinct alignments. This is illustrated in the Au plot in Figure 10.3. Because the posterior topology distribution also is diffuse, it is possible that the diffuseness of the posterior alignment distribution results from the diffuseness of the posterior topology distribution. Therefore, we selected the MAP topology from the joint analysis with heterogeneity and reran the analysis with the topology fixed to this

TABLE 10.1. PARAMETER ESTIMATES FROM THE JOINT MODEL WITH AND WITHOUT RATE HETEROGENEITY

Parameter	Model			
	No Indel Model / Clustal	No Indel Model / MUSCLE	No Rate Variation	GTR+gwF+log-Normal$_8$
log λ	—	—	−3.41 (−3.77, −3.07)	−4.24 (−4.83, −3.74)
log ε	—	—	−0.716 (−1.00, −0.492)	−0.634 (−0.881, −0.437)
f	0.382 (0.0170, 0.956)	0.412 (0.0218, 0.962)	0.407 (0.181, 0.962)	0.250 (0.00994, 0.892)
μ	0.244 (0.172, 0.359)	0.285 (0.193, 0.434)	0.165 (0.124, 0.226)	0.365 (0.226, 0.650)
|T|	11.7 (9.13, 14.4)	12.9 (10.3, 17.9)	7.57 (6.90, 8.28)	16.7 (11.4, 27.8)
V	0.999 (0.750, 1.37)	1.26 (0.935, 1.83)	—	2.20 (1.44, 3.83)
# indels	—	—	58 (49, 70)	57 (50, 66)
|indels|	—	—	111 (93, 137)	118 (101, 145)
# subst	745 (737, 756)	755 (746, 768)	701 (681, 719)	711 (693, 729)
|A|	135	135	166 (155, 181)	172 (160, 194)

NOTE: The posterior median and a 95% Bayesian credible interval are reported for each parameter. Each column contains parameter estimates under a different model. The first and second columns refer to models in which indel information is not used, and the alignment is fixed to the ClustalW and MUSCLE alignment estimates, respectively. The third and fourth columns refer to models in which alignment is allowed to vary, and substitution rate heterogeneity is absent or present, respectively. Note the differences in alignment length and in the parsimony score when the alignment is allowed to vary. This may indicate that ClustalW and MUSCLE align residues too readily, resulting in shorter alignments with more mismatches. In addition to model parameters, we also report ranges for several properties of the tree and alignment. |T| represents the sum of branch lengths on the tree; #indels represents the number of insertions and deletions; |indels| represents the sum of the lengths of the indels; #subst represents the parsimony score for the alignment and tree; |A| represents the number of columns in the alignment.

MAP topology. However, fixing the topology increased the fraction of residues aligned at the 0.5, 0.75, 0.95, and 0.99 levels only slightly (data not shown). Therefore, we conclude that alignment uncertainty in the posterior distribution is not primarily a result of topological uncertainty.

Second, the alignment is ambiguous enough to be biased by the guide tree used during the progressive alignment procedure, so that the use of a single alignment estimate constructed using progressive alignment leads to substantial bias in phylogeny construction. For example, when the alignment is fixed to the ClustalW alignment estimate, the posterior probability (PP) that Eubacteria are monophyletic is only 0.283 with a posterior log odds (PLODs) score of −0.405. This is because the Eubacterium taxon *Campylobacter* is placed among the Archaea, in accordance with the ClustalW guide tree (Figure 10.6). In contrast, when the alignment is allowed to vary during phylogeny estimation, the posterior support for monophyly of the Eubacteria rises to 0.998, with a PLODs score of PLOD = 2.74. This indicates that the support for clustering *Campylobacter* with the Archaea is an artifact of the ClustalW and MUSCLE guide trees. This observation is further evidence for the view that sequential estimation does not yield robust inferences in the presence of alignment uncertainty.

The bias in phylogeny inference that results from conditioning on an alignment is not limited to the placement of the single taxon *Campylobacter*. In Figure 10.5, we compare the posterior topology distributions when the alignment is fixed to the ClustalW estimate and when it is allowed to vary. Splits that are present in the ClustalW guide tree tend to have higher support when the alignment is constructed using the guide tree. Because some taxa can plausibly attach at several places on the tree under both distributions, not many full splits are strongly supported (Figure 10.5a). The exception is that the monophyly of bacteria is strongly supported when the alignment is not fixed to the ClustalW estimate. However, many *partial* splits do show strong support (Figure 10.5b), and partial splits that occur in the guide tree tend to be strongly supported only when the fixed alignment is used that was created using the guide tree.

Rate Heterogeneity

Substitution rate heterogeneity was strongly supported by the data set and substantially decreased the amount of posterior alignment uncertainty. To determine whether rate heterogeneity was supported by the

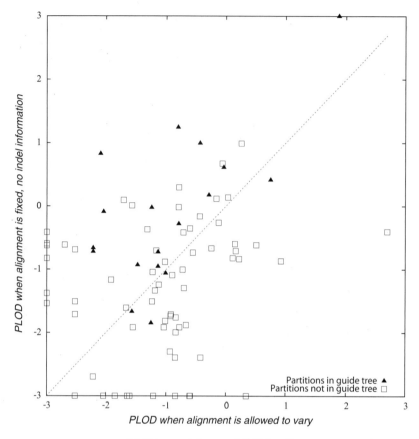

A) Bi-partitions of full leaf taxa

Figure 10.5. Fixed alignment leads to bias toward guide tree. The data set consists of 5S ribosomal RNA sequences from 25 taxa across the Tree of Life. In both panels (A) and (B), each point represents the support for a bipartition when near-optimal alignments are considered using joint Bayesian estimation of phylogeny and alignments (x-axis) and when the alignment is fixed and indel information is ignored (y-axis). Support is indicated by the posterior $\log_{10}$ odds (PLOD). Filled triangles represent bipartitions that are implied by the guide tree used by ClustalW to estimate the fixed alignment; open squares represent bipartitions that contradict this guide tree. Note that triangles tend to fall above the line $y = x$, while open squares tend to fall below it. This indicates that under the fixed alignment, partitions have increased support if they are in the guide tree and decreased support otherwise. This illustrates bias toward the guide tree of a progressive alignment estimate when the alignment is fixed. Points in panel (A) represent bipartitions of the leaf taxa, but points in panel (B) represent bipartitions of subsets of the leaf taxa. Because the posterior topology distribution contains many wandering taxa, bipartitions of the full leaf taxon set do not reveal the full amount of information in the distribution.

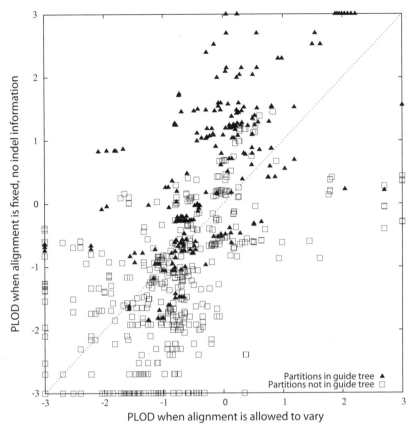

B) Bi-partitions of subsets of leaf taxa

Figure 10.5. *(continued)*

data, we estimated the marginal likelihood of the data set under the GTR+gwF and GTR+gwF+logNormal$_8$ models using the stabilized harmonic mean estimator (Newton and Raftery 1994; Suchard et al. 2003). The marginal likelihood estimates for these two models are -2829.2 ± 0.2 and $-2739:6 \pm 0.4$ on the log$_e$ scale. Placing equal prior weight on both models yields a log Bayes factor of 89.6 in favor of the model with rate heterogeneity. The strength of evidence here is surprisingly high, given that the longest sequence length is only 131 nucleotides. However, this can be explained by the extreme degree of rate heterogeneity, since V is estimated at 1.93 (1.28, 3.24) far from $V = 0$.

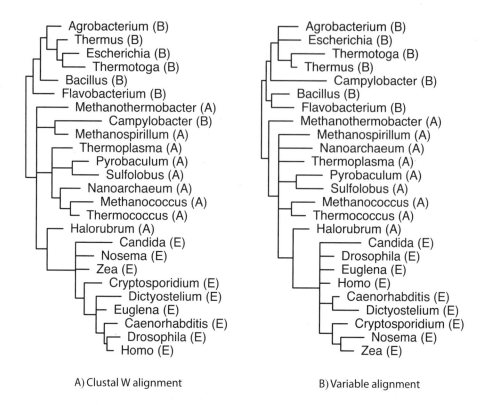

Figure 10.6. Majority *consensu*s trees summarize the posterior phylogeny distribution for the 5S rRNA data set. Partitions with posterior probability greater than 0.5 are displayed in each tree, and branch lengths are posterior mean branch lengths conditional on the topology shown. (A) The tree on the left summarizes the posterior distribution when the alignment is fixed to the ClustalW alignment estimate. (B) The tree on the right summarizes the posterior distribution under the RS07 model when the alignment is allowed to vary. Note that the taxon *Campylobacter* is incorrectly placed among the Archaea when the ClustalW is used, in accordance with the ClustalW guide tree. Each taxon is labeled B, A, or E to indicate the bacteria, archaea, or eukaryotes, respectively.

Including rate heterogeneity in the substitution model substantially decreases the degree of alignment ambiguity in the posterior distribution. This is indicated in comparison of the Au plots of the two models (not shown) and in Figure 10.7. Figure 10.7 shows that the fraction of aligned residues in each pair of sequences tends to be higher under the GTR+gwF+log-normal$_8$ model than under the GTR+gwF model with no rate variation. This can be explained by

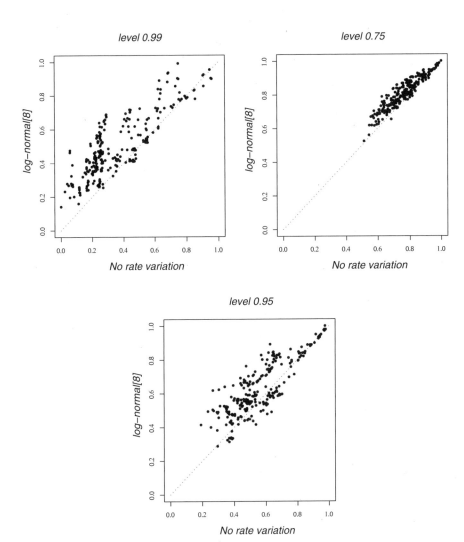

Figure 10.7. Modeling site-to-site variation in substitution rate leads to decreased alignment ambiguity. Each point represents a pair of sequences. Both axes represent the fraction of residues in the shorter sequence that are aligned to a gap or specific residue in the other sequence with a posterior probability greater than the cutoff value. The horizontal axis represents the aligned fraction under the GTR+gwF model, and the vertical axis represents the aligned fraction under the GTR+gwF+log-normal$_8$ model. In each plot, the aligned fraction is substantially greater under the model with rate heterogeneity; this trend increases as the threshold level of posterior probability increases. Additionally, the separate plots illustrate the fact that as the threshold increases, fewer residues in each pair are aligned with the required posterior probability.

the observation that the first half of the 5S rRNA sequence exhibits greater conservation than the second half. By allowing different substitution rates in each site, the GTR+gwF+logNormal$_8$ can improve the likelihood by increasing the observed substitutions in this region. This explanation is within the increased posterior median parsimony score (714 versus 702) under the model with rate heterogeneity. Since Table 10.1 indicates that the substitution rate does indeed vary across the alignment, this trade-off should be biologically realistic and lead to more accurate alignments and trees.

DISCUSSION

In this chapter, we described how alignment ambiguity can undermine bioinformatic inference methods based on sequential estimation, preventing the robust inference of phylogenies and other evolutionary parameters from collections of distantly related sequences. We then described bioinformatic inference methods that make use of homology information in multiple sequence alignments and yet are robust to alignment uncertainty. In doing so, we emphasized that to yield robust inferences and accurate measures of confidence, an inference method must take into account both of the two sources of alignment uncertainty: parameter uncertainty and near-optimal alignments. We also emphasized the importance of jointly estimating the alignment and any other parameters that are mutually dependent on the alignment. When the alignment is ambiguous, joint estimation is necessary to prevent circular reasoning. Bayesian inference methods that fulfill these conditions have long been available for pairwise alignments, but they have only recently become feasible for multiple alignments. In addition to robust inference, these Bayesian methods make it possible to assess alignment uncertainty in an objective and repeatable fashion and can safely make use of information in ambiguous regions of the alignment. Bayesian methods naturally asses fine gradations in support of homologies and allow the characterization of support for homology statements that range in size from an entire alignment to a single pair of residues.

Speed and Sequential Estimation

Despite the many benefits of the joint estimation approach, consideration of near-optimal alignments during estimation of other parameters comes at the cost of increased run time when the data set contains multiple sequences. For example, when the software BAli-Phy is used, considering near-optimal alignments increases the computation time for each iteration by about a factor of 2.2 over a standard Bayesian analysis for the foregoing example, where the sequence lengths are about 120 letters. For longer sequences of about 500 letters, joint estimation is about 10 times slower than the use of a fixed alignment per iteration. In addition, more iterations are often required when the alignment is allowed to vary. Therefore, the benefits of increased power through the use of indel information and the use of substitution information in ambiguous regions must be balanced against the limited size of data sets that currently can be analyzed in the joint estimation paradigm. Because of the computational cost, we predict that sequential estimation based on the use of a single censored alignment will continue to be used commonly by biologists for at least a decade, whether this is advisable or not. Of course, when a rough estimate of the alignment is required for the purpose of visualization, speed will always be the predominant concern. In these cases, progressive alignment and other heuristic techniques can be preferred indefinitely. However, it is still important in such cases to indicate when the alignment should be trusted and when it may be unreliable. Therefore, it is worth considering how much the speed of joint estimation can be increased and what improvements in robustness can be made for methods that attempt to handle alignment uncertainty through censoring.

Speeding Up Joint Estimation of Alignment and Other Parameters Joint Bayesian estimation requires additional computation time compared to traditional Bayesian analysis for two primary reasons: first, because the alignment must continually be resampled and, second, because the Markov chain requires more iterations to give equally precise estimates. The time required for resampling using DP is linear in the number of genes used, but it is quadratic in the length of each gene. Therefore, just as in other applications of DP, the cost of resampling the alignment can be decreased by ignoring the corners of the DP matrix or by constraining the alignment to contain specific homologies that are considered certain. The second cause for slowness can be improved by new transition kernels that increase mixing efficiency, but it is difficult to characterize how much mixing efficiency is decreased by allowing the alignment to vary.

A number of factors can combine to make joint Bayesian estimation quicker in several common cases. First, note that if a rough estimate of the alignment without measures of confidence is all that is needed, then the MCMC analysis may require many fewer iterations, since the analysis can be stopped when burn-in is reached. For example, we estimate that the analysis example in this chapter would require only about an hour or two to produce an estimate, instead of about 2 weeks to produce detailed measures of confidence as well. This amount of time can be further decreased by using a starting tree and alignment that are computed from a quicker method. However, one problem with stopping directly after burn-in is complete is that it is difficult to know the burn-in time without running a longer analysis.

Secondly, in many cases, the branching order of the taxa in the data set is already known. In such cases, the use of a fixed topology should greatly decrease the number of iterations required for convergence of the Markov chain, as well as decreasing the number of samples that need to be collected after convergence. Furthermore, in many such cases, each gene can be analyzed independently to infer substitution rates, insertion and deletion rates, the degree of positive selection, or other parameters. In this case, a large computing cluster should enable whole-genome analysis of coding regions.

Improvements to Traditional Alignment Methods Several new consistency-based methods for multiple sequence alignment hold out the possibility of improving the accuracy of multiple sequence alignments when evolutionary parameters are not known in advance. For example, the ProbCons (Do et al. 2005) approach mitigates the bias induced by the use of a guide tree through iterative refinement and the 3-way consistency transformation, while sequence annealing as implemented in AMAP (Schwartz and Pachter 2007) does not rely on progressive alignment and does not need a guide tree. While these methods can be slower than ClustalW, they are still quite fast.

Many of these methods additionally compute some measure of reliability for homologies in a multiple alignment. For example, ProbCons provides a column reliability score based on posterior probabilities, as do several other consistency-based methods. The program AMAP takes a different approach and provides a series of alignments at various levels of specificity. However, this specificity applies only to homologies and not to gaps. Thus, AMAP leaves more residues unaligned, which requires increased reliability.

Testing Alignment Reliability Measures

Alignment estimation methods are often ranked by the number of correctly predicted homologous pairs or columns as measured against curated alignment databases, such as BAliBase (Thompson et al. 1999a). However, even if an alignment method correctly predicts 40% of aligned residues, this degree of accuracy may not be very useful unless the program also indicates which residues in the alignment estimate are correct aligned. In addition to statistical alignment programs, which indicate confidence in alignment regions using posterior probabilities, a number of other programs—such as T-Coffee (Notredame et al. 2000), ProbCons, and AMAP—also provide estimates of reliability. Thus, to improve or rank these approaches, improved alignment benchmarks that compare sensitivity at each level of specificity are needed.

Verifying Bioinformatic Inferences by Simulation

When developing new bioinformatic inference methods, researchers are hampered by the fact that the accuracy of such methods is often difficult to check. This is because phylogenies and many other evolutionary parameters cannot be directly measured. One way around this problem is to simulate a collection of sequences conditional on a specific set of evolutionary parameters so that the true values for the parameters are known (Ogden and Rosenberg 2006). However, this approach does have a few difficulties. First, it is important that conclusions and generalizations about the accuracy of a method as a whole are carefully based on results for a large collection of different parameter settings. Second, inference methods that are based on a specific evolutionary model often perform well on data sets simulated from that model, but they may perform worse on real data if the model is inaccurate. Inaccuracies on real data often stem from the fact that inference under a model that captures all relevant biological phenomena is computationally prohibitive. For example, all current alignment algorithms fail to account for slipped-strand mispairing that causes insertions and deletions of tandem repeats.

However, it is often computationally feasible to simulate under biologically realistic models that contain complex phenomena, even when it is not feasible to use such models for inference. By using data simulated from biologically realistic models, it should be possible to conduct more tests of bioinformatic inference methods whose results can be extrapolated to real data with greater confidence. In addition, it would be possible to ascertain which aspects of the more complex simulation model must be included in the inference model to improve accuracy. Thus, we believe that the development of more complex and realistic models of sequence evolution would well repay any effort that is put into developing them. Models for simulation would ideally include both substitution and indel rate heterogeneity, and this heterogeneity would have biologically realistic spatial patterns, perhaps based on structures of known proteins.

Acknowledgments

The authors wish to thank Jeff Thorne and Eric Stone for helpful comments and stimulating discussions. MAS is an Alfred P. Sloan Research Fellow and a John Simon Guggenheim Memorial Foundation Fellow. This research was supported by NIH R01 GM086887, NIH GM070806, and NSF DEB-044180.

NOTES

1. The flip side of ambiguity is the strength of evidence. Ambiguity applies to a variable (e.g., the alignment) that has more than one plausible value. Strength of evidence applies not to a variable, but to a hypothesis, such as the hypothesis that the variable takes on a specific value.

2. This observation has practical consequences for sensitivity analysis, which attempts to gauge alignment ambiguity by using a range of parameter values for uncertain parameters. Despite the fact that the topology is unknown, it seems uncommon for researchers to include a range of values for the guide tree when progressive alignment is used.

3. We note that alignments also may be sensitive to which taxa are used in the analysis. This kind of sensitivity checking is oriented toward finding results that are robust to internal inconsistencies of the method and is not discussed here.

CHAPTER 11

Strategies for Efficient Exploitation of the Informational Content of Protein Multiple Alignments

ANNE FRIEDRICH
OLIVIER POCH
LUC MOULINIER
Institut de Génétique et de Biologie Moléculaire et Cellulaire, France

MSA: Central Role in Protein Analysis 273
    Evolutionary Studies. 274
    Functional Assignments 275
    Structural Studies 277
    Mutation Studies 278
    Interaction Networks 279
    Comparative Genomics. 280
Efficient Construction of MSA 282
    Detection of Similar Sequences 283
    Alignment Software 283
    MSA Quality 286
    Sequence Selection for More Accurate MSA 287
MSA Integrative Exploitation Tools 290
Conclusions and Perspectives 294

Life can now be considered as a complex system in which molecular agents are interconnected in space and time. The information for life is mainly stored and organized by stretched chains of chemical building blocks: four nucleotides for genes and, after gene translation, 20 amino acids to form proteins. Despite the recent developments highlighting the existence of codes and higher orders of organization at the genome level (Kepes 2003; Segal et al. 2006), proteins still represent the major mediator for life information management. Various levels of organization can be considered. First, the primary structure describes the number and arrangement of amino acids characterizing a particular protein. The secondary structure then represents the assemblage of the polypeptide backbone into local regions of alpha-helices, beta-sheets, coils, and turns. The tertiary structure refers to the entire three-dimensional (3D) structure of the protein, and finally the quaternary structure describes interactions between separate polypeptide chains, called subunits, arranged in protein complexes. Furthermore, proteins and macromolecular complexes are interconnected in specialized cellular pathways or processes to perform most of the cellular functions and events. These processes are strongly regulated at the gene, protein, and macromolecular levels by expression, regulation, or chemical modification controls which are highly dependent on a number of different factors, such as the cellular localization, tissue type, developmental stage, or environment.

These interwoven informational levels are clearly essential for the understanding of the functions of a gene or protein system. The widespread use of high-throughput technologies, the so-called "omics" (e.g., structural genomics, transcriptomics, interactomics . . .) has not only increased the total number of DNA and protein sequences available but has also given us access to many information levels that were not accessible before.

In this context, sequence comparisons facilitate the identification of both major and minor genetic events (e.g., adaptation; selection; speciation; genetic drift; flows, duplications, mutations, or recombination of gene and genome) and play a central role in biological studies by introducing the evolutionary dimension to the analysis of the complex gene and protein relationships (Lecompte et al. 2001). Sequence comparison is therefore one of the most common tasks in bioinformatics, increasing our understanding of the relationships between sequence/structure/function and evolution. The purpose of multiple sequence comparisons, or alignments, is to elucidate the complex relationships that exist within a set of related sequences, in terms of conserved (evolutionarily fixed) residues, substitutions, or insertion or deletion events.

Sequence alignments can be classified in two main categories: pairwise alignments, that is, the alignment of two sequences, and multiple alignments, that is, from three to thousands of sequences. For both categories, the sequences can be aligned across their entire length (global alignment) or only across the most similar segments of the sequences (local alignment). Pairwise alignments are commonly used in database search programs (Altschul et al. 1990; Pearson and Lipman 1988) in order to detect sequences that have evolved from the same ancestor, that is, homologous sequences, and multiple sequence alignments (MSAs) are mainly used to study the effect of evolution across the sequence family and have a wide range of applications.

Here, we will concentrate on the MSA of complete proteins, which represents an ideal workbench to study all of the information related to a set of homologous sequences. Indeed, placing the sequence in the context of its overall family permits not only a "horizontal" analysis of an event (residue, sequence motif, domain insertion or fusion, etc.) along the complete length of the proteins but also a "vertical" view of its evolution among different clades or organisms. In systems-level studies, proteins are considered as interacting multidimensional entities, and in this case, multiple sequence alignment can also contribute to our understanding of the majority of these dimensions.

In this chapter, we will first give an overview of the central role of multiple sequence alignments and their main applications in current biological studies. We will then discuss the problem of how to efficiently construct MSAs, depending on the biological question being studied. Finally, we will consider how the standard applications can be integrated for wider exploitation.

MSA: CENTRAL ROLE IN PROTEIN ANALYSIS

Since their introduction in the early seventies (Needleman and Wunsch 1970), sequence comparisons and alignments have been used in a wide range of biological applications, and multiple alignments now play a fundamental role in most of the computational methods used in genomic or proteomic projects. This central role is further emphasized by the numerous studies and analyses that are directly or indirectly associated to MSA, and especially to multiple alignment of complete sequences, ranging from evolutionary, functional, or structural studies up to comparative genomics (Figure 11.1).

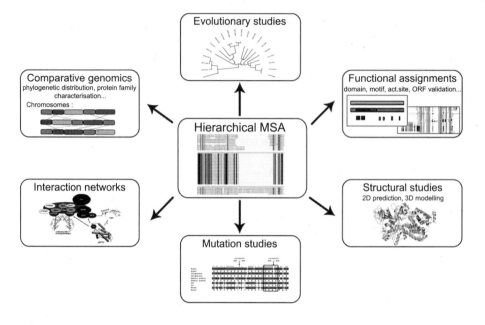

Figure 11.1. The central role of multiple protein sequence alignments in biology. The main applications directly related to hierarchical multiple sequence alignment (MSA) are represented. These applications are mostly interrelated.

Evolutionary Studies

One of the earliest applications of multiple sequence alignments was to define the phylogenetic relationships between organisms (Phillips et al. 2000) through phylogenetic tree construction. Briefly, the objective is to reflect the evolutionary relationships between organisms in a tree. The methods for calculating phylogenetic trees fall into two general categories (Page and Holmes 1998): distance–matrix methods, such as UPGMA and neighbor-joining (also known as clustering or algorithmic methods), which provide an estimation of the distance between two organisms by computing each pairwise sequence distance; and discrete data methods, such as parsimony, maximum-likelihood, or Bayesian methods, also known as tree searching methods.

Strikingly, it has been shown that variation in resulting phylogeny is more dependent on the mode of alignment than on the method of phylogenetic reconstruction (Morrison and Ellis 1997). This emphasizes the fact that generating high-quality sequence alignments is clearly a critical step in the path from raw sequence to phylogeny. Multiple alignments are also essential for the estimation of the reliability of each branch of the constructed tree, using bootstrapping methods which involve resampling of the alignment columns and subsequent tree reconstruction.

The information from several multiple alignments can be used for the estimation of major genetic events, and even of the Tree of Life (Wolf et al. 2002), by concatenating various alignments of distinct protein families (Brown et al. 2001; Clarke et al. 2002). The concatenation increases the number of informative sites for the phylogeny reconstruction and overcomes the discrepancies introduced by lateral gene transfers or by the existence of divergent evolutionary rates. Another commonly used approach to combine phylogenetic information is based on multiple, independently reconstructed trees (Sicheritz-Ponten and Andersson 2001; Wolf et al. 2001). These trees can be calculated using different types of multiple alignments involving protein, DNA, or RNA sequences.

Phylogenetic studies can also be exploited for the reconstruction of ancestral protein sequences, which may improve the sensitivity of remote homolog searches and the prediction of functional sites. The ancestral reconstruction has to take into account the observed variation of evolutionary rates among specific positions of the alignment that more precisely describes the evolution of protein families (Cai et al. 2004). Finally, proteins may evolve at different rates, and evolutionary studies can be used to determine family-specific evolutionary rates. Indeed, the mutational process which acts on a protein is sometimes subject to a specific selective pressure, and MSA represents the basis for the estimation of this evolutionary rate. As an example, the analysis of Luz and Vingron (2006) revealed that the evolution of indispensable proteins is constrained by selection while extracellular proteins are evolving faster.

Functional Assignments

In the context of protein functional characterization, the first levels of analysis cover the detection of sequence errors, the identification of functional domains, and the study of domain organization.

In most genome annotation projects, the standard approach to characterize a novel protein is to search the sequence databases for homologs and to propagate the structural/functional annotation from the known to the unknown protein. Most automatic annotation methods use information from the top best hits in database searches, but it is now well documented that best hits do not necessarily represent the closest sequence relatives (Koski and Golding 2001), thereby casting doubt on the reliability of this approach and leading to a certain number of annotation errors (Bork and Koonin 1998). Two main error types can be met (Sasson et al. 2006): underprediction, when a function is not transferred although it should be, for example, when all of the top blast hits are uncharacterized, or overprediction, when similarity is restricted to a region of the sequence and all annotations are transferred. Another approach is to look for similarities to known domains in precompiled databases, such as Interpro (Mulder et al. 2005). These databases contain representations, such as profiles or hidden Markov models (HMMs), of individual protein domains based on multiple alignments of known sequences.

Recent developments in database search methods have exploited multiple alignments to detect more distant sequences (Altschul et al. 1997; Eddy 1998; Yona and Levitt 2002): database similarity searches are performed with a profile constructed by considering the whole protein family. This allows a more sensitive detection of new family members.

MSA can alleviate some of the problems due to sequencing errors (Dandekar et al. 2000) or the presence of annotation errors in the source database, which are frequently ignored in automatic annotation systems. By integrating a sequence in the context of its close relatives, multiple alignment of complete sequences allows the detection of inconsistencies in open reading frame determination. The vALId (Bianchetti et al. 2005) program has been developed to take advantage of this feature of MSA. It warns about the occurrence of suspicious insertions, deletions, and divergent segments absent of close relatives, and proposes corrections based on mRNA or genome-based translation of alternative frame-shifted regions.

Wrong annotation (Devos and Valencia 2001) due to human or automatic assignments can be introduced in the public sequence databases, and some studies have shown that once an erroneous annotation is introduced into a database, it tends to propagate via automatic annotation inference methods that are based on sequence similarity (Linial 2003). Therefore, existing functional assignments need to be cross-validated before they can be used as a basis for reliable annotations. This has lead to the introduction of knowledge-based systems for automated multiple alignment annotation (e.g., Thompson et al. 2006). The major advantage of these integrated systems is that the sequence information mined from the public databases can be cross-validated within the MSA to distinguish between reliable, consistent information and spurious predictions. The validated data then provide a basis for the accurate propagation of information from known to unknown sequences.

MSAs can also be used to study the domain organization of proteins (Sonnhammer and Kahn 1994): in the course of evolution, the fusion of functional/structural domains and the splitting of multi-domain proteins have occurred frequently, and as a result, some protein families present a complex mosaic picture. Furthermore, insights into cellular localization information can be obtained with the prediction of transmembrane helices, for example, for which MSA improves the reliability (Chen et al. 2002).

MSAs also reveal evolutionary constraints existing at particular sequence positions that conserve residues or their physico-chemical properties, and thus allow a straightforward detection of potential functional sites (e.g., those involved in catalytic sites, protein interactions, ligand binding) that characterize a protein family or subfamily (Ison et al. 2000; Nevill-Manning et al. 1998). Methods for the prediction of these functional sites have been developed that use as the primary indicator of potential sites the conservation of amino acid in MSA (e.g., Lichtarge et al. 1996; Valdar 2002). Other prediction methods exploit structural information, but it has now become clear that neither sequence nor structure alone is sufficient for accurate predictions. Efforts are now being concentrated on the combined use of both sequence conservation and structural information (e.g., Chelliah et al. 2004; Cheng et al. 2005).

Structural Studies

MSAs play an important role in a number of aspects of the characterization of protein structures, in particular in the improvement of secondary and tertiary structure prediction, aimed at estimating the role a residue plays in a protein structure (e.g., helix or strand; buried or exposed).

The current most successful secondary structure prediction methods all employ positional information from multiple alignments (Heringa 2000; Lee et al. 2006; Rost 2001). Prediction based on MSA allows the increase of the prediction accuracy from about 60% (Garnier et al. 1996) to 75% (Jones 1999; Rost et al. 1994), compared to prediction based on single sequences. Indeed, the use of aligned sequences allows better application of the propensities of particular residues for particular secondary structures and improved identification of patterns of hydrophobicity (King and Sternberg 1996). The rationale behind such improvements is that the pattern of substitutions observed in a column directly reflects the type of constraints imposed on that position in the course of evolution.

Protein tertiary structures are classified into families based on a limited set of folding patterns, defined as the three-dimensional fitting of the secondary structure elements; for example, see SCOP (Lo Conte et al. 2002) and CATH (Pearl et al. 2003). In general, proteins sharing the same function have similar structure and functional sites; that is, the residues that are involved in the protein molecular function maintain identical structural positions. Such observations became the foundation of homology modeling, which is the most accurate *in silico* method for determining the tertiary structure of an unknown protein. Sequence similarity between proteins usually indicates a structural resemblance, and accurate sequence alignments provide a practical approach for structure modeling, when a 3D structural prototype is available. MODELLER (Sali et al. 1995), Geno3D (Combet et al. 2002), SWISS-MODEL (Schwede et al. 2003), and 3D-JIGSAW (Bates et al. 2001) are accurate tools for protein modeling. For models based on distant evolutionary relationships, it has been shown that multiple sequence alignments often improve the accuracy of the structural prediction (Moult 2005).

Mutation Studies

A considerable effort is now underway to relate human phenotypes to variation at the DNA level. Most human genetic variation is represented by single nucleotide polymorphisms (SNPs), and many of them are believed to cause phenotypic differences between individuals (Ramensky et al. 2002). One of the main goals of SNP research is therefore to understand the genetics of human phenotype variation and especially the genetic basis of complex diseases, thus providing a basis for assessing susceptibility to diseases and designing individual therapy. Whereas a large number of SNPs may be functionally neutral, others may have deleterious effects on the regulation or the functional activity of specific gene products. Nonsynonymous single-nucleotide polymorphisms (nsSNPs) that lead to an amino acid change in the protein product are of particular interest because they account for nearly half of the known genetic variations related to human inherited disease

(Stenson et al. 2003). There is a surprisingly high fraction of nsSNPs that affect the structure and probably the functional sites of proteins (Sunyaev et al. 2000), which implies a probably negative effect on the associated phenotype.

Given the huge number of nsSNPs already discovered (Fredman et al. 2004; Irizarry et al. 2000), it has become imperative to predict the phenotype of an nsSNP *in silico*. Computational tools are therefore being developed, for example, nsSNP Analyzer (Bao et al. 2005), SIFT (Ng and Henikoff 2003), and PolyPhen (Ramensky et al. 2002), which use structural information or evolutionary information from MSA to predict an nsSNP's phenotypic effect and to identify disease-associated nsSNPs. Recent studies have shown that combining information obtained from multiple sequence alignment and three-dimensional protein structure can increase the prediction accuracy (Bao et al. 2005; Saunders and Baker 2002).

Interaction Networks

Biological entities are more and more considered as part of complex networks of interactions. Powerful experimental techniques, such as the yeast two-hybrid system or tandem-affinity purification and mass spectrometry, are used to determine protein–protein interactions systematically, and protein interaction maps have been built for several organisms, such as *Saccharomyces cerevisiae* (Uetz et al. 2000), *Escherichia coli* (Butland et al. 2005), and *Drosophila melanogaster* (Formstecher et al. 2005; Giot et al. 2003). In parallel with these experimental developments, a number of computational techniques that give rise to the term comparative interactomics (Cesareni et al. 2005) have been designed for predicting protein interactions. The basic assumption of these techniques is that interactions may be conserved during evolution and can thus be inferred by studying the orthology of proteins. The "Interaction Domain Profile Pairs" (IDPP) (Wojcik and Schachter 2001) approach uses a high-quality reference protein interaction map to predict an interaction map in another organism. This method combines sequence similarity searches and alignments with clustering based on interaction patterns and interaction domain information. Some other methods explore genome organization: one implies that proteins that are consistently present or absent in different proteome sets are likely to interact functionally (Pellegrini et al. 1999). The Rosetta stone method (Marcotte et al. 1999) is based on the observation that some interacting proteins have homologues in other organisms fused into a single protein chain. Pairs of interacting proteins have also been predicted using a measure of the similarity between phylogenetic trees of protein families (Pazos and Valencia 2001). This method was adapted to consider the multi-domain nature of proteins by breaking the sequence into a set of segments of predetermined size, and constructing a separate profile for each segment (Kim and Subramaniam 2006). Another approach involves quantifying the degree of covariation (based on the MSA) between residues from pairs of interacting proteins (correlated mutations), known as the "in silico two-hybrid" method. For some proteins that are known to interact, correlated mutations have been demonstrated to be able to select the correct structural arrangement of two proteins based on the accumulation of signals in the proximity of interacting surfaces (Pazos and Valencia 2002). This relationship between correlated residues and interacting surfaces has been extended to the prediction of interacting protein pairs based on the differential accumulation of correlated mutations between the interacting partners (inter-protein correlated mutations) and within the individual proteins (intra-protein correlated mutations) (Pazos et al. 1997). Comparison of predicted data with experimental data has demonstrated that the predictive power is increased when several independent sources of data and different algorithms are combined (Eisenberg et al. 2000).

Comparative Genomics

Comparative genomics is a discipline based on whole genome comparison that has become more and more informative as genomic and proteomic sequence data accumulate (Hardison 2003). This discipline is generally based on DNA sequences, but is clearly essential in the context of gene and protein characterization, mainly at the evolutionary and functional levels. At the end of 2006, over 1000 genomes (from bacteria, archaea, and eukaryota, as well as viruses and organelles) were either complete or being determined. In this era of complete genome sequences, it has become possible to perform comparative multiple sequence analysis at the genome level and also at a proteome level. A number of software tools have been developed for use in comparative genomics to explore the similarities and differences between genomes at different levels. Because of the volume and nature of the data involved, almost all of the visualization tools in this field use a web interface to access large databases of precomputed sequence comparisons and annotations, for example, VISTA (Frazer et al. 2004), Ensembl (Curwen et al. 2004), and UCSC (Hsu et al. 2005). Platforms dedicated to comparative genomics have also been made available, such as Phytome (Hartmann et al. 2006), which provides online resources that can be applied to functional plant genomics, molecular breeding, and evolutionary studies. Phytome contains predicted protein sequences, protein family assignments, multiple sequence alignments, phylogenies, and functional annotations for proteins from a large, phylogenetically diverse set of plant taxa.

At the protein analysis level, comparative genomics take essentially a medium-resolution view. By identifying all of the known proteins from one genome and finding their protein homologues, if they exist, in another genome, one can determine which proteins have been conserved between species, and which are unique. Studies are mainly based on the determination of protein presence/absence, to deduce phylogenetic profiles. A comparison of the genomes of yeast, worms, and flies revealed that these eukaryotes encode many of the same proteins, but different gene families are expanded in each genome (Rubin et al. 2000). Genes and gene products that are responsible for species-specific phenotypes can also be identified with this method. By comparing the *Helicobacter pylori* genome with its closest relatives for which complete genome sequences are available, the pathogenic *Haemophilus influenzae* and the benign *Escherichia coli* K12 (Huynen et al. 1998), it has been possible to identify a set of species-specific genes that are responsible for pathogen-specific features (Figure 11.2).

Determining protein functions from genomic sequences can also be achieved using the phylogenetic profile method (Pellegrini et al. 1999). This method is based on the assumption that proteins that function together in a pathway or structural complex, that is, functionally linked proteins, are likely to evolve in a correlated fashion. The phylogenetic profile of a protein describes the presence or absence of homologs in organisms. Proteins that make up multimeric structural complexes are likely to have similar profiles. Also, proteins that are known to participate in given biochemical pathways are likely to have close phylogenetic profiles. Comparing profiles is therefore a useful tool for identifying the macromolecular complex or pathway in which a protein participates. It is possible to assign the function of unchar-

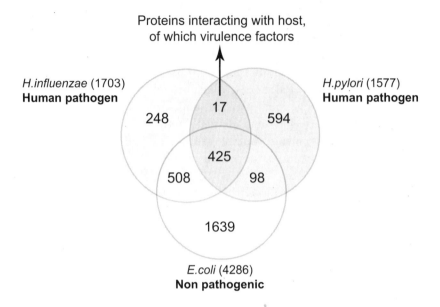

Figure 11.2. Representation of the comparison of the *H. pylori*, *H. influenzae*, and *E. coli* genomes. The total number of proteins for each species is shown between brackets. Some 425 orthologs are shared by all three species. The 17 orthologs shared by *H. influenzae* and *H. pylori* are responsible for pathogen-specific features.

acterized proteins by examining the function of proteins with identical phylogenetic profiles.

EFFICIENT CONSTRUCTION OF MSA

Genomics and proteomics technologies, together with the new system biology strategies, have led to a paradigm shift in bioinformatics. The traditional reductionist approach has been replaced by a more global, integrated view. MSA is one example where a shift of thinking or focus is now leading to the development of new methods.

The construction of an MSA traditionally begins with a sequence database search to find homologous sequences. These sequences are then aligned with appropriate software and automatically and/or manually refined to obtain a high-quality multiple alignment. With the exponential increase in the size of the sequence databases, it is no longer possible to align all of the homologous sequences available, and therefore biologists have to develop novel strategies to select the pertinent sequences, according to their analysis goals.

Detection of Similar Sequences

The alignment methods make sense only if they are assumed to be dealing with homologous sequences, that is, sequences sharing a common ancestor. Several programs, such as BLAST (Altschul et al. 1997) and FASTA (Pearson and Lipman 1988), search for local similarity between the query and database sequences, to detect homologous sequences. These similarities might lead to evolutionary clues about the structure and/or function of the query sequence. Each comparison is given a score reflecting the degree of similarity between the query and the sequence being compared: the higher the score, the greater the degree of similarity. In both BLAST and FASTA, in addition to the alignment scores, the significance of each alignment is computed as a p-value or an E-value, based on the alignment scores expected by chance in the total sequence space. Although these methods have proven their ability to detect closely related sequences (those with sequence identity larger than 30%), their performance decreases quickly when one tries to find more distant homologs (Brenner et al. 1998).

For remotely related proteins, conservation is weak and is often restricted to short segments of the sequences. To overcome this limitation, researchers have developed various homology search methods based on the common features of a group of related sequences. These methods, for example, HMMER (Eddy 1998), based on HMMs, and PSI-BLAST (Altschul et al. 1997), exploit multiple alignments to detect more and more distant homologues. With PSI-BLAST, a profile is automatically constructed from a multiple alignment of the highest scoring hits in an initial BLAST search. Highly conserved positions receive high scores, and weakly conserved positions receive scores near zero, and the profile is used to perform a second BLAST search (etc.). The results of each iteration are used to refine the profile, which increases the sensitivity for homologues search.

Alignment Software

Once the homologous sequences to be aligned have been determined, the biologist has to choose appropriate alignment software. Three main considerations have to be taken into account in choosing a program: biological accuracy, execution time and memory usage, biological accuracy being the most important concern. It has been shown that the best choice of an alignment program depends on the sequences set to be aligned and that no single "best" program exists (Lassmann and Sonnhammer 2002; Nuin et al. 2006; Thompson et al. 1999b): no automatic alignment method can be expected to produce biologically meaningful alignments, whatever the sequences to be aligned. There is therefore a continuous effort to improve the biological accuracy of MSA tools. Until recently, the choice for building MSAs was limited to a handful of packages, but a recent increase in genomic data has fuelled the development of many novel methods that are more accurate and faster than the older ones (Nuin et al. 2006).

The next section briefly describes the evolution of multiple alignment methods from their beginnings in the seventies to the recent introduction of new integrative and cooperative strategies (Figure 11.3).

Although a dynamic programming algorithm that guarantees a mathematically optimal alignment exists, the method is limited to a small number of short sequences, since the computing power required for

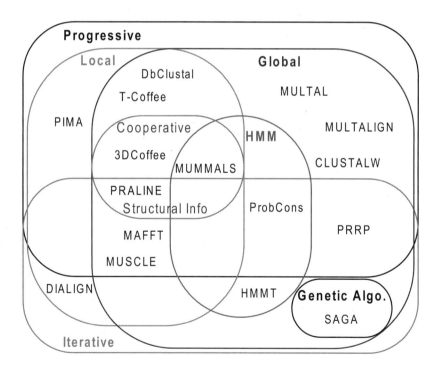

Figure 11.3. Overview of different alignment algorithms.

larger alignments becomes too prohibitive (Carrillo and Lipman 1988). To overcome this problem, various heuristic approaches have been developed, leading to a huge quantity of programs using fundamentally different strategies based on very different algorithms.

Traditionally, the most popular method has been the progressive alignments (Feng and Doolittle 1987) that exploit the fact that homologous sequences are evolutionarily related. A multiple alignment is built up gradually by aligning the closest sequences first and successively adding in the more distant ones. PIMA (Smith and Smith 1992), MULTALIGN (Barton and Sternberg 1987b), MULTAL (Taylor 1987), and CLUSTALW (Thompson et al. 1994), for example, are based on this method, and mainly differ in the method used to determine the order of alignment of the sequences. Progressive methods have the great advantage of speed and simplicity combined with reasonable sensitivity, even if it is by nature a heuristic that does not guarantee any level of optimization.

Later, numerous alignment algorithms that applied iterative strategies were developed, introducing the capacity to refine previously aligned sequences at each iteration with the goal of improving the overall alignment quality, using an objective function (see MSA quality). A local alignment approach is implemented in the DIALIGN (Morgenstern et al. 1998a) program to construct multiple alignments based on segment-to-segment comparison rather than the residue-to-residue comparison used previously. The PRRP (Gotoh 1996) program optimizes a progressive, global alignment by iteratively dividing the sequences into two groups, which are subsequently realigned using a global group-to-group alignment algorithm. SAGA (Notredame and Higgins 1996) uses a genetic algorithm to select from an evolving population the alignment that optimizes the COFFEE (Notredame et al. 1998) objective function (see MSA quality). The program HMMT (Eddy 1998) uses a simulated annealing method to maximize the probability that an HMM represents the sequences to be aligned. Iterative methods often produce more accurate alignments, although these methods are generally slower than the progressive ones (Thompson et al. 1999b).

The complexity of the multiple alignment challenge has led to the combination of different sequence-based alignment algorithms, and even, more recently, to the incorporation of biological information other than the sequence itself (Edgar and Batzoglou 2006). DbClustal (Thompson et al. 2000), for example, uses information about locally conserved segments detected during a database search to guide or "anchor" a global multiple alignment. T-Coffee (Notredame et al. 2000) uses information from a precompiled library of pairwise alignments (local, global, and structural alignments), and constructs a multiple alignment with the highest possible level of consistency with the alignments within the library. MAFFT (Katoh et al. 2005) and MUSCLE (Edgar 2004a) both use fast algorithms to detect local sequence similarities, and then use a progressive alignment algorithm together with an iterative refinement step to produce an accurate alignment of large sets of sequences. ProbCons (Do et al. 2005) uses HMM-derived posterior probabilities and three-way alignment consistency in a global, progressive alignment, together with an iterative refinement step. The most recent multiple alignment methods that are being developed can be considered as knowledge-based strategies: 3DCoffee (O'Sullivan et al. 2004) integrates structural information, PRALINE (Simossis and Heringa 2005) exploits protein secondary structure information either from 3D structures or from computational predictions, and MUMMALS (Pei and Grishin 2006) uses HMM structural information (Figure 11.3).

Other recent developments address the problems associated with sequences having distinct domain organization, such as RAlign (Sammeth and Heringa 2006), POA (Lee et al. 2002), and AliWABA (Jones et al. 2006) that employ partially ordered graphs. These three programs use local algorithms suitable for multi-domain proteins that may contain repeated or shuffled elements.

Despite the enormous progress achieved, current implementations of multiple alignment algorithms are still heuristic, and it is necessary to examine the alignment to check that there are no obvious errors. It is therefore common practice to use automatic refinement programs such as RASCAL (Thompson et al. 2003) or Refiner (Chakrabarti et al. 2006), and then manually improve the alignments, using an alignment editor such as Jalview (Clamp et al. 2004), SEAVIEW (Galtier et al. 1996), ANTHEPROT (Deleage et al. 2001), or CINEMA (Parry-Smith et al. 1998).

MSA Quality

Since multiple alignments are usually employed at the beginning of a series of bioinformatics and/or biological studies, they have to be of high quality. Errors in the alignment will lead to further errors in the subsequent analyses and might generate misleading patterns and result in false hypotheses. It is therefore critical to be able to distinguish "good" alignments from bad ones. A good alignment is one that corresponds to the biologically correct alignment, accurately reflecting the evolutionary, structural, and functional relationships between the sequences.

A number of studies have been realized to determine which alignment programs work best in which situations. These studies are usually based on benchmarks that have been specifically designed for this purpose (see Chapter 8. Constructing Alignment Benchmarks), taking the form of databases of precompiled alignments to which the alignments generated by test algorithms are compared.

None of the current implementations of MSA guarantees a full alignment optimization. Given a particular set of sequences, an objective score that describes the optimal or "biologically correct" multiple alignment is needed. Suboptimal or incorrect alignments would then score less than this maximal score. Such measures, also known as objective functions, are currently used to evaluate and compare multiple alignments from different sources and to detect low-quality alignments. They are also used in iterative alignment methods to improve the alignment by seeking to maximize the objective function. One of the first scoring systems was the sum-of-pairs score (Carrillo and Lipman 1988), based on pairwise scores. Pairwise scores are also used in the COFFEE objective function (Notredame et al. 1998), which reflects the level of consistency between a multiple sequence alignment and a library containing pairwise alignments of the same sequences. Column statistics have also been used to score alignments. One approach uses an information content statistic (relative entropy) (Hertz and Stormo 1999), but it considers only the frequencies of identical residues in each column and does not take into account similarities between residues. Another column-based measure, norMD (Thompson et al. 2001), is based on the mean distance (MD) column scores implemented in ClustalX (Thompson et al. 1997), and its main advantage is incorporation of *ab initio* sequence information, such as the number, length, and similarity of the sequences to be aligned. Other developments that measure local reliability or column conservation have also been recently reported (e.g., Cline et al. 2002; Pei and Grishin 2001; Schlosshauer and Ohlsson 2002).

Sequence Selection for More Accurate MSA

We are currently facing an explosion of the data available in databases, due to the generalization of high-throughput experimental technologies: genome sequencing as well as structural genomics or transcriptomics have increased the number of DNA and protein sequences available, and the rate at which genome and sequence products are being functionally characterized.

Some biological sequence databases are highly redundant, for two main reasons: some databases keep redundant sequences with many identical and nearly identical sequences, and natural sequences often have high sequence identities due to gene duplication.

As a direct consequence of the recent database growth, homology searches frequently lead to the identification of hundreds to thousands of potential homologues for a single query sequence. Dealing with so much data can be detrimental in terms of computational and human analysis time; indeed, for the majority of alignment programs, the number of sequences to be aligned is limited, and even if all sequences can be aligned, the more sequences an alignment contains, the more difficult it would be for the biologist to analyze it. Moreover, this vast quantity of available sequences can also be detrimental in terms of the accuracy of the significance of the results. Apart from inherent database noise linked to sequencing or intron/exon prediction errors, problems, such as redundancy or the presence of partial sequences, may represent a significant source of noise, depending on the biological question under study. Redundancy is now a major problem, notably if we consider that aligning a defined number of sequences according to user-defined or program limitation may hinder access to significant information present in the additional non-aligned sequences.

It is now essential to develop new complementary approaches for an efficient exploitation of this large sequence space. As a consequence, the current challenges are shifting from the multiple alignment to the choice of sequences to align that will yield the most biologically correct and informative alignments. Even though the alignment-based methods are ubiquitous and have been much discussed (Edgar and Batzoglou 2006; Notredame 2002), very little has been written about the actual selection of sequences the alignment is built on. Strategies for sequence selection are essentially based on the assumption that, at the structural and functional level, closely related sequences may not add relevant information, whereas diversity is usually more informative (Przybylski and Rost 2002).

It is clearly advantageous in terms of processing time to reduce the sequence space at the earliest possible stage of an analysis, that is, during the initial homology search step. There are clearly two possibilities in this case: an *a priori* reduction of the sequence database search space and an *a posteriori* sampling of the sequences detected by the database search program. The *a priori* reduction of the search space can sometimes easily be done by choosing an appropriate sequence database to run the homology searches according to the biological question under study:

- For comparative genomic studies, the search space is usually restricted to complete genome or proteome databases.
- For some evolutionary studies, the sequence search space can be restricted to specific clades, such as mammals or vertebrates, and numerous Blast servers, such as the Blast NCBI site, propose the restriction to specific organisms.
- To annotate an unknown sequence, biologists can choose to work with a reduced database with higher-quality annotation such as SWISS-PROT (Bairoch and Apweiler 2000), a curated protein sequence database containing a minimal level of redundancy.
- For the majority of structural or functional assignments, and in the case of information propagation, preprocessed nonredundant databases, where sequences are clustered by means of their percentage of identity, such as the UniRef database series (Wu et al. 2006), can be used.
- For protein homology modeling, the database search can be done in the Protein Data Bank (PDB) (Berman et al. 2000), a protein 3D structure database that includes structures experimentally determined by X-ray crystallography or by NMR.

Obviously, the set of sequences detected by homology searches can also be *a posteriori* sampled. The current most widely used methods for detecting homologous sequences is the BLAST suite of programs, and software for BLAST results post-treatment have been developed. BLAST Filter (Spalding and Lammers 2004) generates smaller sequence sets by filtering the BLAST results: the filter is based on 15 user-configurable predefined rules. UniqueProt (Mika and Rost 2003) creates a representative and unbiased dataset by removing overrepresented sequences: the user has to define an identity threshold above which sequences will be clustered. These two programs are based on sequence similarity criteria and need at least a brief analysis of the BLAST search results to be powerful.

The MyHits (Pagni et al. 2004) web server, which is an integrated service dedicated to the annotation of protein sequences and to the analysis of their domains and signatures, also provides a tool for the automated clustering of identical and highly similar protein sequences. Typically, a representative subset out of a large set of sequences is extracted, based on a taxonomic filter and a procedure to group the matched sequences with an identity level equal or superior to a user-defined threshold.

A recent study has concentrated on how far the homologous sequences detected by BlastP can be reduced while maintaining the relevant information concerning the active site residues, as annotated in the PDB structural database (Friedrich et al. 2007). This study has demonstrated that, on average, it is sufficient to align 30% of the detected sequences to efficiently maintain the relevant functional information when more than 100 sequences are detected, but the sequence selection cannot be performed randomly. The most suitable method tested is based on the BlastP E-values, and allows the conservation of the potential structural and functional information in the sampled set while restricting the alignment computation time to a reasonable limit.

The PhyloGenie (Frickey and Lupas 2004) suite of programs, dedicated to automated phylogenetic studies, integrates the java program Blammer, which consists of five Blast post-processing steps that convert sets of high-scoring segment pairs (HSPs) to multiple alignments and perform a sequence selection optimization.

In some specific cases, a filtering step can be integrated once the MSA has been computed to ease and optimize its analysis. QR factorization (O'Donoghue and Luthey-Schulten 2005) produces minimally redundant sets of protein sequences, based on column pivoting of a matrix encoding the sequence alignment. This software has been designed to choose automatically a linearly independent subset of sequences, and has been integrated in MultiSeq (Roberts et al. 2006), a unified bioinformatics analysis environment that allows one to organize, display, align, and analyze both sequence and structure data for proteins and nucleic acids.

Mihalek et al. (2006) propose, in the case of proteins with known structure, a heuristic Metropolis Monte Carlo strategy to select sequences from a large set of homologues, to improve detection of functional surfaces. The optimization is based on the clustering of residues that are under increased evolutionary pressure, according to the sample of sequences under consideration.

MSA INTEGRATIVE EXPLOITATION TOOLS

Bioinformatics has to respond to the challenges represented by the vast amounts of heterogeneous data available in the databases. Studying sequence/structure/function/evolution relationships within a protein family can be achieved by integrating the results of many different analyses in the context of MSAs.

Thanks to the increasing adoption of common data standards and exchange formats, a number of systems have been described recently that allow the integration and visualization of heterogeneous information in the context of a family of proteins. Pfaat (Johnson et al. 2003) is a semiautomatic protein family annotation tool incorporated in a multiple alignment viewer. The application merges display features such as dendrograms and secondary and tertiary protein structure with Sequence Retrieval System (SRS) (Etzold et al. 1996), subgroup comparison, and extensive user-annotation capabilities. MyHits (Pagni et al. 2004), an integrated web server dedicated to the annotation of protein sequences based on MSA, helps the analysis of the protein domains and signatures. A dictionary-driven annotation has also been developed, based on Bio-dictionary (Rigoutsos et al. 2002), a collection of amino acid patterns that completely covers the natural sequence space and can capture functional and structural signals that have been reused during evolution, within and across protein families.

A tool combining knowledge-based methods with complementary *ab initio* sequence-based predictions for protein family analysis has also been recently developed. MACSIMS (Thompson et al. 2006) integrates different types of data in the framework of the multiple alignment, taking advantage of the recently developed Multiple Alignment Ontology (MAO) (Thompson et al. 2005a). A wide range of information, from taxonomic data and functional descriptions to individual sequence features, such as structural domains and active site residues, is mined from the public databases using SRS. MACSIMS also includes a number of algorithms developed for reliable data validation, con*sensu*s predictions, and rational propagation of information from the known to the unknown sequences. Jalview (Clamp et al. 2004) has been adapted to be able to visualize all of the mined and predicted information in MACSIMS. Another tool, OrdAlie (Ordered Alignment Information Explorer, manuscript in preparation) is available that is designed to allow interactive analysis and exploration of protein sequence, structure, function, and evolution relationships. With OrdAlie, the multiple sequence alignment is displayed in a graphical window, together with user-selected sequence features. Sequences can be clustered automatically into subfamilies, and a detailed, hierarchical analysis of residue conservation can be performed at the family or subfamily level. Conserved residues can also be visualized in the context of

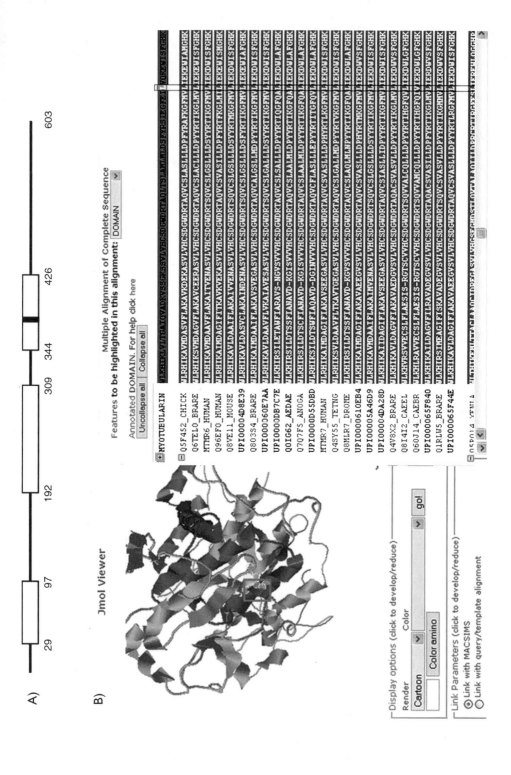

their 3D structural environment, using the RASMOL structure viewer (Sayle and Milner-White 1995).

Several recent programs, such as VISSA (Li and Godzik 2006), ViTO (Catherinot and Labesse 2004) and Friend (Abyzov et al. 2005), also introduced functions to simultaneously visualize multiple alignments and other information, mainly structural. These bioinformatics applications are mainly designed for simultaneous analysis and visualization of multiple structures and sequences, for refinement of sequence/structure alignment for homology modeling, and to assist researchers in handling many practical issues involving simultaneous analysis of both sequence alignments and 3D structures, for example, checking what the conservation of residues around the active site is.

A recent web server, MAGOS (Garnier et al. 2006) integrates, starting with a single query sequence, the construction of a multiple alignment, its annotation with MACSIMS, and the construction of a model of the query sequence. The model and the query protein in the context of the annotated multiple alignment are then interconnected via a user-friendly interface. MAGOS data integration capabilities can be illustrated by the analysis of the known point mutations and their relative position according to functional and structural features of a protein implicated in genetic disease, such as myotubularin (Figure 11.4). Myotubularin is implicated in the X-linked myotubular myopathy, a severe muscular disease characterized by a marked hypotonia and generalized muscle weakness. Missense mutations that directly affect residues implicated in the active site pocket are usually associated with severe phenotypes (early postnatal death). The L406P mutation affects a highly conserved residue located in the tyrosine–protein phosphatase domain, located in the middle of a 10-residue helix. The severe phenotype associated with

---

Figure 11.4. Illustration of MAGOS capabilities. A. Myotubularin domain organization. The protein, which is 603 residues long, is divided in three distinct domains: the PH-GRAM domain from residue 29 to 97, the myotubularin-related domain from residue 192 to 309, and the tyrosine–protein phosphatase domain from residue 344 to 426, which encompasses the active site pocket $C(X)_5R$ in black. B. MAGOS results interface. The annotated alignment is shown on the right, the first sequence being the myotubularin, which is interconnected to its model on the left side. The tyrosine–protein phosphatase domain is highlighted in dark grey in the alignment and on the model. The leucine situated in position 406 in the myotubularin sequence is also highlighted both on the myotubularin sequence in the context of the MSA and on the constructed 3D model.

this mutation may be linked to the helix break due to the presence of the proline.

CONCLUSIONS AND PERSPECTIVES

Bioinformatics techniques are constantly evolving in response to the challenges presented by the mass of heterogeneous and complex information made available thanks to the new high-throughput technologies. In the context of sequence-based biology, there is no doubt that multiple alignment of protein sequences will remain a central application in the foreseeable future.

The MSA construction process can be divided into five crucial steps (Figure 11.5). However, it is not possible to define a single process that is effective in all study cases, and efficient MSA construction should take the biological question under study into account to choose an appropriate method for sequence selection, as well as the most suitable alignment program.

Protein multiple alignment tools are constantly evolving and improving in terms of scalability and accuracy. It is now clear that primary sequence data alone are not sufficient to construct accurate multiple alignments. Improvements in multiple alignment construction are likely to come by combining sequence alignment with other information, such as known structures, results of database searches for local similarities, experimental data, and, in general, anything that may come from expression data and proteomic analysis. As mentioned in an earlier section, recently developed software already integrate structural and/or local similarity information, and usually increase the quality of the resulting alignment compared to methods based solely on primary sequences. Moreover, with the exponential increase of the amount of sequence data available, it becomes fundamental to develop strategies aimed at reducing the number of sequences to be aligned without losing any relevant information.

As multiple sequence alignments provide an ideal environment for the integration of structural and functional information in the context of the protein family, new information management systems will continue to develop for the collection, validation, and analysis of the vast amount of heterogeneous data available. A major challenge will be the selection of the most descriptive and useful information and its presentation in a suitable format for the biologist. The ideal tool will allow the validation, visualization, integration, and interpretation, in a biological

Information Content of Protein MSA

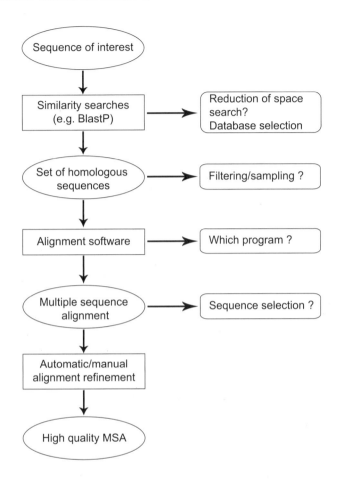

Figure 11.5. MSA efficient construction. Overview of the stages where the user can intervene to reduce the size of his alignment and/or increase its quality.

context, of the vast amount of diverse data generated by the application of proteomic and genomic discovery science tools.

Acknowledgments

We are thankful for the support from the Institut National de la Santé et de la Recherche Médicale (INSERM), the Centre National de la Recherche Scientifique (CNRS), and the Université Louis Pasteur (ULP) of Strasbourg.

References

Aagesen, L., G. Petersen, and O. Seberg. 2005. Sequence length variation, indel costs, and congruence in sensitivity analysis. *Cladistics* 21:15–30.

Abyzov, A., M. Errami, C. M. Leslin, and V. A. Ilyin. 2005. Friend, an integrated analytical front-end application for bioinformatics. *Bioinformatics* 21:3677–3678.

Alberts, B., D. Bray, J. Lewis, M. Raff, K. Roberts, and J. D. Watson. 1994. *Molecular Biology of the Cell.* 3rd Edition. New York: Garland.

Alexandersson, M., S. Cawley, and L. Pachter. 2003. SLAM: Cross-species gene finding and alignment with a generalized pair hidden Markov model. *Genome Research* 13:496–502.

Alkan, C., J. A. Bailey, E. E. Eichler, S. C. Sahinalp, and E. Tuzan. 2002. An algorithmic analysis of the role of unequal crossover in alpha-satellite DNA evolution. *Genome Informatics* 13:93–102.

Alkemar, G., and O. Nygård. 2003. A possible tertiary rRNA interaction between expansion segments ES3 and ES6 in eukaryotic 40S ribosomal subunits. *RNA* 9:20–24.

———. 2004. Secondary structure of two regions in expansion segments ES3 and ES6 with the potential of forming a tertiary interaction in eukaryotic 40S ribosomal subunits. *RNA* 10:403–411.

Allison, L., and C. S. Wallace. 1994. The posterior probability distribution of alignments and its application to parameter estimation of evolutionary trees and to optimization of multiple alignments. *Journal of Molecular Evolution* 39:418–430.

Allison, L., C. S. Wallace, and C. N. Yee. 1992a. Finite-state models in the alignment of macromolecules. *Journal of Molecular Evolution* 35:77–90.

———. 1992b. Minimum message length encoding, evolutionary trees and multiple alignment. Pp. 663–674 in *Proceedings of the 25th Annual Hawaii International Conference on System Sciences.*

Allison, L., and C. N. Yee. 1990. Minimum message length encoding and the comparison of macromolecules. *Bulletin of Mathematical Biology* 52:431–453.

Altschul, S. F. 1998. Generalized affine gap costs for protein sequence alignment. *Proteins* 32:88–96.

Altschul, S. F., and B. W. Erickson. 1986. Optimal sequence alignment using affine gap costs. *Bulletin of Mathematical Biology* 48:603–616.

Altschul, S. F., W. Gish, W. Miller, E. W. Myers, and D. J. Lipman. 1990. Basic Local Alignment Search Tool. *Journal of Molecular Biology* 215:403–410.

Altschul, S. F., T. L. Madden, A. A. Schaffer, J. Zhang, Z. Zhang, W. Miller, and D. J. Lipman. 1997. Gapped BLAST and PSI-BLAST: A new generation of protein database search programs. *Nucleic Acids Research* 25:3389–3402.

Amrine-Madsen, H., K. P. Koepfli, R. K. Wayne, and M. S. Springer. 2003. A new phylogenetic marker, apolipoprotein B, provides compelling evidence for eutherian relationships. *Molecular Phylogenetics and Evolution* 28:225–240.

Armougom, F., S. Moretti, V. Keduas, and C. Notredame. 2006. The iRMSD: A local measure of sequence alignment accuracy using structural information. *Bioinformatics* 22:e35–e39.

Armougom, F., S. Moretti, O. Poirot, S. Audic, P. Dumas, B. Schaeli, V. Keduas, and C. Notredame. 2006. Expresso: Automatic incorporation of structural information in multiple sequence alignments using 3D-Coffee. *Nucleic Acids Research* 34:W604–W608.

Arribas-Gil, A., D. Metzler, and J.-L. Plouhinec. 2007. Statistical alignment with a sequence evolution model allowing rate heterogeneity along the sequence. *IEEE/ACM Transactions on Computational Biology and Bioinformatics August 29, 2007*, IEEE Computer Society Digital Library. http://doi.ieeecomputersociety.org/10.1109/TCBB.2007.70246.

Bafna, V., and D. H. Huson. 2000. The conserved exon method for gene finding. *Bioinformatics* 16:190–202.

Bahr, A., J. D. Thompson, J. C. Thierry, and O. Poch. 2001. BAliBASE (Benchmark Alignment dataBASE): Enhancements for repeats, transmembrane sequences and circular permutations. *Nucleic Acids Research* 29:323–326.

Bailey, T. L., and C. Elkan. 1994. Fitting a mixture model by expectation maximization to discover motifs in biopolymers. Pp. 28–36 in *Proceedings of the Second International Conference on Intelligent Systems for Molecular Biology*, R. Altman, D. Brutlag, P. Karp, R. Lanthrop, and D. Searls, eds. Menlo Park, CA: AAAI Press.

Bairoch, A., and R. Apweiler. 2000. The SWISS-PROT protein sequence database and its supplement TrEMBL in 2000. *Nucleic Acids Research* 28:45–48.

Baker, D., and A. Sali. 2001. Protein structure prediction and structural genomics. *Science* 294:93–96.

Baldi, P., Y. Chauvin, T. Hunkapiller, and M. A. McClure. 1993. Hidden Markov models in molecular biology: New algorithms and applications. Pp. 747–754 in *Advances in Neural Information Processing Systems*, S. J. Hanson, J. D. Cowan, and C. Lee Giles, eds. San Mateo, CA: Kaufmann.

———. 1994. Hidden Markov models of biological primary sequence information. *Proceedings of the National Academy of Sciences USA* 91:1059–1063.

Ban, N., P. Nissen, J. Hansen, P. B. Moore, and T. A. Steitz. 2000. The complete atomic structure of the large ribosomal subunit at 2.4 Å resolution. *Science* 289:905–920.

Bao, L., M. Zhou, and Y. Cui. 2005. nsSNPAnalyzer: Identifying disease-associated nonsynonymous single nucleotide polymorphisms. *Nucleic Acids Research* 33:W480–W482.

Barton, G. J., and M. J. E. Sternberg. 1987a. Evaluation and improvements in the automatic alignment of protein sequences. *Protein Engineering* 1:89–94.

———. 1987b. A strategy for the rapid multiple alignment of protein sequences. *Journal of Molecular Biology* 198:327–337.

Bates, P. A., L. A. Kelley, R. M. MacCallum, and M. J. Sternberg. 2001. Enhancement of protein modeling by human intervention in applying the automatic programs 3D-JIGSAW and 3D-PSSM. *Proteins* Supplement 5:39–46.

Batzoglou, S., L. Pachter, J. P. Mesirov, B. Berger, and E. S. Lander. 2000. Human and mouse gene structure: Comparative analysis and application to exon prediction. *Genome Research* 10:950–958.

Bauer, M., G. Klau, and K. Reinert. 2005. Multiple structural RNA alignment with Lagrangian relaxation. *Lecture Notes in Computer Science* 3692:303–314.

Baumer, A., F. Dutly, D. Balmer, M. Riegel, T. Tukel, M. Krajewska–Walasek, and A. A. Schinzel. 1998. High level of unequal meiotic crossovers at the origin of the 22q11. 2 and 7q11.23 deletions. *Human Molecular Genetics* 7:887–894.

Benavides, E., R. Baum, D. McClellan, and J. W. Sites, Jr. 2007. Molecular phylogenetics of the lizard genus *Microlophus* (Squamata: Tropiduridae): Aligning and retrieving indel signal from nuclear introns. *Systematic Biology* 56:776–797.

Benner, S. A., M. A. Cohen, and G. H. Gonnet. 1993. Empirical and structural models for insertions and deletions in the divergent evolution of proteins. *Journal of Molecular Biology* 229:1065–1082.

Berman, H. M., T. Battistuz, T. N. Bhat, W. F. Bluhm, P. E. Bourne, K. Burkhardt, Z. Feng, G. L. Gilliland, L. Iype, S. Jain, P. Fagan, J. Marvin, D. Padilla, V. Ravichandran, B. Schneider, N. Thanki, H. Weissig, J. D. Westbrook, and C. Zardecki. 2002. The Protein Data Bank. *Acta Crystallographica D Biological Crystallography* 58:899–907.

Berman, H. M., J. Westbrook, Z. Feng, G. Gilliland, T. N. Bhat, H. Weissig, I. N. Shindyalov, and P. E. Bourne. 2000. The Protein Data Bank. *Nucleic Acids Research* 28:235–242.

Bianchetti, L., J. D. Thompson, O. Lecompte, F. Plewniak, and O. Poch. 2005. vALId: Validation of protein sequence quality based on multiple alignment data. *Journal of Bioinformatics and Computational Biology* 3:929–947.

Bishop, M. J., and E. A. Thompson. 1986. Maximum likelihood alignment of DNA sequences. *Journal of Molecular Biology* 190:159–165.

Blackshields, G., I. M. Wallace, M. Larkin, and D. G. Higgins. 2006. Analysis and comparison of benchmarks for multiple sequence alignment. *In Silico Biology* 6:321–339.

Blanchette, M., E. D. Green, W. Miller, and D. Haussler. 2004. Reconstructing large regions of an ancestral mammalian genome *in silico*. *Genome Research* 14:2412–2423.

Blanchette, M., W. J. Kent, C. Riemer, L. Elnitski, A. F. A. Smit, K. M. Roskin, R. Baertsch, K. Rosenbloom, H. Clawson, E. D. Green, D. Haussler, and W. Miller. 2004. Aligning multiple genomic sequences with the threaded blockset aligner. *Genome Research* 14:708–715.

Borg, I., and P. Groenen. 1997. *Modern Multidimensional Scaling*, 2nd ed. New York: Springer Verlag.

Bork, P., and E. V. Koonin. 1998. Predicting functions from protein sequences–Where are the bottlenecks? *Nature Genetics* 18:313–318.

Boutonnet, N. S., M. J. Rooman, M. E. Ochagavia, J. Richelle, and S. J. Wodak. 1995. Optimal protein structure alignments by multiple linkage clustering: Application to distantly related proteins. *Protein Engineering* 8:647–662.

Bradley, P., K. M. Misura, and D. Baker. 2005. Toward high-resolution de novo structure prediction for small proteins. *Science* 309:1868–1871.

Bray, N., I. Dubchak, and L. Pachter. 2003. AVID: A global alignment program. *Genome Research* 13:97–102.

Breitbart, R. E., A. Andreadis, and B. Nadal-Ginard. 1987. Alternative splicing: A ubiquitous mechanism for the generation of multiple protein isoforms from single genes. *Annual Review of Biochemistry* 56:467–495.

Bremer, K. 1988. The limits of amino acids sequence data in angiosperm phylogenetic reconstruction. *Evolution* 42:795–803.

Brenner, S. E., C. Chothia, and T. J. Hubbard. 1997. Population statistics of protein structures: Lessons from structural classifications. *Current Opinion in Structural Biology* 7:369–376.

———. 1998. Assessing sequence comparison methods with reliable structurally identified distant evolutionary relationships. *Proceedings of the National Academy of Sciences USA* 95:6073–6078.

Briffeuil, P., G. Baudoux, C. Lambert, X. De Bolle, C. Vinals, E. Feytmans, and E. Depiereux. 1998. Comparative analysis of seven multiple protein sequence alignment servers: Clues to enhances reliability of predictions. *Bioinformatics* 14:357–366.

Brodersen, D. E., W. M. Clemons, Jr., A. P. Carter, B. T. Wimberly, and V. Ramakrishnan. 2002. Crystal structure of the 30S ribosomal subunit from *Thermus thermophilus*: Structure of the proteins and their interactions with 16S RNA. *Journal of Molecular Biology* 316:725–768.

Brooks, D. R., and D. A. McLennan. 1994. Historical ecology as a research programme: Scope, limitations, and the future. Pp. 1–27 in *Phylogenetics and Ecology*, P. Eggleton and R. I. Vanne-Write, eds. London: Academic Press.

Brown, J. R., C. J. Douady, M. J. Italia, W. E. Marshall, and M. J. Stanhope. 2001. Universal trees based on large combined protein sequence data sets. *Nature Genetics* 28:281–285.

Brudno, M., M. Chapman, B. Göttgens, S. Batzoglou, and B. Morgenstern. 2003. Fast and sensitive multiple alignment of large genomic sequences. *BMC Bioinformatics* 4:66.

Brudno, M., C. B. Do, G. M. Cooper, M. F. Kim, E. Davydov, E. D. Green, A. Sidow, and S. Batzoglou. 2003. LAGAN and Multi-LAGAN: Efficient tools for large-scale multiple alignment of genomic DNA. *Genome Research* 13:721–731.

References

Brudno, M., S. Malde, A. Poliakov, C. B. Do, O. Couronne, I. Dubchak, and S. Batzoglou. 2003. Glocal alignment: Finding rearrangements during alignment. *Bioinformatics* 19:i54–i62.
Brudno, M., and B. Morgenstern. 2002. Fast and sensitive alignment of large genomic sequences. Pp. 138–147 in *Proceedings of the First IEEE Computer Society Conference on Bioinformatics*. Washington, DC: IEEE Computer Society Press.
Brudno, M., R. Steinkamp, and B. Morgenstern. 2004. The CHAOS/DIALIGN WWW server for multiple alignment of genomic sequences. *Nucleic Acids Research* 32:W41–W44.
Buckley, T. R., C. Simon, P. K. Flook, and B. Misof. 2000. Secondary structure and conserved motifs of the frequently sequenced domains IV and V of the insect mitochondrial large subunit rRNA gene. *Insect Molecular Biology* 9:565–580.
Burge, C., and S. Karlin. 1997. Prediction of complete gene structures in human genomic DNA. *Journal of Molecular Biology* 268:78–94.
Butland, G., J. M. Peregrin-Alvarez, J. Li, W. H. Yang, X. C. Yang, V. Canadien, A. Starostine, D. Richards, B. Beattie, N. Krogan, M. Davey, J. Parkinson, J. Greenblatt, and A. Emili. 2005. Interaction network containing conserved and essential protein complexes in *Escherichia coli*. *Nature* 433:531–537.
Cai, W., J. Pei, and N. V. Grishin. 2004. Reconstruction of ancestral protein sequences and its applications. *BMC Evolutionary Biology* 4:33.
Cameron, S. L., and M. F. Whiting. 2007. Mitochondrial genome comparisons of the subterranean termites from the genus *Reticulitermes* (Insecta: Isoptera: Rhinotermitidae). *Genome* 50:188–202.
Cannone, J. J., S. Subramanian, M. N. Schnare, J. R. Collett, L. M. D'Souza, Y. Du, B. Feng, N. Lin, L. V. Madabusi, K. M. Muller, N. Pande, Z. Shang, N. Yu, and R. R. Gutell. 2002. The Comparative RNA Web (CRW) Site: An online database of comparative sequence and structure information for ribosomal, intron and other RNAs. *BMC Bioinformatics* 3:2.
Cantarel, B. L., H. G. Morrison, and W. Pearson. 2006. Exploring the relationship between sequence similarity and accurate phylogenetic trees. *Molecular Biology and Evolution* 23:2090–2100.
Cao, Y., J. Adachi, A. Janke, S. Pääbo, and M. Hasegawa. 1994. Phylogenetic relationships among eutherian orders estimated from inferred sequences of mitochondrial proteins: Instability of a tree based on a single gene. *Journal of Molecular Evolution* 39:519–527.
Carrillo, H., and D. Lipman. 1988. The multiple sequence alignment problem in biology. *SIAM Journal on Applied Mathematics* 48:1073–1082.
Carter, D., and R. Durbin. 2006. Vertebrate gene finding from multiple-species alignments using a two-level strategy. *Genome Biology* 7:1–12.
Cartwright, R. A. 2005. DNA Assembly With Gaps (DAWG): Simulating sequence evolution. *Bioinformatics* 21:iii31–iii38.
Cartwright, R. A. 2006. Logarithmic gap costs decrease alignment accuracy. *BMC Bioinformatics* 7:527.
Castresana, J. 2000. Selection of conserved blocks from multiple alignments for their use in phylogenetic analysis. *Molecular Biology and Evolution* 17:540–552.

Cate, J. H., M. M. Yusupov, G. Z. Yusupova, T. N. Earnest, and H. F. Noller. 1999. X-ray crystal structures of 70S ribosome functional complexes. *Science* 285:2095–2104.

Catherinot, V., and G. Labesse. 2004. ViTO: Tool for refinement of protein sequence-structure alignments. *Bioinformatics* 20:3694–3696.

Cesareni, G., A. Ceol, C. Gavrila, L. M. Palazzi, M. Persico, and M. V. Schneider. 2005. Comparative interactomics. *FEBS Letters* 579:1828–1833.

Chain, P., S. Kurtz, E. Ohlebusch, and T. R. Slezak. 2003. An applications-focused review of comparative genomics tools: Capabilities, limitations, and future challenges. *Briefings in Bioinformatics* 4:105–123.

Chakrabarti, S., C. J. Lanczycki, A. R. Panchenko, T. M. Przytycka, P. A. Thiessen, and S. H. Bryant. 2006. Refining multiple sequence alignments with conserved core regions. *Nucleic Acids Research* 34:2598–2606.

Chang, M. S. S., and S. A. Benner. 2004. Empirical analysis of protein insertions and deletions determining parameters for the correct placement of gaps in protein sequence alignments. *Journal of Molecular Biology* 341:617–631.

Chapman, M. A., F. J. Charchar, S. Kinston, C. P. Bird, D. Grafham, J. Rogers, F. Grützner, J. A. Marshall Graves, A. R. Green, and B. Göttgens. 2003. Comparative and functional analysis of LYL1 loci establish marsupial sequences as a model for phylogenetic footprinting. *Genomics* 81:249–259.

Chelliah, V., L. Chen, T. L. Blundell, and S. C. Lovell. 2004. Distinguishing structural and functional restraints in evolution in order to identify interaction sites. *Journal of Molecular Biology* 342:1487–1504.

Chen, C. P., A. Kernytsky, and B. Rost. 2002. Transmembrane helix predictions revisited. *Protein Science* 11:2774–2791.

Cheng, G., B. Qian, R. Samudrala, and D. Baker. 2005. Improvement in protein functional site prediction by distinguishing structural and functional constraints on protein family evolution using computational design. *Nucleic Acids Research* 33:5861–5867.

Clamp, M., J. Cuff, S. M. Searle, and G. J. Barton. 2004. The Jalview Java alignment editor. *Bioinformatics* 20:426–427.

Clark, C. G., B. W. Tague, V. C. Ware, and S. A. Gerbi. 1984. *Xenopus laevis* 28S ribosomal RNA: A secondary structural model and its evolutionary and functional implications. *Nucleic Acids Research* 12:6197–6220.

Clarke, G. D., R. G. Beiko, M. A. Ragan, and R. L. Charlebois. 2002. Inferring genome trees by using a filter to eliminate phylogenetically discordant sequences and a distance matrix based on mean normalized BLASTP scores. *Journal of Bacteriology* 184:2072–2080.

Claude, J. B., K. Suhre, C. Notredame, J. M. Claverie, and C. Abergel. 2004. CaspR: A web server for automated molecular replacement using homology modelling. *Nucleic Acids Research* 32:W606–W609.

Cline, M., R. Hughey, and K. Karplus. 2002. Predicting reliable regions in protein sequence alignments. *Bioinformatics* 18:306–314.

Collins, T. M., P. H. Wimberger, and G. Naylor. 1994. Compositional bias, character-state bias, and character-state reconstruction using parsimony. *Systematic Biology* 43:482–496.

Combet, C., M. Jambon, G. Deleage, and C. Geourjon. 2002. Geno3D: Automatic comparative molecular modelling of protein. *Bioinformatics* 18:213–214.

Corpet, F. 1988. Multiple sequence alignment with hierarchical clustering. *Nucleic Acids Research* 16:10881–10890.

Coventry, A., D. J. Kleitman, and B. Berger. 2004. MSARI: Multiple sequence alignments for statistical detection of RNA secondary structure. *Proceedings of the National Academy of Sciences USA* 101:12102–12107.

Cox, M. F., and M. A. A. Cox. 2001. *Multidimensional Scaling.* New York: Chapman and Hall.

Csuros, M., and I. Miklós. 2005. Statistical alignment of retropseudogenes and their functional paralogs. *Molecular Biology and Evolution* 22:2457–2471.

Curwen, V., E. Eyras, T. D. Andrews, L. Clarke, E. Mongin, S. M. Searle, and M. Clamp. 2004. The Ensembl automatic gene annotation system. *Genome Research* 14:942–950.

Dandekar, T., M. Huynen, J. T. Regula, B. Ueberle, C. U. Zimmermann, M. A. Andrade, T. Doerks, L. Sanchez-Pulido, B. Snel, M. Suyama, Y. P. Yuan, R. Herrmann, and P. Bork. 2000. Reannotating the *Mycoplasma pneumoniae* genome sequence: Adding value, function and reading frames. *Nucleic Acids Research* 28:3278–3288.

Darling, A. C. E., B. Mau, F. R. Blattner, and N. T. Perna. 2004. MAUVE: Multiple alignment of conserved genomic sequence with rearrangements. *Genome Research* 14:1394–1403.

Daubin, V., N. A. Moran, and H. Ochman. 2003. Phylogenetics and the cohesion of bacterial genomes. *Science* 301:829–832.

Dayhoff, M. O., and R. V. Eck. 1968. *Atlas of Protein Sequence and Structure.* Volume 3. Silver Spring, MD: National Biomedical Research Foundation.

Dayhoff, M. O., R. M. Schwartz, and B. C. Orcut. 1978. A model of evoutionary change in proteins. Pp. 345–352 in *Atlas of Protein Sequence and Structure*, M. O. Dayhoff, ed., Volume 5. Silver Spring, MD: National Biomedical Research Foundation.

De Laet, J., and W. C. Wheeler. 2003. POY version 3.0.11 command line documentation. New York: distributed by the authors.

Delcher, A. L., S. Kasif, R. D. Fleischmann, J. Peterson, O. White, and S. L. Salzberg. 1999. Alignment of whole genomes. *Nucleic Acids Research* 27:2369–2376.

Delcher, A. L., A. Phillippy, J. Carlton, and S. L. Salzberg. 2002. Fast algorithms for large-scale genome alignment and comparison. *Nucleic Acids Research* 30:2478–2483.

Deleage, G., C. Combet, C. Blanchet, and C. Geourjon. 2001. ANTHEPROT: An integrated protein sequence analysis software with client/server capabilities. *Computers in Biology and Medicine* 31:259–267.

Delsuc, F., M. Scally, O. Madsen, M. J. Stanhope, W. W. de Jong, F. M. Catzeflis, M. S. Springer, and E. J. P. Douzery. 2002. Molecular phylogeny of living xenarthrans and the impact of character and taxon sampling on the placental tree rooting. *Molecular Biology and Evolution* 19:1656–1671.

Denduangboripant, J., and Q. C. B. Cronk. 2001. Evolution and alignment of the hypervariable arm 1 of *Aeschynanthus* (Gesneriaceae) ITS2 nuclear ribosomal DNA. *Molecular Phylogenetics and Evolution* 20:163–172.

Depiereux, E., G. Baudoux, P. Briffeuil, I. Regnister, X. De Boll, C. Vinals, and E. Feytmans. 1997. Match-Box server: A multiple sequence alignment tool placing emphasis on reliability. *Computer Applications in Bioscience* 13:249–256.

Depiereux, E., and E. Feytmans. 1992. MATCH-BOX: A fundamentally new algorithm for the simultaneous alignment of several protein sequences. *Computer Applications in Bioscience* 8:501–509.
Devos, D., and A. Valencia. 2001. Intrinsic errors in genome annotation. *Trends in Genetics* 17:429–431.
di Bernardo, D., T. Down, and T. Hubbard. 2004. ddbRNA: Detection of conserved secondary structures in multiple alignments. *Bioinformatics* 19:1606–1611.
Didier, G., and C. Guziolowski. 2007. Mapping sequences by parts. *Algorithms for Molecular Biology* 2:11.
Dixon, M. T., and D. M. Hillis. 1993. Ribosomal RNA secondary structure: Compensatory mutations and implications for phylogenetic analysis. *Molecular Biology and Evolution* 10:256–267.
Do, C. B., M. S. P. Mahabhashyam, M. Brudno, and S. Batzoglou. 2005. PROBCONS: Probabilistic consistency-based multiple sequence alignment. *Genome Research* 15:330–340.
Donoghue, M. J., R. G. Olmstead, J. F. Smith, and J. D. Palmer. 1992. Phylogenetic relationships of *Dipsacales* based on *rbcl* sequences. *Annals of the Missouri Botanical Garden* 79:333–345.
Douzery, E., and F. M. Catzeflis. 1995. Molecular evolution of the mitochondrial 12S rRNA in Ungulata (Mammalia). *Journal of Molecular Evolution* 41:622–636.
Dowell, R., and S. Eddy. 2004. Evaluation of several lightweight stochastic context-free grammars for RNA secondary structure prediction. *BMC Bioinformatics* 5:71.
Doyle, J. J., and J. I. Davis. 1998. Homology in molecular phylogenetics: A parsimony perspective. Pp. 101–131 in *Molecular Systematics of Plants II: DNA Sequencing*, D. E. Soltis, P. S. Soltis, and J. J. Doyle, eds. Boston, MA: Kluwer Academic.
Dubchak, I., M. Brudno, G. G. Loots, L. Pachter, C. Mayor, E. M. Rubin, and K. A. Frazer. 2000. Active conservation of noncoding sequences revealed by three-way species comparisons. *Genome Research* 10:1304–1306.
Durbin, R., S. Eddy, A. Krogh, and G. Mitchison. 1998. *Biological Sequence Analysis*. Cambridge, UK: Cambridge University Press.
Duret, L., and S. Abdeddaim. 2000. Multiple alignments for structrual, functional, or phylogenetic analyses of homologous sequences. Pp. 51–76 in *Bioinformatics: Sequence, Structure, and Databanks*, D. Higgins and W. Taylor, eds. Oxford: Oxford University Press.
Eddy, S. R. 1995. Multiple alignments using hidden Markov models. Pp. 114–120 in *Proceedings of the Third International Conference on Intelligent Systems for Molecular Biology*, Volume 3, C. Rawlings, D. Clark, R. Altman, L. Hunter, T. Lengauer and S. Wodak, eds. Menlo Park, CA: AAAI Press.
———. 1998. Profile hidden Markov models. *Bioinformatics* 14:755–763.
———. 2004. What is dynamic programming? *Nature Biotechnology* 22:909–910.
Edgar, R. C. 2004a. MUSCLE: A multiple sequence alignment method with reduced time and space complexity. *BMC Bioinformatics* 5:113.
———. 2004b. MUSCLE: Multiple sequence alignment with high accuracy and high throughput. *Nucleic Acids Research* 32:1792–1797.
Edgar, R. C., and S. Batzoglou. 2006. Multiple sequence alignment. *Current Opinion in Structural Biology* 16:368–73.

Eisen, J. A., and C. M. Fraser. 2003. Phylogenomics: Intersection of evolution and genomics. *Science* 300:1706–1707.

Eisenberg, D., E. M. Marcotte, I. Xenarios, and T. O. Yeates. 2000. Protein function in the post-genomic era. *Nature* 405:823–826.

Ellis, J., and D. Morrison. 1995. Effects of sequence alignment on the phylogeny of Sarcocystis deduced from 18S rDNA sequences. *Parasitology Research* 81:696–699.

Etzold, T., A. Ulyanov, and P. Argos. 1996. SRS: Information retrieval system for molecular biology data banks. *Methods in Enzymology* 266:114–128.

Eyre-Walker, A. 1998. Problems with parsimony in sequences of biased base composition. *Journal of Molecular Evolution* 47:686–690.

Farris, J. S., M. Källersjö, A. G. Kluge, and C. Bult. 1994. Testing significance of incongruence. *Cladistics* 10:315–319.

Felsenstein, J. 1981. Evolutionary trees from DNA sequences: A maximum likelihood approach. *Journal of Molecular Evolution* 17:368–376.

———. 1985. Confidence-limits on phylogenies: An approach using the bootstrap. *Evolution* 39:783–791.

———. 2004. *Inferring Phylogenies*. Sunderland, MA: Sinauer.

Feng, D.-F., and R. F. Doolittle. 1987. Progressive sequence alignment as a prerequisite to correct phylogenetic trees. *Journal of Molecular Evolution* 25:351–360.

———. 1990. Progressive alignment and phylogenetic tree construction of protein sequences. *Methods in Enzymology* 183:375–387.

Fitch, J. P., S. N. Gardner, T. A. Kuczmarski, S. Kurtz, R. Myers, L. L. Ott, T. R. Slezak, E. A. Vitalis, A. T. Zemla, and P. M. McCready. 2002. Rapid development of nucleic acid diagnostics. *Proceedings of the IEEE* 90:1708–1721.

Fitch, W. M. 1966. An improved method of testing for evolutionary homology. *Journal of Molecular Biology* 16:9–16.

———. 1971. Toward defining the course of evolution: Minimum change for a specific tree topology. *Systematic Zoology* 20:406–416.

Fleißner, R. 2004. *Sequence alignment and phylogenetic inference*. Berlin: Logos Verlag.

Fleißner, R., D. Metzler, and A. von Haeseler. 2000. Can one estimate distances from pairwise sequence alignments? Pp. 89–95 in *Proceedings of the German Conference on Bioinformatics*, E. Bornberg-Bauer, U. Rost, J. Stoye, and M. Vingron, eds. Berlin: Logos Verlag.

———. 2005. Simultaneous statistical multiple alignment and phylogeny reconstruction. *Systematic Biology* 54:548–561.

Flicek, P., E. Keibler, P. Hu, I. Korf, and M. R. Brent. 2003. Leveraging the mouse genome for gene prediction in human: From whole-genome shotgun reads to a global synteny map. *Genome Research* 13:46–54.

Formstecher, E., S. Aresta, V. Collura, A. Hamburger, A. Meil, A. Trehin, C. Reverdy, V. Betin, S. Maire, C. Brun, B. Jacq, M. Arpin, Y. Bellaiche, S. Bellusci, P. Benaroch, M. Bornens, R. Chanet, P. Chavrier, O. Delattre, V. Doye, R. Fehon, G. Faye, T. Galli, J. A. Girault, B. Goud, J. de Gunzburg, L. Johannes, M. P. Junier, V. Mirouse, A. Mukherjee, D. Papadopoulo, F. Perez, A. Plessis, C. Rosse, S. Saule, D. Stoppa-Lyonnet, A. Vincent, M. White, P. Legrain, J. Wojcik, J. Camonis, and L. Daviet. 2005. Protein interaction mapping: A *Drosophila* case study. *Genome Research* 15:376–384.

Frank, J., and R. K. Agrawal. 2000. A ratchet-like inter-subunit reorganization of the ribosome during translocation. *Nature* 406:318–322.

———. 2001. Ratchet-like movements between the two ribosomal subunits: Their implications in elongation factor recognition and tRNA translocation. *Cold Spring Harbor Symposia on Quantitative Biology* 66:67–75.

Frank, J., P. Penczek, R. A. Grassucci, A. B. Heagle, C. M. T. Spahn, and R. K. Agrawal. 2000. Cryo-electron microscopy of the translational apparatus: Experimental evidence for the paths of mRNA, tRNA, and the polypeptide chain. Pp. 45–51 in *The Ribosome: Structure, Function, Antibiotics, and Cellular Interactions*, R. A. Garrett, S. R. Douthwaite, A. Liljas, A. T. Matheson, P. B. Moore, and H. F. Noller, eds. Washington, DC: ASM Press.

Frazer, K. A., L. Pachter, A. Poliakov, E. M. Rubin, and I. Dubchak. 2004. VISTA: Computational tools for comparative genomics. *Nucleic Acids Research* 32:W273–W279.

Fredman, D., G. Munns, D. Rios, F. Sjoholm, M. Siegfried, B. Lenhard, H. Lehvaslaiho, and A. J. Brookes. 2004. HGVbase: A curated resource describing human DNA variation and phenotype relationships. *Nucleic Acids Research* 32:D516–D519.

Freudenstein, J. V., and M. W. Chase. 2001. Analysis of mitochondrial nad1b-c intron sequences in Orchidaceae: Utility and coding of length-change characters. *Systematic Botany* 26:643–657.

Frickey, T., and A. N. Lupas. 2004. PhyloGenie: Automated phylome generation and analysis. *Nucleic Acids Research* 32:5231–5238.

Friedrich, A., R. Ripp, N. Garnier, E. Bettler, G. Deleage, O. Poch, and L. Moulinier. 2007. Blast sampling for structural and functional analyses. *BMC Bioinformatics* 8:62.

Fuellen, G., J. W. Wagele, and R. Giegerich. 2002. Minimum conflict: A divide-and-conquer approach to phylogeny estimation. *Bioinformatics* 17:1168–1178.

Galtier, N., M. Gouy, and C. Gautier. 1996. SEAVIEW and PHYLO_WIN: Two graphic tools for sequence alignment and molecular phylogeny. *Computer Applications in Bioscience* 12:543–548.

Gardner, P. P., A. Wilm, and S. Washietl. 2005. A benchmark of multiple sequence alignment programs upon structural RNAs. *Nucleic Acids Research* 33:2433–2439.

Garnier, J., J. F. Gibrat, and B. Robson. 1996. GOR method for predicting protein secondary structure from amino acid sequence. *Methods in Enzymology* 266:540–553.

Garnier, N., A. Friedrich, R. Bolze, E. Bettler, L. Moulinier, C. Geourjon, J. D. Thompson, G. Deleage, and O. Poch. 2006. MAGOS: Multiple alignment and modelling server. *Bioinformatics* 22:2164–2165.

Gatesy, J., R. DeSalle, and W. C. Wheeler. 1993. Alignment-ambiguous nucleotide sites and the exclusion of systematic data. *Molecular Phylogenetics and Evolution* 2:152–157.

Gatesy, J., C. Hayashi, R. DeSalle, and E. Vrba. 1994. Rate limits for pairing and compensatory change: The mitochondrial ribosomal DNA of antelopes. *Evolution* 48:188–196.

Geman, S., and D. Geman. 1984. Stochastic relaxation, Gibbs distributions and the Bayesian restoration of images. *IEEE Transactions on Pattern Analysis and Machine Intelligence* 6:721–741.

Gibson, A., V. Gowri-Shankar, P. G. Higgs, and M. Rattray. 2005. A comprehensive analysis of mammalian mitochondrial genome base composition and improved phylogenetic methods. *Molecular Biology and Evolution* 22:251–264.

Gillespie, J. J. 2004. Characterizing regions of ambiguous alignment caused by the expansion and contraction of hairpin-stem loops in ribosomal RNA molecules. *Molecular Phylogenetics and Evolution* 33:936–943.

———. 2005. Structure-based methods for the phylogenetic analysis of ribosomal RNA molecules. Ph.D. dissertation, Texas A&M University, College Station, TX.

Gillespie, J. J., J. J. Cannone, R. R. Gutell, and A. I. Cognato. 2004. A secondary structural model of the 28S rRNA expansion segments D2 and D3 from rootworms and related leaf beetles (Coleoptera: Chrysomelidae: Galerucinae). *Insect Molecular Biology* 13:495–518.

Gillespie, J. J., J. S. Johnston, J. J. Cannone, and R. R. Gutell. 2006. Characteristics of the nuclear (18S, 5.8S, 28S, and 5S) and mitochondrial (16S and 12S) rRNA genes of *Apis mellifera* (Insecta: Hymenoptera): Structure, organization and retrotransposable elements. *Insect Molecular Biology* 15:657–686.

Gillespie, J. J., C. H. McKenna, M. J. Yoder, R. R. Gutell, J. S. Johnston, J. Kathirithamby, and A. I. Cognato. 2005. Assessing the odd secondary structural properties of nuclear small subunit ribosomal RNA sequences (18S) of the twisted-wing parasites (Insecta: Strepsiptera). *Insect Molecular Biology* 14:625–643.

Gillespie, J. J., J. B. Munro, J. M. Heraty, M. J. Yoder, A. K. Owen, and A. E. Carmichael. 2005. A secondary structural model of the 28S rRNA expansion segments D2 and D3 for chalcidoid wasps (Hymenoptera: Chalcidoidea). *Molecular Biology and Evolution* 22:1593–1608.

Gillespie, J. J., M. J. Yoder, and R. A. Wharton. 2005. Predicted secondary structures for expansion segments D2–D10 of the 28S large subunit ribosomal RNA from ichneumonoid Hymenoptera: Homology assignment and phylogenetic implications. *Journal of Molecular Evolution* 61:114–137.

Ginalski, K., N. V. Grishin, A. Godzik, and L. Rychlewski. 2005. Practical lessons from protein structure prediction. *Nucleic Acids Research* 33:1874–1891.

Giot, L., J. S. Bader, C. Brouwer, A. Chaudhuri, B. Kuang, Y. Li, Y. L. Hao, C. E. Ooi, B. Godwin, E. Vitols, G. Vijayadamodar, P. Pochart, H. Machineni, M. Welsh, Y. Kong, B. Zerhusen, R. Malcolm, Z. Varrone, A. Collis, M. Minto, S. Burgess, L. McDaniel, E. Stimpson, F. Spriggs, J. Williams, K. Neurath, N. Ioime, M. Agee, E. Voss, K. Furtak, R. Renzulli, N. Aanensen, S. Carrolla, E. Bickelhaupt, Y. Lazovatsky, A. DaSilva, J. Zhong, C. A. Stanyon, R. L. Finley, K. P. White, M. Braverman, T. Jarvie, S. Gold, M. Leach, J. Knight, R. A. Shimkets, M. P. McKenna, J. Chant, and J. M. Rothberg. 2003. A protein interaction map of *Drosophila melanogaster*. *Science* 302:1727–1736.

Giribet, G. 2005. Review. TNT: Tree analysis using new technology. *Systematic Biology* 54:176–178.

Giribet, G., and W. C. Wheeler. 2001. Some unusual small-subunit ribosomal RNA sequences of metazoans. *American Museum Novitates* 3337:1–14.

———. 2007. The case for sensitivity: A response to Grant and Kluge. *Cladistics* 23:1–3.

Gladstein, D. S., and W. C. Wheeler. 1999. POY: Phylogeny reconstruction via direct optimization of DNA data. Ver. 2.0. New York: American Museum of Natural History.

Godzik, A. 1996. The structural alignment between two proteins: Is there a unique answer? *Protein Science* 5:1325–1338.

Gogarten, P., W. F. Doolittle, and J. G. Lawrence. 2002. Prokaryotic evolution in light of gene transfer. *Molecular Biology and Evolution* 19:2226–2238.

Goldman, N. 1993. Statistical tests of models of DNA substitution. *Journal of Molecular Evolution* 36:182–198.

———. 1998. Effects of sequence alignment procedures on estimates of phylogeny. *BioEssays* 20:287–290.

Goldman, N., and S. Whelan. 2002. A novel use of equilibrium frequencies in models of sequence evolution. *Molecular Biology and Evolution* 19:1821–1831.

Gonnet, G. H., M. A. Cohen, and S. A. Brenner. 1992. Exhaustive matching of the entire protein database. *Science* 256:1443–1445.

Goonesekere, N. C., and B. Lee. 2004. Frequency of gaps observed in a structurally aligned protein pair database suggests a simple gap penalty function. *Nucleic Acids Research* 32:2838–2843.

Gotoh, O. 1982. An improved algorithm for matching biological sequences. *Journal of Molecular Biology* 162:705–708.

———. 1986. Alignment of three biological sequences with an efficient traceback procedure. *Journal of Theoretical Biology* 121:327–337.

———. 1996. Significant improvement in accuracy of multiple protein sequence alignments by iterative refinement as assessed by referent to structural alignments. *Journal of Molecular Biology* 264:823–838.

———. 1999. Multiple sequence alignments: Algorithms and applications. *Advances in Biophysics* 36:159–206.

Göttgens, B., L. M. Barton, J. G. R. Gilbert, A. J. Bench, M. J. Sanchez, S. Bahn, S. Mistry, D. Grafham, A. McMurray, M. Vaudin, E. Amaya, D. R. Bentley, and A. R. Green. 2000. Analysis of vertebrate SCL loci identifies conserved enhancers. *Nature Biotechnology* 18:181–186.

Grant, T., and A. G. Kluge. 2005. Stability, sensitivity, science, and heurism. *Cladistics* 21:597–604.

Grasso, C., and C. Lee. 2004. Combining partial order alignment and progressive multiple sequence alignment increases alignment speed and scalability to very large alignment problems. *Bioinformatics* 20:1546–1556.

Griffiths-Jones, S., and A. Bateman. 2002. The use of structure information to increase alignment accuracy does not aid homologue detection with profile HMMs. *Bioinformatics* 18:1243–1249.

Griffiths-Jones, S., S. Moxon, M. Marshall, A. Khanna, S. R. Eddy, and A. Bateman. 2005. Rfam: Annotating non-coding RNAs in complete genomes. *Nucleic Acids Research* 33:D121–D124.

References

Gusfield, D. 1997. *Algorithms on Strings, Trees, and Sequences: Computer Science and Computational Biology.* Cambridge, UK: Cambridge University Press.

Gutell, R. R. 1996. Comparative sequence analysis and the structure of 16S and 23S rRNA. Pp. 111–128 in *Ribosomal RNA: Structure, Evolution, Processing, and Function in Protein Biosynthesis*, A. E. Dahlberg and R. A. Zimmermann, eds. Boca Raton, FL: CRC Press.

Gutell, R. R., J. C. Lee, and J. J. Cannone. 2002. The accuracy of ribosomal RNA comparative structure models. *Current Opinion in Structural Biology* 12:301–310.

Hadjiolov, A. A., O. I. Georgiev, V. V. Nosikov, and L. P. Yarachev. 1984. Primary and secondary structure of rat 28S ribosomal RNA. *Nucleic Acids Research* 12:3677–3693.

Hall, B. G. 2005. Comparison of the accuracies of several phylogenetic methods using protein and DNA sequences. *Molecular Biology and Evolution* 22:792–802.

———. 2006. Simple and accurate estimation of ancestral protein sequences. *Proceedings of the National Academy of Sciences USA* 103:5431–5436.

———. 2008. Simulating DNA coding sequence evolution with EvolveA Gene3. *Molecular Biology and Evolution* 25:688–695.

Handt, O., S. Meyer, and A. von Haeseler. 1998. Compilation of human mtDNA control region sequences. *Nucleic Acids Research* 26:126–130.

Harding, R. M., A. J. Boyce, and J. B. Clegg. 1992. The evolution of tandemly repetitive DNA: Recombination rules. *Genetics* 132:847–859.

Hardison, R. C. 2003. Comparative genomics. *PLoS Biology* 1:E58.

Harms, J., F. Schluenzen, R. Zarivach, A. Bashan, S. Gat, I. Agmon, H. Bartels, F. Franceschi, and A. Yonath. 2001. High resolution structure of the large ribosomal subunit from a mesophilic eubacterium. *Cell* 107:679–688.

Hartmann, S., D. Lu, J. Phillips, and T. J. Vision. 2006. Phytome: A platform for plant comparative genomics. *Nucleic Acids Research* 34:D724–D730.

Hasegawa, M., H. Kishino, and T. Yano. 1985. Dating of the human-ape splitting by a molecular clock of mitochondrial DNA. *Journal of Molecular Evolution* 22:160–174.

Hastings, W. K. 1970. Monte Carlo sampling methods using Markov chains and their applications. *Biometrika* 57:97–109.

Havgaard, J. H., R. Lyngsø, G. D. Stormo, and J. Gorodkin. 2005. Pairwise local structural alignment of RNA sequences with sequence similarity less than 40%. *Bioinformatics* 21:1815–1824.

Hein, J. 2001. An algorithm for statistical alignment of sequences related by a binary tree. *Pacific Symposium on Biocomputing* 6:179–190.

Hein, J., J. L. Jensen, and C. N. S. Pedersen. 2003. Recursions for statistical multiple alignment. *Proceedings of the National Academy of Sciences USA* 100:14960–14965.

Hein, J., C. Wiuf, B. Knudsen, M. B. Møller, and G. Wibling. 2000. Statistical alignment: Computational properties, homology testing and goodness-of-fit. *Journal of Molecular Biology* 302:265–279.

Henikoff, S., and J. G. Henikoff. 1992. Amino acid substitution matrices from protein blocks. *Proceedings of the National Academy of Sciences USA* 89:10915–10919.

Heringa, J. 2000. Computational methods for protein secondary structure prediction using multiple sequence alignments. *Current Protein and Peptide Science* 1:273–301.

Hertz, G. Z., and G. D. Stormo. 1999. Identifying DNA and protein patterns with statistically significant alignments of multiple sequences. *Bioinformatics* 15:563–577.

Hickson, R. E., C. Simon, A. Cooper, G. S. Spicer, J. Sullivan, and D. Penny. 1996. Conserved sequence motifs, alignment, and secondary structure for the third domain of animal 12S rRNA. *Molecular Biology and Evolution* 13:150–169.

Hickson, R. E., C. Simon, and S. W. Perrey. 2000. The performance of several multiple-sequence alignment programs in relation to secondary-structure features for an rRNA sequence. *Molecular Biology and Evolution* 17:530–539.

Higgins, D. G., and P. M. Sharp. 1988. CLUSTAL: A package for performing multiple sequence alignment on a microcomputer. *Gene* 73:237–244.

Hillis, D. M. 1995. Approaches for assessing phylogenetic accuracy. *Systematic Biology* 44:3–16.

Hillis, D. M., and J. J. Bull. 1993. An empirical testing of bootstrapping as a method for assessing confidence in phylogenetic analysis. *Systematic Biology* 42:182–192.

Hillis, D. M., T. A. Heath, and K. St John. 2005. Analysis and visualization of tree space. *Systematic Biology* 54:471–482.

Hofacker, I., M. Fekete, and P. Stadler. 2002. Secondary structure prediction for aligned RNA sequences. *Journal of Molecular Biology* 319:1059–1066.

Hofacker, I. L., S. H. Bernhart, and P. F. Stadler. 2004. Alignment of RNA base pairing probability matrices. *Bioinformatics* 20:2222–2227.

Hogan, J. J., R. R. Gutell, and H. F. Noller. 1984. Probing the conformation of 26S rRNA in yeast 60S ribosomal subunits with kethoxal. *Biochemistry* 23:3330–3335.

Hogeweg, P., and B. Hesper. 1984. The alignment of sets of sequences and the construction of phyletic trees: An integrated method. *Journal of Molecular Evolution* 20:175–186.

Höhl, M., S. E. Kurtz, and E. Ohlebusch. 2002. Efficient multiple genome alignment. *Bioinformatics* 18:S312–S320.

Hollich, V., L. Milchert, L. Arvestad, and E. L. L. Sonnhammer. 2005. Assessment of protein distance measures and tree-building methods for phylogenetic tree reconstruction. *Molecular Biology and Evolution* 22:2257–2264.

Holm, L., and C. Sander. 1993. Protein structure comparison by alignment of distance matrices. *Journal of Molecular Biology* 233:123–138.

———. 1998. Touring protein fold space with Dali/FSSP. *Nucleic Acids Research* 26:316–319.

Holmes, I. 2003. Using guide tree to construct multiple-sequence evolutionary HMMS. Pp. 147–157 in *Proceedings of the Eleventh International Conference on Intelligent Systems for Molecular Biology*. Menlo Park, CA: AAAI Press.

———. 2005. Accelerated probabilistic inference of RNA structure evolution. *Bioinformatics* 6:73.

Holmes, I., and W. J. Bruno. 2001. Evolutionary HMMs: A Bayesian approach to multiple alignment. *Bioinformatics* 17:803–820.
Hsu, F., T. H. Pringle, R. M. Kuhn, D. Karolchik, M. Diekhans, D. Haussler, and W. J. Kent. 2005. The UCSC Proteome Browser. *Nucleic Acids Research* 33:D454–D458.
Huang, X., R. C. Hardison, and W. Miller. 1990. A space-efficient algorithm for local similarities. *Computer Applications in Bioscience* 6:373–381.
Huang, X., and W. Miller. 1991. A time-efficient, linear-space local similarity algorithm. *Advances in Applied Mathematics* 12:337–357.
Hubbard, T. J., B. Ailey, S. E. Brenner, A. G. Murzin, and C. Chothia. 1999. SCOP: A structural classification of proteins database. *Nucleic Acids Research* 27:254–256.
Hudek, A., and D. Brown. 2005. Ancestral sequence alignment under optimal conditions. *BMC Bioinformatics* 6:273.
Huelsenbeck, J. P., and B. Rannala. 2004. Frequentist properties of Bayesian posterior probabilities of phylogenetic trees under simple and complex substitution models. *Systematic Biology* 53:904–913.
Hung, G. C., N. B. Chilton, I. Beveridge, and R. B. Gasser. 1999. Secondary structure model for the ITS-2 precursor rRNA of strongyloid nematodes of equids: Implications for phylogenetic inference. *International Journal for Parasitology* 29:1949–1964.
Huynen, M., T. Dandekar, and P. Bork. 1998. Differential genome analysis applied to the species-specific features of *Helicobacter pylori*. *FEBS Letters* 426:1–5.
Hypša, V. 2006. Parasite histories and novel phylogenetic tools: Alternative approaches to inferring parasite evolution from molecular markers. *International Journal for Parasitology* 36:141–155.
Hypša, V., A. Škeříková, and T. Scholz. 2005. Phylogeny, evolution and host-parasite relationships of the order Proteocephalidea (Eucestoda) as revealed by combined analysis and secondary structure characters. *Parasitology* 130:359–371.
Innis, C. A., J. Shi, and T. L. Blundell. 2000. Evolutionary trace analysis of TGF-beta and related growth factors: Implications for site-directed mutagenesis. *Protein Engineering* 13:839–847.
International Human Genome Sequencing Consortium. 2001. Initial sequencing and analysis of the human genome. *Nature* 409:860–892.
Irizarry, K., V. Kustanovich, C. Li, N. Brown, S. Nelson, W. Wong, and C. J. Lee. 2000. Genome-wide analysis of single-nucleotide polymorphisms in human expressed sequences. *Nature Genetics* 26:233–236.
Ison, J. C., M. J. Blades, A. J. Bleasby, S. C. Daniel, J. H. Parish, and J. B. Findlay. 2000. Key residues approach to the definition of protein families and analysis of sparse family signatures. *Proteins* 40:330–341.
Jackson, S. A., J. J. Cannone, J. C. Lee, R. R. Gutell, and S. A. Woodson. 2002. Distribution of rRNA introns in the three-dimensional structure of the ribosome. *Journal of Molecular Biology* 323:35–52.
Jeanmougin, F., J. D. Thompson, M. Gouy, D. G. Higgins, and T. J. Gibson. 1998. Multiple sequence alignment with Clustal X. *Trends in Biochemical Sciences* 23:403–405.

Jensen, J. L., and J. Hein. 2005. Gibbs sampler for statistical multiple alignment. *Statistica Sinica* 15:889–907.
Johnson, J. M., K. Mason, C. Moallemi, H. Xi, S. Somaroo, and E. S. Huang. 2003. Protein family annotation in a multiple alignment viewer. *Bioinformatics* 19:544–545.
Jones, D. T. 1999. Protein secondary structure prediction based on position-specific scoring matrices. *Journal of Molecular Biology* 292:195–202.
Jones, D. T., W. R. Taylor, and J. M. Thornton. 1992. The rapid generation of mutation data matrices from protein sequences. *Computer Applications in Bioscience* 8:275–282.
Jones, N. C., D. Zhi, and B. J. Raphael. 2006. AliWABA: Alignment on the web through an A-Bruijn approach. *Nucleic Acids Research* 34:W613–W616.
Jow, H., V. Gowri-Shankar, and B. Guillard. 2005. *PHASE*: A software package for *Ph*ylogenetics *A*nd *S*equence *E*volution: Program and documentation available at http://www.cs.man.ac.uk/~gowrishv/beta-release/).
Jow, H., C. Hudelot, M. Rattay, and P. G. Higgs. 2002. Bayesian phylogenetics using an RNA substitution model applied to early mammalian evolution. *Molecular Biology and Evolution* 19:1591–1601.
Jue, R. A., N. W. Woodbury, and R. F. Doolittle. 1980. Sequence homologies among *E. coli* ribosomal proteins: Evidence for evolutionarily related groupings and internal duplications. *Journal of Molecular Evolution* 15:129–148.
Jukes, T. H., and C. R. Cantor. 1969. Evolution of protein molecules. Pp. 21–132 in *Mammalian Protein Metabolism*, H. N. Munro, ed. New York: Academic Press.
Kann, M., B. Qian, and R. A. Goldstein. 2000. Optimization of a new score function for the detection of remote homologues. *Proteins* 41:498–503.
Karplus, K., C. Barrett, and R. Hughey. 1998. Hidden Markov models for detecting remote protein homologies. *Bioinformatics* 10:846–856.
Karplus, K., and B. Hu. 2001. Evaluation of protein multiple alignments by SAM-T99 using the BAliBASE multiple alignment test set. *Bioinformatics* 17:713–720.
Katoh, K., K. Kuma, H. Toh, and T. Miyata. 2005. MAFFT version 5: Improvement in accuracy of multiple sequence alignment. *Nucleic Acids Research* 33:511–518.
Katoh, K., K. Misawa, K.-I. Kuma, and T. Miyata. 2002. MAFFT: A novel method for rapid multiple sequence alignment based on fast Fourier transform. *Nucleic Acids Research* 30:3059–3066.
Keightley, P. D., and T. Johnson. 2004. MCALIGN: Stochastic alignment of noncoding DNA sequences based on an evolutionary model of sequence evolution. *Genome Research* 14:442–450.
Kent, W. J., and A. M. Zahler. 2000. Conservation, regulation, synteny, and introns in a large-scale *C. briggsae-C. elegans* genomic alignment. *Genome Research* 10:1115–1125.
Kepes, F. 2003. Periodic epi-organization of the yeast genome revealed by the distribution of promoter sites. *Journal of Molecular Biology* 329:859–865.
Kim, Y., and S. Subramaniam. 2006. Locally defined protein phylogenetic profiles reveal previously missed protein interactions and functional relationships. *Proteins* 62:1115–1124.

Kimura, M. 1980. A simple method for estimating evolutionary rates of base substitutions through comparative studies of nucleotide sequences. *Journal of Molecular Evolution* 16:111–120.

———. 1983. *The Neutral Theory of Molecular Evolution.* New York: Cambridge University Press.

Kindlund, E., M. T. Tammi, E. Arner, D. Nilsson, and B. Andersson. 2007. GRAT—genome-scale rapid alignment tool. *Computer Methods and Programs in Biomedicine* 86:87–92.

King, R. D., and M. J. Sternberg. 1996. Identification and application of the concepts important for accurate and reliable protein secondary structure prediction. *Protein Science* 5:2298–2310.

Kirkness, E. F., V. Bafna, A. L. Halpern, S. Levy, K. Remington, D. B. Rusch, A. L. Delcher, M. Pop, W. Wang, C. M. Fraser, and J. C. Venter. 2003. The dog genome: Survey sequencing and comparative analysis. *Science* 301:1898–1903.

Kjer, K. M. 1995. Use of rRNA secondary structure in phylogenetic studies to identify homologous positions: An example of alignment and data presentation from the frogs. *Molecular Phylogenetics and Evolution* 4:314–330.

———. 1997. An alignment template for amphibian 12S rRNA, domain III: Conserved primary and secondary structural motifs. *Journal of Herpetology* 31:599–604.

———. 2004. Aligned 18S and insect phylogeny. *Systematic Biology* 53:506–514.

Kjer, K. M., G. D. Baldridge, and A. M. Fallon. 1994. Mosquito large subunit ribosomal RNA: Simultaneous alignment of primary and secondary structure. *Biochimica et Biophysica Acta* 1217:147–155.

Kjer, K. M., R. J. Blahnik, and R. W. Holzenthal. 2001. Phylogeny of Trichoptera (caddisflies): Characterization of signal and noise within multiple datasets. *Systematic Biology* 50:781–816.

Kjer, K. M., J. J. Gillespie, and K. A. Ober. 2006. Structural homology in ribosomal RNA, and a deliberation on POY. *Arthropod Systematics and Phylogeny* 64:71–76.

———. 2007. Opinions on multiple sequence alignment, and an empirical comparison of repeatability and accuracy between POY and structural alignment. *Systematic Biology* 56:133–146.

Kjer, K. M., and R. L. Honeycutt. 2007. Site specific rates of mitochondrial genomes and the phylogeny of Eutheria. *BMC Evolutionary Biology* 7:8.

Klein, R. J., and S. R. Eddy. 2003. RSEARCH: Finding homologs of single structured RNA sequences. *BMC Bioinformatics* 4:44.

Knudsen, B., and J. Hein. 1999. RNA secondary structure prediction using stochastic context-free grammars and evolutionary history. *Bioinformatics* 15:446–454.

Knudsen, B., and M. M. Miyamoto. 2003. Sequence alignments and pair hidden Markov models. *Journal of Molecular Biology* 333:453–465.

Koehl, P. 2001. Protein structure similarities. *Current Opinion in Structural Biology* 11:348–353.

Kondrashov, F. A., and E. V. Koonin. 2003. Evolution of alternative splicing: Deletions, insertions and origin of functional parts of proteins from intron sequences. *Trends in Genetics* 19:115–119.

Korf, I., P. Flicek, D. Duan, and M. R. Brent. 2001. Integrating genomic homology into gene structure prediction. *Bioinformatics* 17:S140–S148.

Korn, L. J., C. L. Queen, and M. N. Wegman. 1977. Computer analysis of nucleic acid regulatory sequences. *Proceedings of the National Academy of Sciences USA* 74:4401–4405.

Koski, L. B., and G. B. Golding. 2001. The closest BLAST hit is often not the nearest neighbor. *Journal of Molecular Evolution* 52:540–542.

Krane, D. E., and M. L. Raymer. 2003. *Fundamental Concepts of Bioinformatics*. San Francisco: Benjamin Cummings.

Kraus, F., L. Jarecki, M. Miyamoto, S. Tanhauser, and P. Laipis. 1992. Mispairing and compensational changes during the evolution of mitochondrial ribosomal RNA. *Molecular Biology and Evolution* 9:770–774.

Krogh, A., M. Brown, I. S. Mian, K. V. Sjölander, and D. Haussler. 1994. Hidden Markov models in computational biology: Applications to protein modeling. *Journal of Molecular Biology* 235:1501–1531.

Kryshtafovych, A., C. Venclovas, K. Fidelis, and J. Moult. 2005. Progress over the first decade of CASP experiments. *Proteins* 61:225–236.

Kumar, S. 1996. A stepwise algorithm for finding minimum evolution trees. *Molecular Biology and Evolution* 13:584–593.

Kumar, S., and A. Filipski. 2007. Multiple sequence alignment: In pursuit of homologous DNA positions. *Genome Research* 17:127–135.

Kurland, C. G., B. Canback, and O. G. Berg. 2003. Horizontal gene transfer: A critical view. *Proceedings of the National Academy of Sciences USA* 100:9658–9662.

Laamanen, T. R., R. Meier, M. A. Miller, A. Hille, and B. M. Wiegmann. 2005. Phylogenetic analysis of *Themira* (Sepsidae: Diptera): Sensitivity analysis, alignment, and indel treatment in a multigene study. *Cladistics* 21:258–271.

Lackner, P., W. A. Koppensteiner, M. J. Sippl, and F. S. Domingues. 2000. ProSup: A refined tool for protein structure alignment. *Protein Engineering* 13:745–752.

Lake, J. A. 1991. The order of sequence alignment can bias the selection of tree topology. *Molecular Biology and Evolution* 8:378–385.

Lanyon, S. M. 1988. The stochastic mode of molecular evolution: What consequences for systematic investigations? *Auk* 105:563–573.

Lassmann, T., and E. L. L. Sonnhammer. 2002. Quality assessment of multiple alignment programs. *FEBS Letters* 529:126–130.

———. 2005a. Automatic assessment of alignment quality. *Nucleic Acids Research* 33:7120–7128.

———. 2005b. Kalign—an accurate and fast multiple sequence alignment algorithm. *BMC Bioinformatics* 6:298.

———. 2006. Kalign, Kalignvu and Mumsa: Web servers for multiple sequence alignment. *Nucleic Acids Research* 34:W596–W599.

Lecompte, O., J. D. Thompson, F. Plewniak, J.-C. Thierry, and O. Poch. 2001. Multiple alignment of complete sequences (MACS) in the post-genomic era. *Gene* 270:17–30.

Lee, C., C. Grasso, and M. F. Sharlow. 2002. Multiple sequence alignment using partial order graphs. *Bioinformatics* 18:452–464.

Lee, J. C., and R. R. Gutell. 2004. Diversity of base-pair conformations and their occurrence in rRNA structure and RNA structural motifs. *Journal of Molecular Biology* 344:1225–1249.
Lee, S., B. C. Lee, and D. Kim. 2006. Prediction of protein secondary structure content using amino acid composition and evolutionary information. *Proteins* 62:1107–1114.
Lehtonen, S. 2008. Phylogeny estimation and alignment via POY versus Clustal + PAUP*: A response to Ogden and Rosenberg (2007). *Systematic Biology* 57:653–657.
Lesk, A. M. 2002. *Introduction to Bioinformatics.* Oxford, UK: Oxford University Press.
Lesk, A. M., and C. Chothia. 1980. How different amino acid sequences determine similar protein structures: The structure and evolutionary dynamics of the globins. *Journal of Molecular Biology* 136:225–270.
Levinson, G., and G. A. Gutman. 1987. Slipped-strand mispairing: A major mechanism for DNA sequence evolution. *Molecular Biology and Evolution* 4:203–221.
Li, W., and A. Godzik. 2006. VISSA: A program to visualize structural features from structure sequence alignment. *Bioinformatics* 22:887–888.
Lichtarge, O., H. R. Bourne, and F. E. Cohen. 1996. An evolutionary trace method defines binding surfaces common to protein families. *Journal of Molecular Biology* 257:342–358.
Linial, M. 2003. How incorrect annotations evolve—the case of short ORFs. *Trends in Biotechnology* 21:298–300.
Lipman, D. J., and W. R. Pearson. 1985. Rapid and sensitive protein similarity searches. *Science* 227:1435—1441.
Liu, D., X. Xiong, Z. G. Hou, and B. DasGupta. 2005. Identification of motifs with insertions and deletions in protein sequences using self-organizing neural networks. *Neural Networks* 18:835–842.
Liu, J. S. 2001. *Monte Carlo Strategies in Scientific Computing.* New York: Springer.
Lo Conte, L., S. E. Brenner, T. J. Hubbard, C. Chothia, and A. G. Murzin. 2002. SCOP database in 2002: Refinements accommodate structural genomics. *Nucleic Acids Research* 30:264–267.
Lokau, H. 2006. Vergleich von zwei Bayesschen Sampling-Methoden für statistisches multiples Alignment am Beispiel vierblättriger Bäume. Frühere Diplomanden dissertation, University of Frankfurt, Frankfurt, Germany.
Lopez-Correa, C., H. Brems, C. Lazaro, P. Marynen, and L. E. 2000. Unequal meiotic crossover: A frequent cause of NF1 microdeletions. *American Journal of Human Genetics* 66:1969–1974.
Löytynoja, A., and M. C. Milinkovitch. 2003. A hidden Markov model for progressive multiple alignment. *Bioinformatics* 19:1505–1513.
Lunter, G., I. Miklos, A. Drummond, J. L. Jensen, and J. Hein. 2005. Bayesian coestimation of phylogeny and sequence alignment. *BMC Bioinformatics* 6:83.
Lunter, G., I. Miklós, and J. Jensen. 2003. Bayesian phylogenetic inference under a satistical insertion-deletion model. Pp. 228–244 in *Proceedings of the Third InternationalWorkshop on Algorithms in Bioinformatics, Budapest,* G. Benson and R. Page, eds. Berlin: Springer.

Lunter, G. A., I. Miklos, Y. S. Song, and J. Hein. 2003. An efficient algorithm for statistical multiple alignment on arbitrary phylogenetic trees. *Journal of Computational Biology* 10:869–889.

Lunter, G., C. Ponting, and J. Hein. 2006. Genome-wide identication of human functional DNA using a neutral indel model. *PLoS Computational Biology* 13:e5.

Lutzoni, F. M., P. Wagner, V. Reeb, and S. Zoller. 2000. Integrating ambiguously aligned regions of DNA sequences in phylogenetic analyses without violating positional homology. *Systematic Biology* 49:628–651.

Luz, H., and M. Vingron. 2006. Family specific rates of protein evolution. *Bioinformatics* 22:1166–1171.

Madhusudhan, M. S., M. A. Marti-Renom, R. Sanchez, and A. Sali. 2006. Variable gap penalty for protein sequence-structure alignment. *Protein Engineering, Design and Selection* 19:129–133.

Marcotte, E. M., M. Pellegrini, H. L. Ng, D. W. Rice, T. O. Yeates, and D. Eisenberg. 1999. Detecting protein function and protein-protein interactions from genome sequences. *Science* 285:751–753.

Marti-Renom, M. A., A. C. Stuart, A. Fiser, R. Sanchez, F. Melo, and A. Sali. 2000. Comparative protein structure modeling of genes and genomes. *Annual Review of Biophysics and Biomolecular Structure* 29:291–325.

Mathews, D. H., and D. H. Turner. 2002. Dynalign: An algorithm for finding the secondary structure common to two RNA sequences. *Journal of Molecular Biology* 317:191–203.

McClure, M. A., T. K. Vasi, and W. M. Fitch. 1994. Comparative analysis of multiple protein sequence alignment methods. *Molecular Biology and Evolution* 11:571–592.

McGuire, G., M. C. Denham, and D. J. Balding. 2001. Models of sequence evolution for DNA sequences containing gaps. *Molecular Biology and Evolution* 18:481–490.

McKenna, M. C., and S. K. Bell. 1997. *Classification of Mammals: Above the Species Level*. New York: Columbia University Press.

Mclachlan, A. D. 1972. A mathematical procedure for superimposing atomic coordinates of proteins. *Acta Crystallographica A* A28:656–657.

Mears, J. A., M. R. Sharma, R. R. Gutell, A. S. McCook, P. E. Richardson, T. R. Caulfield, R. K. Agrawal, and S. C. Harvey. 2006. A structural model for the large subunit of the mammalian mitochondrial ribosome. *Journal of Molecular Biology* 358:193–212.

Metzler, D. 2003. Statistical alignment based on fragment insertion and deletion models. *Bioinformatics* 19:490–499.

Metzler, D., R. Fleißner, A. Wakolbinger, and A. von Haeseler. 2001. Assessing variability by joint sampling of alignments and mutation rates. *Journal of Molecular Evolution* 53:660–669.

———. 2005. Stochastic insertion-deletion processes and statistical sequence alignment. Pp. 247–264 in *Interacting Stochastic Systems*, J. Deuschel and A. Greven, eds. Berlin: Springer.

Meyer, A. 1994. Shortcomings of the cytochrome b genes as a molecular marker. *Trends in Ecology and Evolution* 9:278–280.

Meyer, I. M., and R. Durbin. 2002. Comparative *ab initio* prediction of gene structures using pair HMMs. *Bioinformatics* 18:1309–1318.

Michot, B., N. Hassouna, and J.-P. Bachelorize. 1984. Secondary structure of mouse 28S rRNA and general model for the folding of the large rRNA in eukaryotes. *Nucleic Acids Research* 12:4259–4279.

Mihalek, I., I. Res, and O. Lichtarge. 2006. A structure and evolution-guided Monte Carlo sequence selection strategy for multiple alignment-based analysis of proteins. *Bioinformatics* 22:149–156.

Mika, S., and B. Rost. 2003. UniqueProt: Creating representative protein sequence sets. *Nucleic Acids Research* 31:3789–3791.

Miklós, I., G. A. Lunter, and I. Holmes. 2004. A "long indel" model for evolutionary sequence alignment. *Molecular Biology and Evolution* 21:529–540.

Miller, M. P., and S. Kumar. 2001. Understanding human disease mutations through the use of interspecific genetic variation. *Human Molecular Genetics* 10:2319–2328.

Miller, M. P., J. D. Parker, S. W. Rissing, and S. Kumar. 2003. Quantifying the intragenic distribution of disease mutations. *Annals of Human Genetics* 67:567–579.

Mills, R. E., E. A. Bennett, R. C. Iskow, C. T. Luttig, C. Tsui, W. S. Pittard, and S. E. Devine. 2006. Recently mobilized transposons in the human and chimpanzee genomes. *American Journal of Human Genetics* 78:671–679.

Mills, R. E., C. T. Luttig, C. E. Larkins, A. Beauchamp, C. W. Tsui, S. Pittard, and S. E. Devine. 2006. An initial map of insertion and deletion (indel) variation in the human genome. *Genome Research* 16:1182–1190.

Mishler, B. D., K. Bremer, C. J. Humphries, and S. P. Churchill. 1988. The use of nucleic acid sequence data in phylogenetic reconstruction. *Taxon* 37:391–395.

Misof, B., and G. Fleck. 2003. Comparative analysis of mt LSU rRNA secondary structures of Odonates: Structural variability and phylogenetic signal. *Insect Molecular Biology* 12:535–547.

Misof, B., and K. Misof. In press. A Monte Carlo approach successfully identifies randomness in multiple sequence alignments: A more objective means of data exclusion. *Systematic Biology*.

Mitchison, G. 1999. A probabilistic treatment of phylogeny and sequence alignment. *Journal of Molecular Evolution* 49:11–22.

Mitchison, G., and R. Durbin. 1995. Tree-based maximal likelihood substitution matrices and hidden Markov models. *Journal of Molecular Evolution* 41:1139–1151.

Mizuguchi, K., C. M. Deane, T. L. Blundell, M. S. Johnson, and J. P. Overington. 1998. JOY: Protein sequence–structure representation and analysis. *Bioinformatics* 14:617–623.

Mizuguchi, K., C. M. Deane, T. L. Blundell, and J. P. Overington. 1998. HOMSTRAD: A database of protein structure alignments for homologous families. *Protein Science* 7:2469–2471.

Morgan, J. A. T., and D. Blair. 1998. Trematode and monogenean rRNA ITS2 secondary structures support a four-domain model. *Journal of Molecular Evolution* 47:406–419.

Morgenstern, B. 1999. DIALIGN 2: Improvement of the segment-to-segment approach to multiple sequence alignment. *Bioinformatics* 15:211–218.

———. 2000. A space-efficient algorithm for aligning large genomic sequences. *Bioinformatics* 16:948–949.

———. 2002. A simple and space-efficient fragment-chaining algorithm for alignment of DNA and protein sequences. *Applied Mathematics Letters* 15:11–16.

———. 2004. DIALIGN: Multiple DNA and protein sequence alignment at BiBiServ. *Nucleic Acids Research* 32:W33–W36.

Morgenstern, B., W. R. Atchley, K. Hahn, and A. Dress. 1998. Segment-based scores for pairwise and multiple sequence alignments. Pp. 115–121 in *Proceedings of the International Conference on Intelligent Systems for Molecular Biology*, Volume 6.

Morgenstern, B., A. Dress, and T. Werner. 1996. Multiple DNA and protein sequence alignment based on segment-to-segment comparison. *Proceedings of the National Academy of Sciences USA* 93:12098–12103.

Morgenstern, B., K. Frech, A. Dress, and T. Werner. 1998. DIALIGN: Finding local similarities by multiple sequence alignment. *Bioinformatics* 14:290–294.

Morgenstern, B., S. J. Prohaska, D. Pöhler, and P. F. Stadler. 2006. Multiple sequence alignment with user-defined anchor points. *Algorithms for Molecular Biology* 1:6.

Morgenstern, B., S. J. Prohaska, N. Werner, J. Weyer-Menkhoff, I. Schneider, A. R. Subramanian, and P. F. Stadler. 2004. Multiple sequence alignment with user-defined constraints. *Proceedings of the German Conference on Bioinformatics 2004* 53:25–36.

Morrison, D. A. 2006. Multiple sequence alignment for phylogenetic purposes. *Australian Systematic Botany* 19:479–539.

Morrison, D. A., and J. T. Ellis. 1997. Effects of nucleotide sequence alignment on phylogeny estimation: A case study of 18S rDNAs of Apicomplexa. *Molecular Biology and Evolution* 14:428–441.

Mott, R. 1999. Local sequence alignments with monotonic gap penalties. *Bioinformatics* 15:455–462.

Moult, J. 2005. A decade of CASP: Progress, bottlenecks and prognosis in protein structure prediction. *Current Opinion in Structural Biology* 15:285–289.

Mugridge, N. B., D. A. Morrison, T. Jakel, A. R. Heckeroth, A. M. Tenter, and A. M. Johnson. 2000. Effects of sequence alignment and structural domains of ribosomal DNA on phylogeny reconstruction for the protozoan family Sarcocystidae. *Molecular Biology and Evolution* 17:1842–1853.

Mulder, N. J., R. Apweiler, T. K. Attwood, A. Bairoch, A. Bateman, D. Binns, P. Bradley, P. Bork, P. Bucher, L. Cerutti, R. Copley, E. Courcelle, U. Das, R. Durbin, W. Fleischmann, J. Gough, D. Haft, N. Harte, N. Hulo, D. Kahn, A. Kanapin, M. Krestyaninova, D. Lonsdale, R. Lopez, I. Letunic, M. Madera, J. Maslen, J. McDowall, A. Mitchell, A. N. Nikolskaya, S. Orchard, M. Pagni, C. P. Ponting, E. Quevillon, J. Selengut, C. J. Sigrist, V. Silventoinen, D. J. Studholme, R. Vaughan, and C. H. Wu. 2005. InterPro, progress and status in 2005. *Nucleic Acids Research* 33:D201–D205.

Müller, T., and M. Vingron. 2000. Modeling amino acid replacement. *Journal of Computational Biology* 7:761–776.

Murata, M., J. S. Richardson, and J. L. Sussman. 1985. Simultaneous comparison of three protein sequences. *Proceedings of the National Academy of Sciences USA* 82:3073–3077.

Myers, E. W., and W. Miller. 1988. Optimal alignments in linear-space. *Computer Applications in Bioscience* 4:11–17.

Naor, D., and D. L. Brutlag. 1994. On near-optimal alignments of biological sequences. *Journal of Computational Biology* 1:349–366.

Needleman, S. B., and T. T. Blair. 1969. Homology of *Pseudomonas* cytochrome c-551 with eukaryotic c-cytochromes. *Proceedings of the National Academy of Sciences USA* 63:1227–1233.

Needleman, S. B., and C. D. Wunsch. 1970. A general method applicable to the search for similarities in the amino acid sequence of two proteins. *Journal of Molecular Biology* 48:443–453.

Nei, M. 1996. Phylogenetic analysis in molecular evolutionary genetics. *Annual Review of Genetics* 30:371–403.

Nei, M., S. Kumar, and K. Takahashi. 1998. The optimization principle in phylogenetic analysis tends to give incorrect topologies when the number of nucleotides or amino acids used is small. *Proceedings of the National Academy of Sciences USA* 95:12390–12397.

Nesbø, C. L., S. L. Haridon, K. O. Stetter, and W. F. Doolittle. 2001. Phylogenetic analysis of two "archaeal" genes in *Thermotoga maritima* reveal multiple transfers between Archaea and Bacteria. *Molecular Biology and Evolution* 18:362–375.

Nevill-Manning, C. G., T. D. Wu, and D. L. Brutlag. 1998. Highly specific protein sequence motifs for genome analysis. *Proceedings of the National Academy of Sciences USA* 95:5865–5871.

Newton, M., and A. Raftery. 1994. Approximate Bayesian inference with the weighted likelihood bootstrap. *Journal of the Royal Statistical Society, Series B* 56:3–48.

Ng, P. C., and S. Henikoff. 2003. SIFT: Predicting amino acid changes that affect protein function. *Nucleic Acids Research* 31:3812–3814.

Noller, H. F. 2005. RNA structure: Reading the ribosome. *Science* 309: 1508–1514.

Norris, J. R. 1997. *Markov Chains*. Cambridge, UK: Cambridge University Press.

Notredame, C. 2002. Recent progress in multiple sequence alignment: A survey. *Pharmacogenomics* 3:131–144.

Notredame, C., and C. Abergel. 2003. Using multiple alignment methods to assess the quality of genomic data analysis. Pp. 30–50 in *Bioinformatics and Genomes: Current Perspectives*, M. A. Andrade, ed. Wymondham, UK: Horizon Scientific Press.

Notredame, C., and D. G. Higgins. 1996. SAGA: Sequence alignment by genetic algorithm. *Nucleic Acids Research* 24:1515–1524.

Notredame, C., D. G. Higgins, and J. Heringa. 2000. T-COFFEE: A novel method for fast and accurate multiple sequence alignment. *Journal of Molecular Biology* 302:205–217.

Notredame, C., L. Holm, and D. G. Higgins. 1998. COFFEE: An objective function for multiple sequence alignments. *Bioinformatics* 14:407–422.

Novacek, M. J. 1992. Mammal phylogenies: Shaking the tree. *Nature* 356: 121–125.

Novacek, M. J., A. R. Wyss, and M. C. McKenna. 1988. The major groups of eutherian mammals. Pp. 31–71 in *The Phylogeny and Classification of the Tetrapods*, M. J. Benton, ed. Oxford, UK: Clarendon Press.

Nuin, P. A., Z. Wang, and E. R. Tillier. 2006. The accuracy of several multiple sequence alignment programs for proteins. *BMC Bioinformatics* 7:471.

Nussinov, R., and A. B. Jacobson. 1980. Fast algorithm for predicting the secondary structure of a single-stranded RNA. *Proceedings of the National Academy of Sciences USA* 78:6309–6313.

O'Donoghue, P., and Z. Luthey-Schulten. 2005. Evolutionary profiles derived from the QR factorization of multiple structural alignments gives an economy of information. *Journal of Molecular Biology* 346:875–894.

O'Sullivan, O., K. Suhre, C. Abergel, D. G. Higgins, and C. Notredame. 2004. 3DCoffee: Combining protein sequences and structures within multiple sequence alignments. *Journal of Molecular Biology* 340:385–395.

O'Sullivan, O., M. Zehnder, D. G. Higgins, P. Bucher, A. Grosdidier, and C. Notredame. 2003. APDB: A novel measure for benchmarking sequence alignment methods without reference alignments. *Bioinformatics* 19:i215–i221.

Ogden, T. H., and M. S. Rosenberg. 2006. Multiple sequence alignment accuracy and phylogenetic inference. *Systematic Biology* 55:314–328.

———. 2007a. Alignment and topological accuracy of the direct optimization approach via POY and traditional phylogenetics via ClustalW + PAUP*. *Systematic Biology* 56:182–193.

———. 2007b. How should gaps be treated in parsimony? A comparison of approaches using simulation. *Molecular Phylogenetics and Evolution* 42:817–826.

Ogden, T. H., and M. F. Whiting. 2003. The problem with "the Paleoptera problem": Sense and sensitivity. *Cladistics* 19:432–442.

Ohta, T. 1973. Slightly deleterious mutant substitutions in evolution. *Nature* 246:96–98.

Ophir, R., and D. Graur. 1997. Patterns and rates of indel evolution in processed pseudogenes from humans and murids. *Gene* 205:191–202.

Ouvrard, D., B. C. Campbell, T. Bourgoin, and K. L. Chan. 2000. 18S rRNA secondary structure and phylogenetic position of Peloridiidae (Insecta: Hemiptera). *Molecular Phylogenetics and Evolution* 16:403–417.

Page, R. D. M. 2000. Comparative analysis of insect mitochondrial small subunit ribosomal RNA using maximum weighted matching. *Nucleic Acids Research* 28:3839–3845.

Page, R. D. M., and E. C. Holmes. 1998. *Molecular Evolution: A Phylogenetic Approach*. Oxford: Blackwell Science.

Pagni, M., V. Ioannidis, L. Cerutti, M. Zahn-Zabal, C. V. Jongeneel, and L. Falquet. 2004. MyHits: A new interactive resource for protein annotation and domain identification. *Nucleic Acids Research* 32:W332–W335.

Park, J., K. Karplus, C. Barrett, R. Hughey, D. Haussler, T. J. Hubbard, and C. Chothia. 1998. Sequence comparisons using multiple sequences detect three times as many remote homologues as pairwise methods. *Journal of Molecular Biology* 284:1201–1210.

Parry-Smith, D. J., A. W. Payne, A. D. Michie, and T. K. Attwood. 1998. CINEMA: A novel colour INteractive editor for multiple alignments. *Gene* 221:GC57–GC63.

Pazos, F., M. Helmer–Citterich, G. Ausiello, and A. Valencia. 1997. Correlated mutations contain information about protein-protein interaction. *Journal of Molecular Biology* 271:511–523.

Pazos, F., and A. Valencia. 2001. Similarity of phylogenetic trees as indicator of protein-protein interaction. *Protein Engineering* 14:609–614.

———. 2002. In silico two-hybrid system for the selection of physically interacting protein pairs. *Proteins* 47:219–227.

Pearl, F., A. Todd, I. Sillitoe, M. Dibley, O. Redfern, T. Lewis, C. Bennett, R. Marsden, A. Grant, D. Lee, A. Akpor, M. Maibaum, A. Harrison, T. Dallman, G. Reeves, I. Diboun, S. Addou, S. Lise, C. Johnston, A. Sillero, J. Thornton, and C. Orengo. 2005. The CATH Domain Structure Database and related resources Gene3D and DHS provide comprehensive domain family information for genome analysis. *Nucleic Acids Research* 33:D247–D251.

Pearl, F. M., C. F. Bennett, J. E. Bray, A. P. Harrison, N. Martin, A. Shepherd, I. Sillitoe, J. Thornton, and C. A. Orengo. 2003. The CATH database: An extended protein family resource for structural and functional genomics. *Nucleic Acids Research* 31:452–455.

Pearson, W. R., and D. J. Lipman. 1988. Improved tools for biological sequence comparison. *Proceedings of the National Academy of Sciences USA* 85:2444–2448.

Pedersen, J. S., G. Bejerano, A. Siepel, K. Rosenbloom, K. Lindblad-Toh, E. S. Lander, J. Kent, W. Miller, and D. Haussler. 2006. Identification and classification of conserved RNA secondary structures in the human genome. *PLoS Computational Biology* 2:e33.

Pei, J., and N. V. Grishin. 2001. AL2CO: Calculation of positional conservation in a protein sequence alignment. *Bioinformatics* 17:700–712.

———. 2006. MUMMALS: Multiple sequence alignment improved by using hidden Markov models with local structural information. *Nucleic Acids Research* 34:4364–4374.

———. 2007. PROMALS: Towards accurate multiple sequence alignments of distantly related proteins. *Bioinformatics* 23:802–808.

Pei, J., R. Sadreyev, and N. V. Grishin. 2003. PCMA: A program for fast and accurate multiple sequence alignment. *Bioinformatics* 19:427–428.

Pellegrini, M., E. M. Marcotte, M. J. Thompson, D. Eisenberg, and T. O. Yeates. 1999. Assigning protein functions by comparative genome analysis: Protein phylogenetic profiles. *Proceedings of the National Academy of Sciences USA* 96:4285–4288.

Phillips, A., D. Janies, and W. C. Wheeler. 2000. Multiple sequence alignment in phylogenetic analysis. *Molecular Phylogenetics and Evolution* 16:317–330.

Phuong, T. M., C. B. Do, R. C. Edgar, and S. Batzoglou. 2006. Multiple alignment of protein sequences with repeats and rearrangements. *Nucleic Acids Research* 34:5932–5942.

Picardi, E., and C. Quagliariello. 2006. EdiPy: A resource to simulate the evolution of plant mitochondrial genes under the RNA editing. *Computational Biology and Chemistry* 30:77–80.

Poirot, O., K. Suhre, C. Abergel, E. O'Toole, and C. Notredame. 2005. 3DCoffee at IGS: A web server for combining sequences and structure into a multiple sequence alignment. *Nucleic Acids Research* 32:W37–W40.

Pollard, D., A. Moses, V. Iyer, and M. Eisen. 2006. Detecting the limits of regulatory element conservation and divergence estimation using pairwise and multiple alignments. *BMC Bioinformatics* 7:376.

Pollard, D. A., C. M. Bergman, J. Stoye, S. E. Celniker, and M. B. Eisen. 2004. Benchmarking tools for the alignment of functional noncoding DNA. *BMC Bioinformatics* 5:6.

Przybylski, D., and B. Rost. 2002. Alignments grow, secondary structure prediction improves. *Proteins* 46:197–205.

Raghava, G. P., S. M. Searle, P. C. Audley, J. D. Barber, and G. J. Barton. 2003. OXBench: A benchmark for evaluation of protein multiple sequence alignment accuracy. *BMC Bioinformatics* 4:47.

Rambaut, A., and N. C. Grassly. 1997. Seq-Gen: An application for the Monte Carlo simulation of DNA sequence evolution along phylogenetic trees. *Computer Applications in Bioscience* 13:235–238.

Ramensky, V., P. Bork, and S. Sunyaev. 2002. Human non-synonymous SNPs: Server and survey. *Nucleic Acids Research* 30:3894–3900.

Raphael, B., D. Zhi, H. Tang, and P. Pevzner. 2004. A novel method for multiple alignment of sequences with repeated and shuffled elements. *Genome Research* 14:2336–2346.

Raymond, J., O. Zhaxybayeva, J. P. Gogarten, S. Y. Gerdes, and R. E. Blankenship. 2002. Whole-genome analysis of photosynthetic prokaryotes. *Science* 298:1616–1620.

Redelings, B. D., and M. A. Suchard. 2005. Joint Bayesian estimation of alignment and phylogeny. *Systematic Biology* 54:401–418.

———. 2007. Incorporating indel information into phylogeny estimation for rapidly emerging pathogens. *BMC Evolutionary Biology* 7:40.

Reese, J. T., and W. R. Pearson. 2002. Empirical determination of effective gap penalties for sequence comparison. *Bioinformatics* 18:1500–1507.

Reyes, A., C. Gissi, F. Catzeflis, E. Nevo, G. Pesole, and C. Saccone. 2004. Congruent mammalian trees from mitochondrial and nuclear genes using Bayesian methods. *Molecular Biology and Evolution* 21:397–403.

Rigoutsos, I., T. Huynh, A. Floratos, L. Parida, and D. Platt. 2002. Dictionary-driven protein annotation. *Nucleic Acids Research* 30:3901–16.

Rivas, E., and S. Eddy. 2001. Noncoding RNA gene detection using comparative sequence analysis. *BMC Bioinformatics* 2:8.

Roberts, E., J. Eargle, D. Wright, and Z. Luthey-Schulten. 2006. MultiSeq: Unifying sequence and structure data for evolutionary analysis. *BMC Bioinformatics* 7:382.

Ronquist, F., and J. P. Huelsenbeck. 2003. MrBayes 3: Bayesian phylogenetic inference under mixed models. *Bioinformatics* 19:1572–1574.

Rosenberg, M. S. 2005a. Evolutionary distance estimation and fidelity of pair wise sequence alignment. *BMC Bioinformatics* 6:102.

———. 2005b. Multiple sequence alignment accuracy and evolutionary distance estimation. *BMC Bioinformatics* 6:278.

———. 2005c. MySSP: Non-stationary evolutionary sequence simulation, including indels. *Evolutionary Bioinformatics Online* 1:51–53.

Rosenberg, M. S., and S. Kumar. 2003. Heterogeneity of nucleotide frequencies among evolutionary lineages and phylogenetic inference. *Molecular Biology and Evolution* 20:610–621.

Roshan, U., and D. R. Livesay. 2006. Probalign: Multiple sequence alignment using partition function posterior probabilities. *Bioinformatics* 22:2715–2721.

Rost, B. 2001. Review: Protein secondary structure prediction continues to rise. *Journal of Structural Biology* 134:204–18.

References

Rost, B., C. Sander, and R. Schneider. 1994. PHD: An automatic mail server for protein secondary structure prediction. *Computer Applications in Bioscience* 10:53–60.

Rousset, F., M. Pelandakis, and M. Solignac. 1991. Evolution of compensatory substitutions through GU intermediate state in *Drosophila* rRNA. *Proceedings of the National Academy of Sciences USA* 88:10032–10036.

Rubin, G. M., M. D. Yandell, J. R. Wortman, G. L. Gabor Miklos, C. R. Nelson, I. K. Hariharan, M. E. Fortini, P. W. Li, R. Apweiler, W. Fleischmann, J. M. Cherry, S. Henikoff, M. P. Skupski, S. Misra, M. Ashburner, E. Birney, M. S. Boguski, T. Brody, P. Brokstein, S. E. Celniker, S. A. Chervitz, D. Coates, A. Cravchik, A. E. Gabrielian, R. F. Galle, W. M. Gelbart, R. A. George, L. S. Goldstein, F. Gong, P. Guan, N. L. Harris, B. A. Hay, R. A. Hoskins, J. Li, Z. Li, R. O. Hynes, S. J. Jones, P. M. Kuehl, B. Lemaitre, J. T. Littleton, D. K. Morrison, C. Mungall, P. H. O'Farrell, O. K. Pickeral, C. Shue, L. B. Vosshall, J. Zhang, Q. Zhao, X. H. Zheng, F. Zhong, W. Zhong, R. Gibbs, J. C. Venter, M. D. Adams, and S. Lewis. 2000. Comparative genomics of the eukaryotes. *Science* 287:2204–2215.

Russell, R. B., and G. J. Barton. 1992. Multiple protein sequence alignment from tertiary structure comparison: Assignment of global and residue confidence levels. *Proteins* 14:309–323.

Saitou, N., and M. Nei. 1987. The neighbor-joining method: A new method for reconstructing phylogenetic trees. *Molecular Biology and Evolution* 4:406–425.

Sali, A., and T. L. Blundell. 1990. Definition of general topological equivalence in protein structures. A procedure involving comparison of properties and relationships through simulated annealing and dynamic programming. *Journal of Molecular Biology* 212:403–428.

Sali, A., L. Potterton, F. Yuan, H. van Vlijmen, and M. Karplus. 1995. Evaluation of comparative protein modeling by MODELLER. *Proteins* 23:318–326.

Sammeth, M., and J. Heringa. 2006. Global multiple-sequence alignment with repeats. *Proteins* 64:263–274.

Sankoff, D. 1972. Matching sequences under deletion/insertion constraints. *Proceedings of the National Academy of Sciences USA* 69:4–6.

———. 1973. A test for nucleotide sequence homology. *Journal of Molecular Biology* 77:159–164.

———. 1975. Minimal mutation trees of sequences. *SIAM Journal on Applied Mathematics* 28:35–42.

———. 1985. Simultaneous solution of the RNA folding, alignment and protosequence problems. *SIAM Journal on Applied Mathematics* 45:810–825.

Sankoff, D., and R. J. Cedergren. 1983. Simultaneous comparison of three or more sequences related by a tree. Pp. 253–263 in *Time Warps, String Edits, and Macromolecules: The Theory and Practice of Sequence Comparison*, D. Sankoff and J. B. Kruskal, eds. London: Addison–Wesley Publishing.

Sankoff, D., R. J. Cedergren, and G. Lapalme. 1976. Frequency of insertion-deletion, transversion, and transition in the evolution of 5S ribosomal RNA. *Journal of Molecular Evolution* 7:133–149.

Sankoff, D., C. Morel, and R. J. Cedergren. 1973. Evolution of 5S RNA and the non-randomness of base replacement. *Nature New Biology* 245:232–234.

Sasson, O., N. Kaplan, and M. Linial. 2006. Functional annotation prediction: All for one and one for all. *Protein Science* 15:1557–1562.

Saunders, C. T., and D. Baker. 2002. Evaluation of structural and evolutionary contributions to deleterious mutation prediction. *Journal of Molecular Biology* 322:891–901.

Savill, N., D. Hoyle, and P. Higgs. 2001. RNA sequence evolution with secondary structure constraints: Comparison of substitution rate models using maximum likelihood methods. *Genetics* 157:399–411.

Sayle, R. A., and E. J. Milner-White. 1995. RASMOL: Biomolecular graphics for all. *Trends in Biochemical Sciences* 20:374.

Schlosshauer, M., and M. Ohlsson. 2002. A novel approach to local reliability of sequence alignments. *Bioinformatics* 18:847–854.

Schluenzen, F., A. Tocilj, R. Zarivach, J. Harms, M. Gluehmann, D. Janell, A. Bashan, H. Bartels, A. I., F. Franceschi, and Y. A. 2000. Structure of functionally activated small ribosomal subunit at 3.3 Å resolution. *Cell* 102:615–623.

Schmidt, H. A., K. Strimmer, M. Vingron, and A. von Haeseler. 2002. TREEPUZZLE: Maximum likelihood phylogenetic analysis using quartets and parallel computing. *Bioinformatics* 18:502–504.

Schmollinger, M., K. Nieselt, M. Kaufmann, and B. Morgenstern. 2004. DIALIGN P: Fast pair-wise and multiple sequence alignment using parallel processors. *BMC Bioinformatics* 5:128.

Schnare, M. N., S. H. Damberger, M. W. Gray, and R. R. Gutell. 1996. Comprehensive comparison of structural characteristics in eukaryotic cytoplasmic large subunit (23S-like) ribosomal RNA. *Journal of Molecular Biology* 256:701–719.

Schöniger, M., and A. von Haeseler. 1994. A stochastic model for the evolution of autocorrelated DNA sequences. *Molecular Phylogenetics and Evolution* 3:240–243.

Schwartz, A. S., E. W. Myers, and L. Pachter. 2005. Alignment metric accuracy. arXiv:q-bio/0510052.

Schwartz, A. S., and L. Pachter. 2007. Multiple alignment by sequence annealing. Bioinformatics. *Bioinformatics* 23:e24–e29.

Schwartz, S., W. J. Kent, A. Smit, Z. Zhang, R. Baerstch, R. C. Hardison, D. Haussler, and W. Miller. 2003. Human-Mouse alignments with BLASTZ. *Genome Research* 13:103–107.

Schwede, T., J. Kopp, N. Guex, and M. C. Peitsch. 2003. SWISS-MODEL: An automated protein homology-modeling server. *Nucleic Acids Research* 31:3381–3385.

Segal, E., Y. Fondufe-Mittendorf, L. Chen, A. Thastrom, Y. Field, I. K. Moore, J. P. Wang, and J. Widom. 2006. A genomic code for nucleosome positioning. *Nature* 442:772–778.

Sellers, P. H. 1974. On the theory and computation of evolutionary distances. *SIAM Journal on Applied Mathematics* 26:787–793.

———. 1979. Pattern recognition in genetic sequences. *Proceedings of the National Academy of Sciences USA* 76:3041.

———. 1980. The theory and computation of evolutionary distances: Pattern recognition. *Journal of Algorithms* 1:359–373.

Shabalina, S., and A. S. Kondrashov. 1999. Pattern of selective constraint in *C. elegans* and *C. briggsae* genomes. *Genetical Research* 74:23–30.

Sharkey, M. J., N. M. Laurenne, B. Sharanowski, D. L. J. Quicke, and D. Murray. 2006. Revision of the Agathidinae (Hymenoptera: Braconidae) with comparisons of static and dynamic alignments. *Cladistics* 22:546–567.

Shi, J., T. L. Blundell, and K. Mizuguchi. 2001. FUGUE: Sequence–structure homology recognition using environment-specific substitution tables and structure-dependent gap penalties. *Journal of Molecular Biology* 310:243–257.

Shindyalov, I. N., and P. E. Bourne. 1998. Protein structure alignment by incremental combinatorial extension (CE) of the optimal path. *Protein Engineering* 11:739–747.

Sicheritz–Ponten, T., and S. G. Andersson. 2001. A phylogenomic approach to microbial evolution. *Nucleic Acids Research* 29:545–552.

Siddiqui, A. S., U. Dengler, and G. J. Barton. 2001. 3Dee: A database of protein structural domains. *Bioinformatics* 17:200–201.

Siebert, S., and R. Backofen. 2005. MARNA: Multiple alignment and consensus structure prediction of RNAs based on sequence structure comparisons. *Bioinformatics* 21:3352–3359.

Sigrist, C. J., L. Cerutti, N. Hulo, A. Gattiker, L. Falquet, M. Pagni, A. Bairoch, and P. Bucher. 2002. PROSITE: A documented database using patterns and profiles as motif descriptors. *Briefings in Bioinformatics* 3:265–274.

Sim, S. E., S. Easterbrook, and R. C. Holt. 2003. Using benchmarking to advance research: A challenge to software engineering. Pp. 74–83 in *25th International Conference on Software Engineering*.

Simmons, M. P. 2004. Independence of alignment and tree search. *Molecular Phylogenetics and Evolution* 31:874–879.

Simmons, M. P., K. Müller, and A. P. Norton. 2007. The relative performance of indel-coding methods in simulations. *Molecular Phylogenetics and Evolution* 44:724–740.

Simmons, M. P., and H. Ochoterena. 2000. Gaps as characters in sequence-based phylogenetic analyses. *Systematic Biology* 49:369–381.

Simmons, M. P., A. Reeves, and J. I. Davis. 2004. Character-state space versus rate of evolution for phylogenetic inference. *Cladistics* 20:191–204.

Simon, C. 1991. Molecular systematics at the species boundary: Exploiting conserved and variable regions of the mitochondrial genome of animals via direct sequencing from enzymatically amplified DNA. Pp. 33–71 in *Molecular Techniques in Taxonomy*, G. M. Hewitt, A. W. B. Johnson, and J. P. W. Young, eds. New York: Springer Verlag.

Simossis, V. A., and J. Heringa. 2005. PRALINE: A multiple sequence alignment toolbox that integrates homology-extended and secondary structure information. *Nucleic Acids Research* 33:W289–W294.

Simpson, G. G. 1945. The principles of classification and a classification of mammals. *American Museum of Natural History Bulletin* 85:1–350.

Slowinski, J. B. 1998. The number of multiple alignments. *Molecular Phylogenetics and Evolution* 10:264–266.

Smith, A. B. 1989. RNA sequence data in phylogenetic reconstruction: Testing the limits of its resolution. *Cladistics* 5:321–344.

Smith, R. F., and T. F. Smith. 1990. Automatic generation of primary sequence patterns from sets of related protein sequences. *Proceedings of the National Academy of Sciences USA* 87:118–122.

———. 1992. Pattern-induced multi-sequence alignment (PIMA) algorithm employing secondary structure–dependent gap penalties for use in comparative protein modeling. *Protein Engineering* 5:35–41.

Smith, T. F., and M. S. Waterman. 1981a. Comparison of biosequences. *Advances in Applied Mathematics* 2:482–489.

———. 1981b. Identification of common molecular subsequences. *Journal of Molecular Biology* 147:195–197.

Smith, T. F., M. S. Waterman, and W. M. Fitch. 1981. Comparative biosequence metrics. *Journal of Molecular Evolution* 18:38–46.

Sonnhammer, E. L., and D. Kahn. 1994. Modular arrangement of proteins as inferred from analysis of homology. *Protein Science* 3:482–492.

Spahn, C. M., R. Beckmann, N. Eswar, P. A. Penczek, A. Sali, G. Blobel, and J. Frank. 2001. Structure of the 80S ribosome from *Saccharomyces cerevisiae*-tRNA-ribosome and subunit-subunit interactions. *Cell* 107:373–386.

Spalding, J. B., and P. J. Lammers. 2004. BLAST Filter and GraphAlign: Rule-based formation and analysis of sets of related DNA and protein sequences. *Nucleic Acids Research* 32:W26–W32.

Springer, M. S., and E. Douzery. 1996. Secondary structure and patterns of evolution among mammalian mitochondrial 12S rRNA molecules. *Journal of Molecular Evolution* 43:357–373.

Springer, M. S., L. J. Hollar, and A. Burk. 1995. Compensatory substitutions and the evolution of the mitochondrial 12S rRNA gene in mammals. *Molecular Biology and Evolution* 12:1138–1150.

Stanke, M., O. Schöffmann, B. Morgenstern, and S. Waack. 2006. Gene prediction in eukaryotes with a generalized hidden Markov model that uses hints from external sources. *BMC Bioinformatics* 7:62.

Stebbings, L. A., and K. Mizuguchi. 2004. HOMSTRAD: Recent developments of the Homologous Protein Structure Alignment Database. *Nucleic Acids Research* 32:D203–D207.

Steel, M., and J. Hein. 2001. Applying Thorne–Kishino–Felsenstein model to sequence evolution on a star-shaped tree. *Applied Mathematics Letters* 14:679–684.

Stenson, P. D., E. V. Ball, M. Mort, A. D. Phillips, J. A. Shiel, N. S. Thomas, S. Abeysinghe, M. Krawczak, and D. N. Cooper. 2003. Human Gene Mutation Database (HGMD): 2003 update. *Human Mutation* 21:577–581.

Storm, C. E. V., and E. L. L. Sonnhammer. 2002. Automated ortholog inference from phylogenetic trees and calculation of orthology reliability. *Bioinformatics* 18:92–99.

Stoye, J. 1998. Multiple sequence alignment with the divide-and-conquer method. *Gene* 211:GC45–GC56.

Stoye, J., D. Evers, and F. Meyer. 1998. Rose: Generating sequence families. *Bioinformatics* 14:157–163.

Stoye, J., V. Moulton, and A. W. M. Dress. 1997. DCA: An efficient implementation of the divide-and-conquer approach to simultaneous multiple sequence alignment. *Computer Applications in Bioscience* 13:625–626.

Streisinger, G., Y. Okada, J. Emrich, J. Newton, A. Tsugita, E. Terzaghi, and I. Inouye. 1966. Frameshift mutations and the genetic code. *Cold Spring Harbor Symposia on Quantitative Biology* 31:77–86.

Streisinger, G., and J. Owen. 1985. Mechanisms of spontaneous and induced frameshift mutation in bacteriophage T4. *Genetics* 109:633–659.

Strope, C. L., S. D. Scott, and E. N. Moriyama. 2007. indel-Seq-Gen: A new protein family simulator incorporating domains, motifs, and indels. *Molecular Biology and Evolution* 24:640–649.

Subramanian, A. R., J. Weyer–Menkhoff, M. Kaufmann, and B. Morgenstern. 2005. DIALIGN-T: An improved algorithm for segment-based multiple sequence alignment. *BMC Bioinformatics* 6:66.

Suchard, M. A., C. M. R. Kitchen, J. S. Sinsheimer, and R. E. Weiss. 2003. Hierarchical phylogenetic models for analyzing multipartite sequence data. *Systematic Biology* 52:649–664.

Suchard, M. A., and B. D. Redelings. 2006. B*Ali-Phy*: Simultaneous Bayesian inference of alignment and phylogeny. *Bioinformatics* 22:2047–2048.

Sundström, H., M. T. Webster, and H. Ellegren. 2003. Is the rate of insertion and deletion mutation male baised?: Molecular evolutionary analysis of avian and primate sex chromosome sequences. *Genetics* 164:259–268.

Sunyaev, S., V. Ramensky, and P. Bork. 2000. Towards a structural basis of human non-synonymous single nucleotide polymorphisms. *Trends in Genetics* 16:198–200.

Svenson, G. J., and M. F. Whiting. 2004. Phylogeny of Mantodea based on molecular data: Evolution of a charismatic predator. *Systematic Entomology* 29:359–370.

Swofford, D. L. 1998. *PAUP**. *Phylogenetic Analysis Using Parsimony (*and Other Methods)*. Sunderland, MA: Sinauer Associates.

Swofford, D. L., G. J. Olsen, P. J. Waddell, and D. M. Hillis. 1996. Phylogenetic inference. Pp. 407–514 in *Molecular Systematics*, D. M. Hillis, C. Moritz and B. K. Mable, eds. Sunderland, MA: Sinauer.

Szklarczyk, R., and J. Heringa. 2006. AuberGene--a sensitive genome alignment tool. *Bioinformatics* 22:1431–1436.

Takahashi, K., and M. Nei. 2000. Efficiencies of fast algorithms of phylogenetic inference under the criteria of maximum parsimony, minimum evolution, and maximum likelihood when a large number of sequences are used. *Molecular Biology and Evolution* 17:1251–1258.

Talavera, G., and J. Castresana. 2007. Improvement of phylogenies after removing divergent and ambiguously aligned blocks from protein sequence alignments. *Systematic Biology* 56:564–577.

Tavakoli, N. P., and K. M. Derbyshire. 2001. Tipping the balance between replicative and simple transposition. *EMBO Journal* 20:2923–2930.

Tavaré, S. 1986. Some probabilistic and statistical problems in the analysis of DNA sequences. Pp. 57–86 in *Some Mathematical Questions in Biology: DNA Sequence Analysis*, M. S. Waterman, ed. Providence, RI: The American Mathematical Society.

Taylor, H. M., and S. Karlin. 1998. *An Introduction to Stochastic Modeling*. New York: Academic Press.

Taylor, P. 1984. A fast homology program for aligning biological sequences. *Nucleic Acids Research* 12:447–455.
Taylor, W. R. 1987. Multiple sequence alignment by a pairwise algorithm. *Computer Applications in Bioscience* 3:81–87.
———. 1988. A flexible method to align large numbers of biological sequences. *Journal of Molecular Evolution* 28:161–169.
———. 2000. Protein structure comparison using SAP. *Methods in Molecular Biology* 143:19–35.
Taylor, W. R., and C. A. Orengo. 1989. Protein structure alignment. *Journal of Molecular Biology* 208:1–22.
Terry, M. D., and M. F. Whiting. 2005. Comparison of two alignment techniques within a single complex data set: POY versus Clustal. *Cladistics* 21:272–281.
Thomas, J. W., J. W. Touchman, R. W. Blakesley, G. G. Bouffard, S. M. Beckstrom-Sternberg, E. H. Margulies, M. Blanchette, A. C. Siepel, P. J. Thomas, J. C. McDowell, B. Maskeri, N. F. Hansen, M. S. Schwartz, R. J. Weber, W. J. Kent, D. Karolchik, T. C. Bruen, R. Bevan, D. J. Cutler, S. Schwartz, L. Elnitski, J. R. Idol, A. B. Prasad, S. Q. Lee-Lin, V. V. B. Maduro, T. J. Summers, M. E. Portnoy, N. L. Dietrich, N. Akhter, K. Ayele, B. Benjamin, K. Cariaga, C. P. Brinkley, S. Y. Brooks, S. Granite, X. Guan, J. Gupta, P. Haghighi, S. L. Ho, M. C. Huang, E. Karlins, P. L. Laric, R. Legaspi, M. J. Lim, Q. L. Maduro, C. A. Masiello, S. D. Mastrian, J. C. McCloskey, R. Pearson, S. Stantripop, E. E. Tiongson, J. T. Tran, C. Tsurgeon, J. L. Vogt, M. A. Walker, K. D. Wetherby, L. S. Wiggins, A. C. Young, L. H. Zhang, K. Osoegawa, B. Zhu, B. Zhao, C. L. Shu, P. J. De Jong, C. E. Lawrence, A. F. Smit, A. Chakravarti, D. Haussler, P. Green, W. Miller, and E. D. Green. 2003. Comparative analyses of multi-species sequences from targeted genomic regions. *Nature* 424:788–793.
Thompson, J. D., T. J. Gibson, F. Plewniak, F. Jeanmougin, and D. G. Higgins. 1997. The CLUSTAL_X windows interface: Flexible strategies for multiple sequence alignment aided by quality analysis tools. *Nucleic Acids Research* 25:4876–4882.
Thompson, J. D., D. G. Higgins, and T. J. Gibson. 1994. CLUSTAL W: Improving the sensitivity of progressive multiple sequence alignment through sequence weighting, positions-specific gap penalties and weight matrix choice. *Nucleic Acids Research* 22:4673–4680.
Thompson, J. D., S. R. Holbrook, K. Katoh, P. Koehl, D. Moras, E. Westhof, and O. Poch. 2005. MAO: A Multiple Alignment Ontology for nucleic acid and protein sequences. *Nucleic Acids Research* 33:4164–4171.
Thompson, J. D., P. Koehl, R. Ripp, and O. Poch. 2005. BAliBASE 3.0: Latest developments of the multiple sequence alignment benchmark. *Proteins* 61:127–136.
Thompson, J. D., A. Muller, A. Waterhouse, J. Procter, G. J. Barton, F. Plewniak, and O. Poch. 2006. MACSIMS: Multiple Alignment of Complete Sequences Information Management System. *BMC Bioinformatics* 7:318.
Thompson, J. D., F. Plewniak, and O. Poch. 1999a. BAliBASE: A benchmark alignment database for the evaluation of multiple sequence alignment programs. *Bioinformatics* 15:87–88.

———. 1999b. A comprehensive comparison of multiple sequence alignment programs. *Nucleic Acids Research* 27:2682–2690.
Thompson, J. D., F. Plewniak, R. Ripp, J.-C. Thierry, and O. Poch. 2001. Towards a reliable objective function for multiple sequence alignments. *Journal of Molecular Biology* 314:937–951.
Thompson, J. D., F. Plewniak, J. C. Thierry, and O. Poch. 2000. DbClustal: Rapid and reliable global multiple alignments of protein sequences detected by database searches. *Nucleic Acids Research* 28:2919–2926.
Thompson, J. D., and O. Poch. 2006. Multiple sequence alignment as a workbench for molecular systems biology. *Current Bioinformatics* 1:95–104.
Thompson, J. D., J.-C. Thierry, and O. Poch. 2003. RASCAL: Rapid scanning and correction of multiple sequence alignments. *Bioinformatics* 19:1155–1161.
Thorne, J. L., and H. Kishino. 1992. Freeing phylogenies from artifacts of alignment. *Molecular Biology and Evolution* 9:1148–1162.
Thorne, J. L., H. Kishino, and J. Felsenstein. 1991. An evolutionary model for maximul likelihood alignment of DNA sequences. *Journal of Molecular Evolution* 33:114–124.
———. 1992. Inching toward reality: An improved likelihood model of sequence evolution. *Journal of Molecular Evolution* 34:3–16.
Titus, T., and D. R. Frost. 1996. Molecular homology assessment and phylogeny in the lizard family Opluridae (Squamata: Iguania). *Molecular Phylogenetics and Evolution* 6:49–62.
Tönges, U., S. W. Perrey, J. Stoye, and A. W. M. Dress. 1996. A general method for fast multiple sequence alignment. *Gene* 172:GC33–GC41.
Uetz, P., L. Giot, G. Cagney, T. A. Mansfield, R. S. Judson, J. R. Knight, D. Lockshon, V. Narayan, M. Srinivasan, P. Pochart, A. Qureshi-Emili, Y. Li, B. Godwin, D. Conover, T. Kalbfleisch, G. Vijayadamodar, M. J. Yang, M. Johnston, S. Fields, and J. M. Rothberg. 2000. A comprehensive analysis of protein-protein interactions in *Saccharomyces cerevisiae*. *Nature* 403:623–627.
Ukkonen, E. 1995. On-line construction of suffix trees. *Algorithmica* 14:249–260.
Valdar, W. S. 2002. Scoring residue conservation. *Proteins* 48:227–241.
van de Peer, Y., J.-M. Neefs, P. de Rijk, and R. de Wachter. 1993. Reconstructing evolution from eukaryotic small-ribosomal-subunit RNA sequences: Calibration of the molecular clock. *Journal of Molecular Evolution* 37:221–232.
Van Walle, I., I. Lasters, and L. Wyns. 2005. SABmark--A benchmark for sequence alignment that covers the entire known fold space. *Bioinformatics* 21:1267–1268.
Varón, A., L. S. Vinh, I. Bomash, and W. C. Wheeler. 2007. *Poy 4.0 beta 2013*. New York: American Museum of Natural History.
Vawter, L., and W. M. Brown. 1993. Rates and patterns of base change in the small subunit ribosomal RNA gene. *Genetics* 134:597–608.
Vingron, M., and A. von Haeseler. 1997. Towards integration of multiple alignment and phylogenetic tree construction. *Journal of Computational Biology* 4:23–34.
Vingron, M., and M. S. Waterman. 1994. Sequence alignment and penalty choice: Review of concepts, case studies and implications. *Journal of Molecular Biology* 235:1–12.

Viterbi, A. 1967. Error bounds for convolutional codes and an asymptotically optimum decoding algorithm. *IEEE Transactions on Information Theory* 13:260–269.

Waddell, P. J., and S. Shelley. 2003. Evaluating placental inter-ordinal phylogenies with novel sequences including RAG1, γ-fibrinogen, ND6, and mt-tRNA, pluse MCMC-driven nucleotide, amino acid, and codon models. *Molecular Phylogenetics and Evolution* 28:197–224.

Wagner, R. A., and M. J. Fisher. 1974. The string-to-string correction problem. *Journal of the ACM* 21:168–173.

Wallace, I. M., G. Blackshields, and D. G. Higgins. 2005. Multiple sequence alignments. *Current Opinion in Structural Biology* 15:261–266.

Wallace, I. M., O. O'Sullivan, and D. G. Higgins. 2005. Evaluation of iterative alignment algorithms for multiple alignment. *Bioinformatics* 21:1408–1414.

Wallace, I. M., O. O'Sullivan, D. G. Higgins, and C. Notredame. 2006. M-Coffee: Combining multiple sequence alignment methods with T-Coffee. *Nucleic Acids Research* 34:1692–1699.

Wang, J., P. D. Keightley, and T. Johnson. 2006. MCALIGN2: Faster, accurate global pairwise alignment of non-coding DNA sequences based on explicit models of indel evolution. *BMC Bioinformatics* 7:292.

Wang, L., and D. Gusfield. 1997. Improved approximation algorithms for tree alignment. *Journal of Algorithms* 25:255–273.

Wang, L., and T. Jiang. 1994. On the complexity of multiple sequence alignment. *Journal of Computational Biology* 1:337–348.

Washietl, S., I. Hofacker, and P. Stadler. 2005. Fast and reliable prediction of noncoding RNAs. *Proceedings of the National Academy of Sciences USA* 102:2454–2459.

Waterman, M. S. 1986. Multiple sequence alignment by consensus. *Nucleic Acids Research* 14:9095–9102.

Waterman, M. S., and M. D. Perlwitz. 1984. Line geometrics for sequence comparisons. *Bulletin of Mathematical Biology* 46:567–577.

Weaver, R. F., and P. W. Hedrick. 1992. *Genetics*, 2nd ed. Dubuque, IA: William C. Brown Publishers.

Weiner, A. M. 2000. Do all SINEs lead to LINEs? *Nature Genetics* 24:332–333.

Weiner, P. 1973. Linear pattern matching algorithms. Pp. 1–11 in *Proceedings of the 14th IEEE Symposium on Switching and Automata Theory*.

Wheeler, W. C. 1996. Optimization alignment: The end of multiple sequence alignment in phylogenetics? *Cladistics* 12:1–9.

———. 1999. Fixed character states and the optimization of molecular sequence data. *Cladistics* 15:379–385.

———. 2001. Homology and the optimization of DNA sequence data. *Cladistics* 17:S3–S11.

———. 2003a. Implied alignment: A synampomorphy-based multiple-sequence alignment method and its use in cladogram search. *Cladistics* 19:261–268.

———. 2003b. Iterative pass optimization of sequence data. *Cladistics* 19:254–260.

———. 2003c. Search-based character optimization. *Cladistics* 19:348–355.

———. 2005. Alignment, dynamic homology, and optimization. Pp. 71–80 in *Parsimony, Phylogeny, and Genomics*, V. A. Albert, ed. Oxford: Oxford University Press.

———. 2006. Dynamic homology and the likelihood criterion. *Cladistics* 22:157–170.

———. 2007. The analysis of molecular sequences in large data sets: Where should we put our effort? Pp. 113–128 in *Reconstructing the Tree of Life: Taxonomy and Systematics of Species Rich Taxa*, T. R. Hodkinson and J. A. N. Parnell, eds. Oxford: Oxford University Press.

Wheeler, W. C., J. Gatesy, and R. DeSalle. 1995. Elision: A method for accommodating multiple molecular sequence alignments with alignment-ambiguous sites. *Molecular Phylogenetics and Evolution* 4:1–9.

Wheeler, W. C., D. S. Gladstein, and J. De Laet. 1996–2006. *POY version 3.0*. New York: American Museum of Natural History.

Wheeler, W. C., and R. L. Honeycutt. 1988. Paired sequence difference in ribosomal RNAs: Evolutionary and phylogenetic implications. *Molecular Biology and Evolution* 5:90–96.

Wheeler, W. C., M. F. Whiting, Q. D. Wheeler, and J. M. Carpenter. 2001. The phylogeny of extant insect orders. *Cladistics* 17:113–169.

Whiting, A. S., J. W. Sites, Jr., K. C. M. Pellegrino, and M. T. Rodrigues. 2006. Comparing alignment methods for inferring the history of the new world lizard genus *Mabuya* (Squamata: Scincidae). *Molecular Phylogenetics and Evolution* 38:719–730.

Whiting, M. F., J. C. Carpenter, Q. D. Wheeler, and W. C. Wheeler. 1997. The Strepsiptera problem: Phylogeny of the holometabolous insect orders inferred from 18S and 28S ribosomal DNA sequences and morphology. *Systematic Biology* 46:1–68.

Wilm, A., I. Mainz, and G. Steger. 2006. An enhanced RNA alignment benchmark for sequence alignment programs. *Algorithms for Molecular Biology* 1:19.

Wimberly, B. T., D. E. Brodersen, W. M. Clemons, Jr., R. J. Morgan–Warren, A. P. Carter, C. Vornhein, T. Hartsch, and V. Ramakrishnan. 2000. Structure of the 30S ribosomal subunit. *Nature* 407:327–339.

Wojcik, J., and V. Schachter. 2001. Protein-protein interaction map inference using interacting domain profile pairs. *Bioinformatics* 17:S296–S305.

Wolf, Y. I., I. B. Rogozin, N. V. Grishin, and E. V. Koonin. 2002. Genome trees and the tree of life. *Trends in Genetics* 18:472–479.

Wolf, Y. I., I. B. Rogozin, N. V. Grishin, R. L. Tatusov, and E. V. Koonin. 2001. Genome trees constructed using five different approaches suggest new major bacterial clades. *BMC Evolutionary Biology* 1:8.

Wu, C. H., R. Apweiler, A. Bairoch, D. A. Natale, W. C. Barker, B. Boeckmann, S. Ferro, E. Gasteiger, H. Huang, R. Lopez, M. Magrane, M. J. Martin, R. Mazumder, C. O'Donovan, N. Redaschi, and B. Suzek. 2006. The Universal Protein Resource (UniProt): An expanding universe of protein information. *Nucleic Acids Research* 34:D187–D191.

Wuyts, J., P. De Rijk, Y. Van de Peer, G. Pison, P. Rousseeuw, and R. De Wachter. 2000. Comparative analysis of more than 3000 sequences reveals the existence of two pseudoknots in area V4 of eukaryotic small subunit ribosomal RNA. *Nucleic Acids Research* 28:4698–4708.

Wuyts, J., Y. Van de Peer, and R. De Wachter. 2001. Distribution of substitution rates and location of insertion sites in the tertiary structure of ribosomal RNA. *Nucleic Acids Research* 29:5017–5028.
Xia, X. H., Z. Xie, and K. M. Kjer. 2003. 18S ribosomal RNA and tetrapod phylogeny. *Systematic Biology* 52:283–295.
Yang, Z. 1994. Maximum likelihood phylogenetic estimation from DNA sequences with variable rates over sites: Approximate methods. *Systematic Biology* 44:384–399.
Yang, S., R. F. Doolittle, and P. E. Bourne. 2005. Phylogeny determined by protein domain content. *Proceedings of the National Academy of Sciences USA* 102:373–378.
Ye, L., and X. Huang. 2005. MAP2: Multiple alignment of syntenic genomic sequences. *Nucleic Acids Research* 33:162–170.
Yee, C. N., and L. Allison. 1993. Reconstruction of strings past. *Computer Applications in Bioscience* 9:1–7.
Yoder, M. J. 2007. Contributions to the supraspecific systematics of the Diapriidae (Insecta: Hymenoptera). Ph.D. dissertation, Texas A&M University, College Station, TX.
Yona, G., and M. Levitt. 2002. Within the twilight zone: A sensitive profile-profile comparison tool based on information theory. *Journal of Molecular Biology* 315:1257–1275.
Young, F. W., and R. M. Hamer. 1987. *Multidimensional Scaling: History, Theory and Applications*. Mahwah, NJ: Lawrence Erlbaum Associates.
Yule, G. U. 1924. A mathematical theory of evolution based on the conclusions of Dr J. C. Willis. *Philosophical Transactions of the Royal Society (London) Series B* 213:21–87.
Yura, K., M. Shionyu, K. Hagino, A. Hijikata, Y. Hirashima, T. Nakahara, T. Eguchi, K. Shinoda, A. Yamaguchi, K. Takahashi, T. Itoh, T. Imanishi, T. Gojobori, and M. Go. 2006. Alternative splicing in human transcriptome: Functional and structural influence on proteins. *Gene* 380:63–71.
Yusupov, M. M., G. Z. Yusupova, A. Baucom, K. Lieberman, T. N. Earnest, J. H. D. Cate, and H. F. Noller. 2001. Crystal structure of the ribosome at 5.5 angstrom resolution. *Science* 292:883–896.
Zhang, Y., and J. Skolnick. 2005. TM-align: A protein structure alignment algorithm based on the TM-score. *Nucleic Acids Research* 33:2302–2309.
Zwickl, D. J., and M. T. Holder. 2004. Model parameterization, prior distributions, and the general time-reversible model in Bayesian phylogenetics. *Systematic Biology* 53:877–888.

Index

3D-JIGSAW, 278
3DCoffee, 50, 284, 286
3Dee database, 161–162
accuracy. See error
affine gap cost. See gap cost, affine
agglomerative approaches, 57
alignment score, 157–158, 165, 169, 183–199. See also column scores; overlap score; sum-of-pairs scores
alignment uncertainty plot, 254–256, 259, 264–265
Aliscore, 138
AliWABA, 286
alternative alignments, 9–10, 22
alternative splicing, 28–29, 32, 34
AMAP, 268–269
ambiguity, 22, 73, 92, 134–135, 138, 147, 210–211, 213–219, 221, 223, 225-236, 241, 249–261, 270. See also error
amino acid sequences. See protein sequences
ANTHEPROT, 286
Au plot. See alignment uncertainty plot
AuberGene, 14
AVID, 53

BAliBase, 50, 62, 136, 158–161, 163, 170–171, 173–177, 197–198, 269
BAli-Phy, 259, 267
Bayesian methods, 19, 61, 65, 71–72, 90–93, 98, 135, 221–229, 236–245, 248, 250, 256–268

benchmarks, xv, 16, 49–51, 56, 65–67, 136, 153–177, 180, 182–183, 185–186, 269, 287. See also BAliBase; BRAliBase; HOMSTRAD; IRMBASE; OXBench; SABmark
Bio-dictionary, 291
biological motivation. See motivation, biological
BLAST, 14, 20, 46, 63, 72, 283, 289–290
BLASTZ, 14
BLOSUM matrix, 10, 25, 136, 224
bootstrapping, 72, 124
BRaliBase, 136, 158, 168–169
Bremer support, 124

CATH, 278
CE, 67, 156, 163–164
CHAOS, 13, 53
character-based alignment. See cost-based alignment
CINEMA, 286
circularity, 22, 72, 219–220, 222
CisEvolver, 183
CLUSTAL, xv, 3, 17, 38, 49–53, 58–60, 62–63, 96, 100–102, 112, 128, 136, 141–142, 166, 169, 171–176, 194, 196–197, 202, 233, 259–264, 268, 284–285, 287
COFFEE, 285, 287
column scores, 51, 157–159, 163, 174, 176, 183, 192, 287. See also alignment score; objective function
COMPARER, 166

333

complexity, 5, 42, 45–46, 53, 56–57, 61, 76, 79, 90, 97–99, 169, 223, 230, 242–247, 254. *See also* speed
compositional bias, 114–116, 121, 181
computational goal. *See* motivation, computational
consistency, 58–63, 68–69, 106, 136, 287
cost-based alignment, 19, 216, 220–227, 233–236. *See also* objective function
cost function. *See* objective function
crossing-over. *See* recombination
CS. *See* column scores
culling, 108–110, 138, 235, 250

DALI, 67, 156
database searching, 14, 20, 46, 273, 276, 283, 288, 294. *See also* BLAST; FASTA
data exclusion. *See* culling
Dawg, 182
Dayhoff matrix. *See* PAM matrix
DBClustal, 171–172, 284–285
DCA, 202–203
deletions. *See* indels
DIALIGN, 13, 18, 48–53, 57, 60, 62, 171–176, 202, 284–285
direct optimization, 98–101, 116, 132–134, 139. *See also* POY; joint estimation of alignment and phylogeny
divide and conquer, 57
DNA sequences, xii, xiv, 10, 15–16, 25–26, 180, 201–202
DO. *See* direct optimization
dynamic homology, 97
Dynalign, 169
dynamic programming, 4–11, 36, 41–42, 45–46, 60, 76–77, 79–80, 87, 91–92, 152–153, 228, 230, 245, 247, 251, 254, 267, 284

EdiPy, 183
edit distance, 3, 42, 44
efficiency. *See* complexity; speed
elision, 234–235, 250
error, xi, 2, 17–18, 20, 22, 58, 65–68, 72–73, 106–107, 128, 151–152, 180, 193, 200–206, 217–218, 258, 283, 286–288. *See also* alignment score
evolutionary distance, 21, 23, 67, 187, 194, 199–202, 205–206, 216
EvolveAGene, 182, 200
exons, 28, 32, 34, 80
EXPRESSO, 59, 63, 65, 171, 177

FASTA, 14, 20, 46, 72, 283
fast Fourier transform, 136
Fibonacci sequence, 5–7
FID. *See* fragment insertion-deletion model

Foldalign, 169
forward algorithm, 76, 87–91, 93
fragment insertion-deletion model, 75–81, 83, 87, 92–93
frameshift mutations. *See* mutations, frameshift
Friend, 293
FSSP, 163
FUGUE, 63, 166

gap. *See* indels
gap cost, 3, 10–11, 16, 25, 36–37, 40–41, 73, 115, 117, 119–121, 128, 136–138, 215, 218–219, 222–224, 228, 233–235
  affine, 11, 16, 25–26, 36–38, 120–121, 247
  linear, 11, 36–37, 41–42, 46, 73, 246
gap-free multiple alignment, 47–48, 53
gap length, 79, 110–111, 122–123, 138, 181, 219, 225, 247
GBLOCKS, 138, 231–232, 250
GCB matrix, 25
general insertion-deletion model, 79, 81
genetic algorithms, 57, 152, 171, 284–285
Geno3D, 278
genome alignment, 13–14, 21, 52–53
Gibbs sampling, 47, 90, 93
GID. *See* general insertion-deletion model
GLASS, 14, 53
global alignment, 3–13, 18, 36, 40–44, 47–53, 60, 158, 175, 196, 273, 284–286
goals. *See* motivation
GRAT, 14
greedy algorithms, 17–18, 48, 58, 202
guide tree, 17–18, 42, 57, 72–73, 106, 138, 215, 218–220, 226, 232–234, 261–264, 268, 270

hairpin-stem-loop, 108–113, 117–118, 123–125, 139
Handel, 19
hidden Markov models, 19, 43, 57, 61, 63, 65, 73, 76–77, 79–82, 86–93, 136, 152, 171, 230, 284–286
HMM. *See* hidden Markov models
HMMER, 19, 283
HMMT, 284–285
homology, xi, 1–2, 11–12, 19–21, 39–40, 107, 119, 128, 135, 180–181, 210, 212–216, 229, 241, 266, 282–283, 288
HOMSTRAD, 136, 158, 166, 170
hypervariable regions, 114, 116–117, 122–123

Index

indels, 11, 16, 19, 21, 23–38, 72–75, 110, 116–117, 121–122, 129–130, 139, 181–182, 212, 214–216, 222, 224–225, 228–229, 232–233, 238, 245–249. *See also* fragment insertion-deletion model; gap length; general insertion-deletion model; long-indel model; TKF91 model; TKF92 model
Indel-Seq-Gen, 182
indel size. *See* gap length
information content, 287
insertions. *See* indels
interaction networks, 279–280
introns, 28, 80, 111, 123
IRMBASE, 50, 158, 166–167, 170, 172–173
iterative methods, 17–18, 42, 48, 58, 72, 81, 138, 152, 171, 175, 219, 268, 284–287

Jalview, 286, 291
joint estimation of alignment and phylogeny, xiii, 19, 22, 71–72, 80–86, 90–93, 98–99, 124, 220–223, 228, 237–249, 258–267. *See also* direct optimization
JOY, 166
JTT matrix, 10, 25, 180

Kalign, 59–60

LAGAN, 53
large datasets, 21, 67–68. *See also* genome alignment
likelihood. *See* maximum likelihood
linear gap cost. *See* gap cost, linear
local alignment, 11–14, 18, 20, 36, 43–53, 60, 158, 172, 175, 196, 273, 284–286
long-indel model, 79, 81, 247–248
loops. *See* hairpin-stem-loop

M-Coffee, 59, 61–62
MACSIMS, 291, 293
MAFFT, xv, 52, 59, 61–63, 136, 171–172, 175–176, 202, 284, 286
MAGOS, 293
manual alignment, 107, 126–137, 139, 141–149
MAO. *See* Multiple Alignment Ontology
Map2, 14
Markov-chain Monte Carlo methods, 90–92, 230, 241–243, 245, 249, 253–254, 259, 268
MARNA, 65
Matchbox, 47
matrix. *See* substitution matrices
MAUVE, 14
maximum likelihood, 3, 19, 71, 92, 98, 124–125, 220–229, 234, 242
MCMC. *See* Markov-chain Monte Carlo methods
MEME, 47
meta-methods, 56, 61–62, 68
Metropolis-Hastings, 91
missense mutations. *See* mutations, missense
MNYFIT, 166
MODELLER, 278
motivation
  biological, xii–xiii, 20, 39, 106, 127, 137–138, 180, 224, 246, 284
  computational, xii–xiii, 3, 20, 60, 97–98, 127, 224, 246
MrBayes, 92, 125
MSA. *See* multiple sequence alignment
MultAl, 171, 284–285
MultAlign, 171, 284–285
MultAlin, 18
Multiple Alignment Ontology, 291
multiple sequence alignment, 16–19, 22, 40, 42–43, 46–50, 55–69, 71–72, 80–86, 90–93, 95–97, 99–103, 138, 151–153, 157–177, 191–199, 202–203, 211–213, 215, 217–218, 223, 228–233, 239–243, 248, 251–256, 272–295
MultiSeq, 290
MUMMALS, 59, 61, 63, 171, 177, 284, 286
MUMMER, 14
MUSCLE, xv, 52, 59, 61–62, 171–176, 202, 259–261, 284, 286
mutation matrices. *See* substitution matrices
mutations, 26–29
  frameshift, 28, 30
  insertions and deletions. See indels
  missense, 27, 278, 293
  point, 16, 23–24, 26–27, 181
  transitions and transversions, 10, 25, 92, 119, 121, 181, 225, 233
MyHits, 289, 291
MySSP, 182

near-optimal alignments, 221, 226, 231, 235, 266–267
Needleman-Wunsch algorithm, 4–10, 12, 16–17, 36, 42, 45, 58, 106, 115–116
norMD, 287
nsSNP Analyzer, 279
number of sequences. *See* taxon sampling

objective function, 3–4, 9–10, 12, 14–17, 19–21, 40–42, 45–46, 48, 56, 58, 71, 73, 98, 218, 220, 223–224, 226, 285, 287. *See also* optimality

objectivity, 56, 61, 107, 117, 128, 139, 141, 147, 211, 231, 233. *See also* subjectivity
OPTIMA matrix, 25
optimality, 4–5, 7, 9, 14–17, 20, 40, 44, 60, 108, 127, 135, 137, 220–221, 224, 229, 242, 284, 287. *See also* motivation, computational; near-optimal alignments; objective function
OrdAlie, 291
overlap score, 187. *See also* alignment score
OXBench, 50, 158, 160–163, 170–172

pairwise sequence alignment, 3–10, 16–18, 40–48, 57–60, 63–65, 158, 162, 165, 184–193, 196–198, 201–202, 211, 223, 228–230, 241, 248, 250–252, 256, 265, 273, 286–287
PAM matrix, 10, 25, 136, 180, 224
parsimony methods, 17, 98–99, 107–109, 111–114, 116, 122, 124–125, 223, 234–235. *See also* cost-based alignment
partial alignment, 214
partial order alignment graph. *See* POA
PAUP*, 100–101, 117
PCMA, 61–62, 202
PDB. *See* Protein Data Bank
Pfaat, 291
Pfam database, 161, 166
PHASE, 125
phylogenetic profile method, 281
phylogenetics, xiii, 17, 21, 55, 68, 71–72, 80–86, 90–93, 95–102, 106–107, 111, 121, 124–130–135, 138–139, 152, 182, 199–200, 203–206, 211, 215, 217–223, 231–233, 249, 258–267, 274–275. *See also* joint estimation of alignment and phylogeny
PhyloGenie, 290
PIMA, 171, 284–285
PMcomp, 169
POA, 57, 62, 169, 173, 286
point mutations. *See* mutations, point
PolyPhen, 279
POY, 96, 100–102, 116, 129, 132–134, 139
PRALINE, 59, 63, 65, 284, 286
precision, 79
PREFAB, 50, 67–68, 158, 163–164, 170–172
probabilistic alignment. *See* statistical alignment
ProbAlign, 136, 169
ProbCons, xv, 43, 52, 57, 59, 61–65, 136, 171–176, 268–269, 284, 286

profile alignment, 136
progressive alignment, 17–18, 22, 42, 49, 57–60, 68, 152, 172, 202, 215, 218–219, 224, 261, 268, 270, 284–286
PROMALS, 57, 59, 63, 65, 136
PROSITE, 166
Protein Data Bank, 63, 156–157, 161–162, 166, 289–290
protein sequences, xiv, 10, 15–16, 20, 25–26, 38, 50, 80, 136, 152, 155, 158–167, 170, 182, 272–295
protein structure prediction, 152, 277–279
PRRN, 52
PRRP, 18, 171–172, 284–285
PSI-BLAST, 63, 65, 136, 163, 283

QR factorization, 290
quality score. *See* alignment score

RAlign, 171, 177, 286
random sequences, xii, 51–52, 115, 201
RASCAL, 171, 177, 286
RASMOL, 293
recombination, 27, 33–35
recursion, 5–7, 42, 46, 79, 89, 91, 93, 247
Refiner, 171, 177, 286
repeatability, 107, 126, 128–135, 141, 211, 233
repeat sequences, 34
replication, 27, 30
reproducibility. *See* repeatability
RIBOSUM matrix, 136
RNAsalsa, 137
RNA sequences, xiv, 20, 50, 65–66, 80, 106–126, 129–132, 136–149, 152, 158, 168–170, 257–266
ROSE, 166–167, 182

SABmark, 158, 164–166, 169, 171–172
SAGA, 171–172, 284–285
SAM, 171
SAP, 63–64, 156, 160
SCI. *See* structure conservation index
SCOP protein domain database, 159, 164–166, 278
scoring function. *See* objective function
SEAVIEW, 286
semi-global alignment, 11
sensitivity, 51, 285
sensitivity analysis, 119–121, 132, 135, 213, 226, 233–236, 270
sequence evolution, 10, 78–79
Sequence Retrieval System, 291
sequence simulation. *See* simulation
sequencing errors, 276
sets of active nodes. *See* soan
SIFT, 279

Index

SIM algorithms, 10, 49
simulated annealing, 57, 81, 152, 241–243, 268
simulated sequences. *See* simulation
simulation, 21, 158, 166–167, 170, 180–183, 196–206, 269–270. *See also* CisEvolver; Dawg; EdiPi; EvolveAGene; Indel-Seq-Gen; MySSP; ROSE
simultaneous estimation. *See* joint estimation of alignment and phylogeny
Smith-Waterman algorithm, 12–13, 36, 45–46
soan, 87–88
SOFI, 164
software, xi, xiv–xv, 2–3, 10, 13–15, 18–19, 52–53, 59–60, 136, 171–177, 283–287, 294. *See also* 3DCoffee; AuberGene; AVID; BAliPhy; BLAST; BLASTZ; CHAOS; CLUSTAL; DCA; DIALIGN; EXPRESSO; FASTA; GLASS; GRAT; Handel; HMMER; Kalign; LAGAN; M-Coffee; MAFFT; Map2; MatchBox; MAUVE; MEME; MultAl; MultAlign; MultAlin; MUMMALS; MUMMER; MUSCLE; PCMA; POA; POY; PRALINE; ProbAlign; ProbCons; PROMALS; PRRP; RAlign; RASCAL; Refiner; RNAsalsa; SIM algorithms; T-Coffee; T-Lara; WABA
specificity, 51
speed, xv, 171, 174–176, 223, 230, 242, 245–246, 249, 254, 259, 266–268, 283, 285, 288. *See also* complexity
SPS. *See* sum-of-pairs scores
SRS. *See* Sequence Retrieval System
SSAP. *See* SAP
STAMP, 161
static homology, 97
statistical alignment, 19, 21, 43, 71, 98, 210, 216, 218, 220–229, 236–249, 257–269. *See also* Bayesian methods; maximum likelihood
stem. *See* hairpin-stem-loop
Stemloc, 169
structural alignment, 20–21, 38, 49–50, 56, 63–67, 106–113, 118–119, 121, 123–128, 132–134, 136–137, 139–149, 155–157, 159–166, 168–172, 177, 184, 212–213, 284, 286, 294
structure conservation index, 169

subjectivity, 107, 128, 138–139, 211, 231, 233. *See also* objectivity
substitution matrices, 15, 25, 27, 36, 58–61, 73, 124–125, 136, 180, 225, 228. *See also* BLOSUM matrix; GCB matrix; JTT matrix; OPTIMA matrix; PAM matrix; RIBOSUM matrix
suffix trees, 53
sum-of-pairs scores, 51, 157–159, 165, 169, 174, 176, 183, 186, 191, 195, 224, 287. *See also* alignment score; objective function
SWISS-MODEL, 278
SWISS-PROT, 289

T-Coffee, 18, 49–52, 59–65, 68, 171–176, 202–203, 269, 284–285
T-Lara, 59, 65
tabu search, 57
tandem repeats. *See* repeat sequences
TAP. *See* tree alignment problem
taxon sampling, 202–203, 249, 270, 288–289, 294
template-based alignment, 56, 62–65, 69
terminal gaps, 11
tihls, 84–89, 91
time. *See* speed
TKF91 model, 19, 74–77, 80–82, 86–87, 92, 229–230, 243, 245–247
TKF92 model, 19, 75, 77–81, 87, 92–93, 229–230, 245–247
trace-back matrix, 8–9, 12–13, 41, 46, 251
TRACE SUIT, 166
transitions. *See* mutations, transitions and transversions
transposable elements, 28–33
transversions. *See* mutations, transitions, and transversions
tree alignment problem, 97-100, 102
tree-indexed heirs lines. *See* tihls
TREE-PUZZLE, 80

uncertainty. *See* ambiguity
Uniprot database, 161
UniqueProt, 289
UniRef database, 289

vALID, 276
VISSA, 293
Viterbi algorithm, 58, 76
ViTO, 293

WABA, 14

Composition: Publication Services
Text: 10/13 Sabon
Display: Sabon
Printer and Binder: Thomson-Shore